HANDBOOK OF DIAGNOSTIC CYTOLOGY

The Authors

HELENA E. HUGHES

M.B., Ch.B.(V.U.Manc.), M.R.C.O.G., M.C.Path.
Consultant in Exfoliative Cytology
Department of Pathology, Glasgow Royal Infirmary
Honorary Lecturer in Exfoliative Cytology, University of Glasgow

T. C. DODDS

F.I.M.L.T., F.I.I.P., F.R.P.S.
Director of The Medical Photography Unit
University of Edinburgh

T. SYMINGTON

B.Sc., M.D., F.R.C.P. & S.(Glasg.)
Professor of Pathology, University of Glasgow

HANDBOOK OF DIAGNOSTIC CYTOLOGY

HELENA E. HUGHES
and
T. C. DODDS

Foreword by
T. SYMINGTON

E. & S. LIVINGSTONE LTD
EDINBURGH AND LONDON
1968

SBN 443 00560 5

Printed in Great Britain

FOREWORD

It is more than a hundred years since the first attempts were made to diagnose gastric malignancy from exfoliated cells. Even today such a cytological diagnosis is fraught with difficulty and it is little wonder that this method of tumour diagnosis was allowed to lapse for many years.

The ease with which satisfactory specimens, in the form of smears, could be prepared from the cervix and the introduction by Papanicolaou of a suitable method of staining them, aroused most gynaecologists, but alas only a few pathologists, to the value of exfoliative cytology as a diagnostic tool for cervical cancer. With experience, the technique has been extended to cytological examination of smears from bronchus, serous cavity and other organs with remarkable and gratifying success, and no pathology department can afford to be without this valuable diagnostic tool. This in itself has created problems. Laboratories have to be established with technicians and medical staff alike trained in the use and limitations of the new service.

This practical manual is set out with this purpose in mind, and the examples illustrated are drawn from patients studied by the author. It is hoped that the book will be a useful guide to anyone charged with the responsibility of establishing a cytology laboratory. It gives full details of the collection and processing of specimens. The small inset photographs illustrate the initial view of the smear, and in each instance, higher magnifications focus attention on some important point which requires attention. These are the appearances which anyone beginning a study of the subject will encounter.

While the manual deals for the most part with gynaecological cytology, all systems from which exfoliated cells may be examined are dealt with and illustrated. Although a number of excellent books are available on this subject, there is an undoubted need for a manual such as this, which contains all the information required, yet is presented in a concise and practical manner.

T. Symington

1968

PREFACE

THIS book is intended to provide an introduction to the practice of diagnostic cytology for medical and technical staff training in this field. The first and major part of the book is concerned with the actual appearance, on microscopic examination, of cytological preparations and the clinical significance of these findings in each system. It includes a consideration of specimens from most of the common sites suitable for cytological investigation. The appendix gives details of the essential steps in the collection, preparation, staining and reporting of specimens together with a brief discussion of the significance of normal and abnormal reports.

Detailed consideration is given to both benign and malignant aspects of gynaecological cytology in view of the widespread application of the method to this field. In non-gynaecological cytology it was decided to include an account of the cytology of those specimens which were regularly requested by clinicians and which did not present such difficulties, either in specimen collection or laboratory technique and interpretation, that they were unsuitable for routine investigation. Certain specimens, such as those from the duodenum and prostate, were omitted because specimen collection presents significant problems, while other fields, such as the cytology of malignant cells in blood, were omitted because the many diagnostic difficulties encountered make them more suitable for research than for routine use.

In order that the material illustrated should be truly representative of the work of a routine laboratory, all specimens were photographed without being restained, and were selected from those received in the cytology laboratory of the University Department of Pathology at Glasgow Royal Infirmary during the years 1962-66. For this reason many kind offers of 'interesting' material were declined and this policy is responsible for the omission of certain rare types of specimen. Unless otherwise stated all preparations were stained by the standard Papanicolaou technique and photographed at constant low and high power magnifications. All the colour illustrations are from unretouched colour photomicrographs.

As the colour photomicrographs were exposed under tungsten lighting and the colour printing balanced and corrected under light of a similar nature, it is strongly advised that the reader should, wherever possible, use illumination of this type. The colour balance will appear inferior in daylight or in fluorescent lighting.

The omission of the anatomy, histology, pathology and surgery of the various systems was necessary to keep the book within the size of a handbook. This was especially regretted in regard to pathology, but it was thought that any short account would be superfluous for medical staff already trained in pathology, while any detailed account, including the many controversial lesions involved, would tend to confuse technical staff. However, all the cytological diagnoses shown in the illustrations were subsequently confirmed by pathological examination.

The bibliography in a small handbook is inevitably incomplete. It was decided that the most comprehensive section should be the one dealing with gynaecological cytology because of the extensive use of the method in this field and because many of the technical and medical staff engaged in working on cervical cytology might be unfamiliar with the literature. In non-gynaecological cytology only specifically cytological references have been included, since pathologists would already be acquainted with the literature of their own subject.

It gives us great pleasure to record our thanks to Professor T. Symington for his advice and encouragement; to Dr Elizabeth Peters for critical reading of the proofs; to Dr J. C. J. Ives and his staff in the Bacteriology Department, Dr G. A. MacDonald and his staff in the Haematology Department, and Dr R. S. Patrick and many other colleagues in the Pathology Department for advice and criticism; to the clinical staff of Glasgow Royal Infirmary for the specimens from their patients and, in particular, to Professor D. F. Anderson, Mr H. Stirling and their staff for access to gynaecological and obstetric patients. We are indebted to Miss Flora E. W. Clouston and other members of the staff of the cytology laboratory at Glasgow Royal Infirmary for the technical work; to Miss Loraine Crozier and Miss Ann Healy for typing the manuscript; and to the staff of the Department of Medical Photography of Edinburgh University, especially Mr J. Paul, for their assistance in the preparation of the book. Throughout the writing of the book we have received help and consideration from Messrs E. & S. Livingstone and would like to acknowledge our indebtedness to Mr Charles Macmillan, Mr J. Parker, Mr W. G. Henderson and Mr A. D. Lewis.

THE AUTHORS

1968

TABLE OF CONTENTS

Page

Introduction 2

SECTION I

FEMALE GENITAL TRACT

Chapter 1	The Normal Smear	6
Chapter 2	Non-malignant Disorders	19
Chapter 3	The Cytological Diagnosis of Cancer	34
Chapter 4	Cancer of the Female Genital Tract	37
Chapter 5	Hormonal Effects on the Vaginal Smear	69

SECTION II

RESPIRATORY TRACT

Chapter 6	The Normal Cellular and Non-cellular Patterns	94
Chapter 7	Non-malignant Disorders	102
Chapter 8	Cancer of the Respiratory Tract	108

SECTION III

GASTRO-INTESTINAL TRACT

Chapter 9	The Normal Cellular Content	122
Chapter 10	Non-malignant Disorders	125
Chapter 11	Megaloblastic Anaemia due to a Deficiency of Vitamin B12 and/or Folic Acid	129
Chapter 12	Cancer of the Gastro-intestinal Tract	132

SECTION IV

URINARY TRACT

Chapter 13	The Normal Cellular Patterns	140
Chapter 14	Non-malignant Disorders	142
Chapter 15	Cancer of the Urinary Tract	144

SECTION V

THE BREAST

Page

Chapter 16 Non-malignant Disorders 150
Chapter 17 Breast Cancer 153

SECTION VI

THE BODY FLUIDS

Chapter 18 Non-malignant Disorders 158
Chapter 19 Malignant Cells in the Body Fluids . . . 165

SECTION VII

THE EFFECTS OF RADIATION

Chapter 20 Radiation Effects 176

SECTION VIII

PRACTICAL PROBLEMS IN DIFFERENTIAL DIAGNOSIS

Chapter 21 Practical Problems in Differential Diagnosis . . 186

APPENDIX

THE ORGANISATION OF A CYTOLOGY SERVICE

Chapter 22 General Principles 198
Chapter 23 The Collection and Preparation of Specimens . . 200
Chapter 24 Staining Techniques 215
Chapter 25 Phase Contrast Microscopy 227
Chapter 26 The Organisation of Work in the Laboratory . . 230
Chapter 27 Diagnostic Cytology and the Care of the Patient . 235
Bibliography 239
Index 249

INTRODUCTION

INTRODUCTION

MANY observers in the nineteenth century and the early part of the twentieth century had noted and described benign and malignant exfoliated cells and malignant tissue fragments. The description of the cellular content of colostrum by Donné (1838) was shortly followed by Pouchet's observations (1847) on the appearance of cells in the vaginal secretion. Malignant tissue fragments in sputum were described by Walshe (1843) and in gastric washings by Rosenbach (1882). Malignant cells were recognised in body fluids by Donaldson (1853) and in sputum by Beale (1860).

However, the practical application of the methods of cytology to diagnostic problems is based on the work of Papanicolaou who, in collaboration with Stockard (1917), reported the cyclical changes seen in cells exfoliated from the vaginal epithelium, laying a foundation for subsequent use of the method for the assessment of hormonal function. In 1928 the role of the vaginal smear in the detection of cancer was described by Papanicolaou and in 1943 Papanicolaou and Traut published their classical work, *The Diagnosis of Uterine Cancer by the Vaginal Smear*. In England notable contributions were made by Dudgeon and his colleagues in 1927 and 1934.

Diagnostic cytology, as a method of investigation, depends on the fact that cells exfoliated or collected from a surface reflect many of the features of the underlying tissue. At present its main application is the early diagnosis of cancer, especially of the female genital tract. The method, however, can also be used for many other purposes, some closely linked to cancer diagnosis and others independent of this problem. In patients who have been treated for cancer, diagnostic cytology can be used to evaluate the response to radiotherapy and to follow post-operative progress after surgery. In patients with endocrine disorders cytological assessment of certain hormonal effects is valuable both for primary diagnosis and for the control of hormone therapy. Closely related to this field is the use of a cytological method for the recognition of sex chromatin. Cytological methods may also be applied in recognising certain infections and allergic reactions and in assessing the response of a tissue to infection.

The advantages of diagnostic cytology are readily apparent. In most cases suitable specimens can be collected with minimal discomfort to the patient, making repeated or serial investigations feasible. At some sites the use of cytology may make biopsy unnecessary. The collection techniques usually require only a small amount of inexpensive equipment. In cancer diagnosis, cytology may reveal cell changes at an early stage in the disease when no abnormality is apparent on careful clinical examination. This may offer excellent prospects of control of the disease and may decrease the extent of surgery or radiotherapy required. In endocrine assessments cytological changes may reflect alterations in hormonal status with considerable sensitivity; this combines with the low cost of the method and the ease of serial examinations to make cytology a useful tool in long-term studies.

Diagnostic cytology unfortunately has limitations and disadvantages and without an understanding of these and an awareness of the possible pitfalls and problems the method may easily be abused.

The first stage in cytological diagnosis is the provision of an adequate clinical history. A cytological diagnosis made in a vacuum may well be misleading. Age, sex and the clinical diagnosis are obviously relevant, but such factors as previous pregnancies, a history of cancer, treatment by radiotherapy, recent surgery or hormone therapy may be equally important.

The next stage is the collection of specimens. Occasionally a failure to make a cytological diagnosis is due to the fact that the lesion is inaccessible to direct collection and does not communicate with a hollow organ from which exfoliated material can be obtained. More often such a failure is attributable to a poor collection technique. It is essential that both clinician and laboratory worker should be aware of the advantages and disadvantages of alternative sites and methods of collection for different purposes, and that the site and method used by the clinician should be made known to the laboratory. Different sites and types of specimens may have different 'normal' ranges. The aims should be the collection of suitable specimens with as little contamination as possible, immediate fixation when smears are made or rapid transport to the laboratory, with accurate specimen identification. Failure to fulfil these conditions can lead to difficulties due to smears being obscured by contaminants, to drying distortion, to cellular degeneration in old specimens and to 'negative' results when specimens are collected from unsuitable areas.

In the laboratory the hazards of poor selection of material from specimens, of delays in specimen processing, of drying distortion, of inadequate fixation, of poor staining techniques and of contamination during staining must be avoided.

Apart from an awareness of the technical problems already mentioned, an essential stage in the diagnosis of significant cellular abnormalities is a sound knowledge of the normal appearance of the relevant cells. This implies a recognition of the types of cell which are usually present in a particular specimen, including a knowledge of the minimal cellular criteria, which must be met before a specimen is considered suitable for diagnosis. This knowledge must include, as a base line for cancer diagnosis, a familiarity both with the changes in individual cells and with the patterns of smears in various non-malignant disorders. As a base line for hormonal assessment, an awareness of physiological changes is necessary. Moreover the observer must realise that here also the 'normal' range of cellular components and cell changes varies widely in specimens from different sites.

Another problem which now arises is the necessity for careful scrutiny of the whole of a specimen. Occasionally an abnormal finding may be so striking that a diagnosis can be made immediately but, in most cases, complete examination of the specimen is necessary for a definitive diagnosis. This is time consuming and frequently monotonous.

Lastly there are certain inherent weaknesses in the method. The observer's final assessment is a combination of information derived from the clinical history, from the pattern of the smear and from the appearance of individual cells and groups of cells. However, there remain several unknown factors. In cancer diagnosis, one can never be certain that the negative report is not due to an inaccessible lesion or an incorrectly collected specimen. Such a report should not outweigh clinical suspicion or exclude further investigation. The positive report, too, has its limitations. Tumour type and differentiation can often be estimated on cytological grounds, but these may be misleading in certain tumours which ultimately show varying degrees of differentiation. The geographical site of the tumour may be fairly readily assessed with direct collection specimens, but may require further investigation with exfoliated specimens. The presence or absence of invasion cannot be determined on cytological evidence, although the distribution patterns of abnormal cells may suggest a pre-invasive or an invasive lesion. A positive diagnosis should therefore be followed by a further confirmatory specimen, by localising investigations and, whenever possible, by biopsy to obtain a histological diagnosis before a patient is subjected to major surgery or radiotherapy.

Different problems arise in a hormonal assessment in which the observer is viewing the end-result of the interplay of several hormones. Serial observations are usually necessary and may be usefully combined with other observations such as basal body temperature recordings, endometrial biopsies and urinary hormone assays. Caution should be exercised in interpreting a single specimen, although occasionally such a specimen may be a useful pointer to further investigation.

Cytology can be safely used alone or in combination with other methods of investigation as a diagnostic technique, provided due regard is paid to the many factors involved in obtaining and reporting a suitable diagnostic specimen. However, without this background knowledge, cytology may be misleading, with the risk that the patient may be subjected to unnecessary surgery or alternatively left with a completely false sense of security.

SECTION I

FEMALE GENITAL TRACT

CHAPTER 1

THE NORMAL SMEAR

THE cellular content of smears, prepared by scraping the cervix or aspirating the posterior fornix of the vagina, can be divided into epithelial and non-epithelial cells. The distribution of the various components of the normal smear varies with the age of the patient, the stage of the menstrual cycle and the site from which the smear is taken.

The epithelial cells present are *squamous cells*, *columnar endocervical* and *endometrial cells*. The non-epithelial cells include *erythrocytes*, *polymorphonuclear leucocytes*, *histiocytes*, *lymphocytes*, *plasma cells* and *spermatozoa*.

EPITHELIAL CELLS

SQUAMOUS CELLS

Superficial, intermediate and parabasal squamous cells are the three types usually seen in smears. Rarely, keratinised anucleated squames and true basal squamous cells may be present.

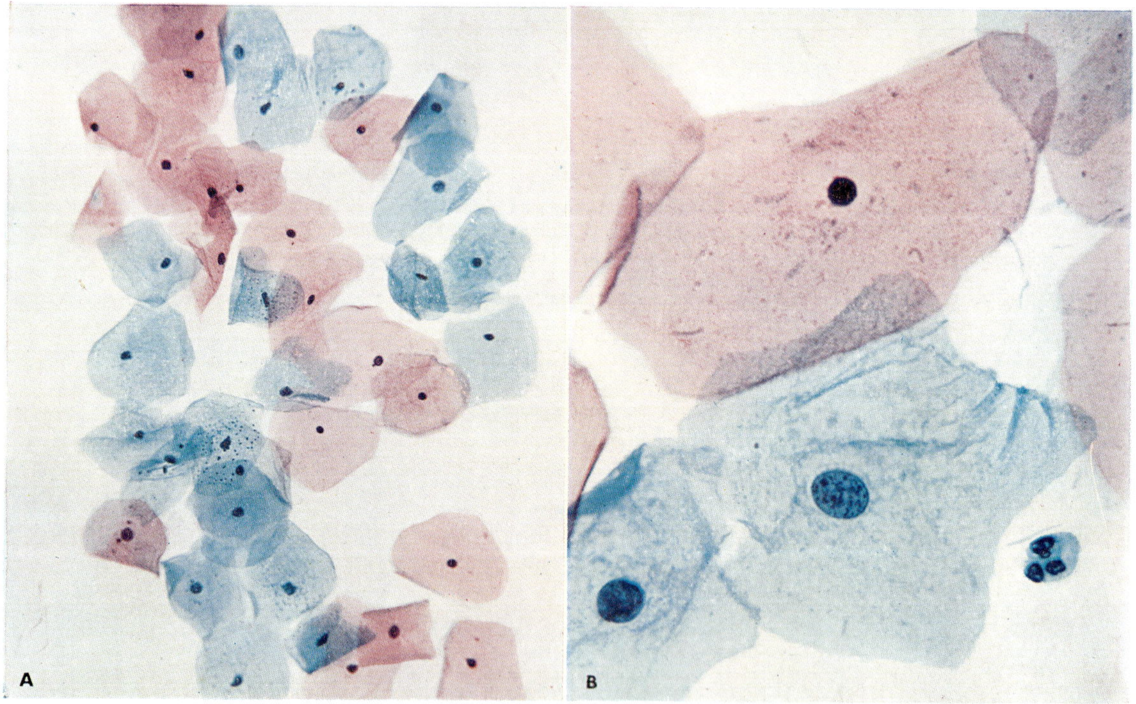

FIG. 1

Benign Superficial and Intermediate Squamous Cells

A. Both the superficial and the intermediate cells have a large amount of translucent cytoplasm. In the superficial cell, the pyknotic nucleus is densely hyperchromatic, small and homogeneous; the cytoplasm is commonly eosinophilic, but not necessarily so. In the intermediate cell, the vesicular nucleus is larger, round or oval in shape, with a fine chromatin network; the cytoplasm is usually basophilic, but may be eosinophilic. Kerato-hyaline granules are seen in the cytoplasm of superficial and intermediate cells. (×200.)

B. A superficial cell with translucent eosinophilic cytoplasm and a pyknotic nucleus is seen above two intermediate cells with translucent basophilic cytoplasm and vesicular nuclei. The size of these cells can be judged against the polymorphonuclear leucocyte seen in the lower right quadrant. (×800.)

Superficial squamous cells. These are large flat, usually polygonal cells, with transparent cytoplasm and small dense homogeneous pyknotic nuclei. The diameter of the cells varies from 40 to 50 μ and the nucleus is 6 μ or less in diameter. The cytoplasm is usually eosinophilic but, occasionally, may be basophilic. Tiny dark granules, *keratohyaline granules*, may be seen in the cytoplasm.

The **intermediate squamous cell** is smaller than the superficial cell, but shares with it a polygonal shape and translucent cytoplasm. It may appear in flat or folded form. The cytoplasmic staining is usually basophilic but occasionally may be eosinophilic. The nucleus is large and vesicular, with a fine chromatin network. In many accounts of squamous cells the larger, more angular intermediate squamous cells are included among the 'superficial' cells and described as 'pre-cornified' cells to distinguish them from so called 'cornified' cells with pyknotic nuclei. The difficulty of this subdivision is the inclusion in one category of 'superficial cells' of two cell types with morphologically different nuclei. This obscures the clear definition of the group and is especially unfortunate because oestrogen activity shows a quantitative correlation with the presence of superficial cells with pyknotic nuclei. It is therefore preferable to include those large cells with vesicular nuclei in the intermediate cell group. These two groups, the superficial and intermediate squamous cells, predominate in smears from women in the child-bearing years, when parabasal squamous cells are seen only occasionally.

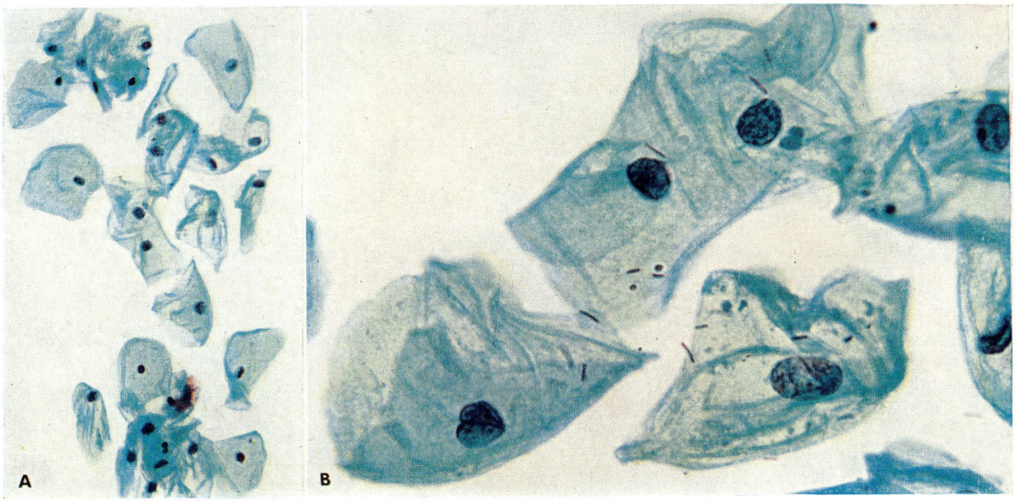

FIG. 2
Intermediate Squamous Cells

A. This group of cells consists mainly of intermediate squamous cells with vesicular nuclei, although an occasional basophilic superficial cell with a pyknotic nucleus is seen. (×200.)

B. The basophilic translucent cytoplasm and the vesicular nuclei of intermediate cells are shown. In this group wrinkling and folding of the cytoplasm is clearly seen. (×800.)

Parabasal squamous cells. These are small, round or oval cells with opaque, usually basophilic, vacuolated cytoplasm, varying in diameter from 15 to 30 μ. As their nuclei are similar in size to those of intermediate cells they have a high nucleocytoplasmic ratio. Their vesicular nuclei show a coarser chromatin pattern than those of intermediate cells. In children and in some post-menopausal women, these cells may be the main epithelial component of a smear.

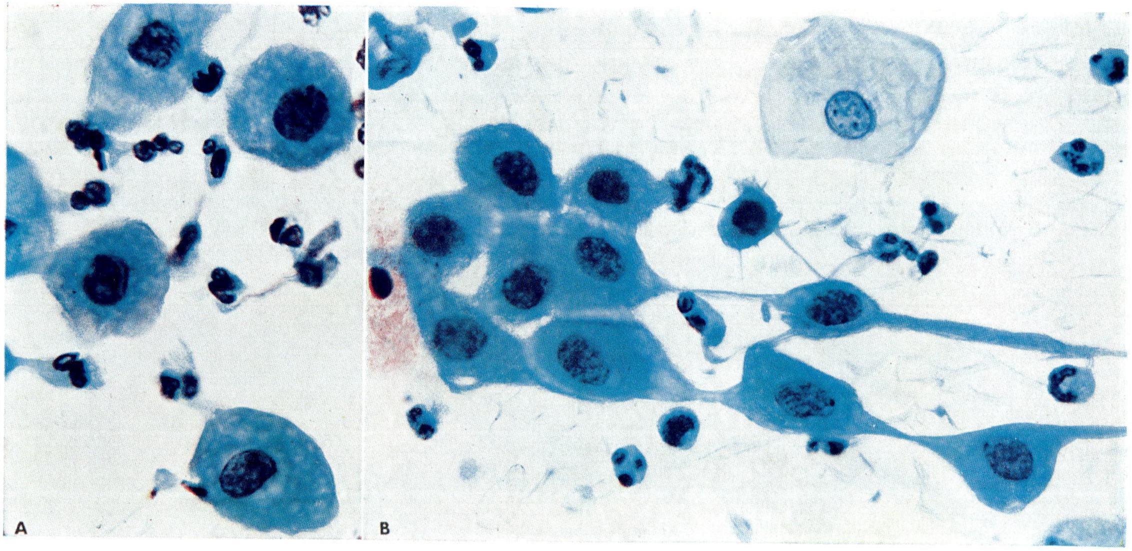

FIG. 3

Parabasal (Outer Layer Basal) Squamous Cells

A. These relatively small epithelial cells with basophilic opaque vacuolated cytoplasm and vesicular nuclei are parabasal squamous cells. Although the shape of a single cell is usually round, some may show an angular outline when in close proximity to others. This is seen in the cell in the centre of the group, where the cytoplasmic process to the adjacent cell on the left border of the field is shown. (×800.)

B. This group again shows the opaque cytoplasm, vesicular nuclei and high nucleocytoplasmic ratio characteristic of parabasal squamous cells. In a group, however, the cells appear more angular and tails of cytoplasm are seen. Comparison with the intermediate cell in the upper right quadrant emphasises the opacity of their cytoplasm. Moreover, it can be seen that the nuclei of the parabasal cells are of similar size or only slightly larger than the intermediate cell nucleus. Their apparent large size is due to the high nucleocytoplasmic ratio. The parabasal nuclei have a coarser chromatin pattern than is seen in the intermediate cell nucleus but their outlines are smoothly oval and their chromatin evenly distributed. (×800.)

Under exceptional circumstances two other types of squamous cell may be seen in normal smears. Keratinisation of upper vaginal or cervical epithelium does not occur unless it is exposed as an external surface. This happens in procidentia, when anucleated keratinised flat translucent orange staining squames may appear in the smear. Exfoliation of basal cells does not normally occur, but they may be present as a result of trauma or severe inflammation. They share the general characteristics of parabasal cells, but are smaller, have densely opaque cytoplasm, a higher nucleocytoplasmic ratio and nuclei showing an even more marked clumping of chromatin to form chromocentres, which are small, round and basophilic.

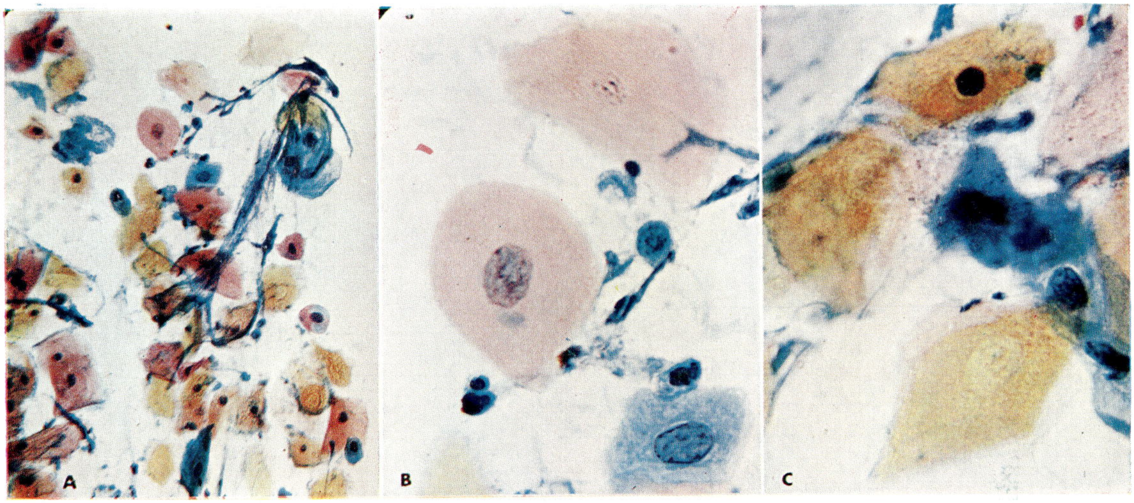

FIG. 4

Keratinised Anucleated Squamous Cells

A. Keratinised anucleated squamous cells (squames) showing characteristic bright orange or yellow staining of cytoplasm. (×200.)

B. The top squamous cell shows a transitional stage with almost complete disappearance of the nucleus by karyolysis. (×800.)

C. Progressive stages in the disappearance of the nucleus. The top squamous cell shows bright orange translucent cytoplasm with retention of a pyknotic nucleus. The squamous cell in the lower part of the field shows yellow cytoplasm with a 'ghost' nucleus. The orange squame on the left shows the final stage with complete disappearance of the nucleus. (×800.)

These keratinised cells are not normally present in cervical smears. They appear when the cervical epithelium is exposed, e.g., in a procidentia. They are also obtained on scraping leukoplakic areas of the cervix. Their appearance in a vaginal smear suggests that it has been taken from the lower third of the vagina. The smear illustrated was taken from the cervix in a case of procidentia.

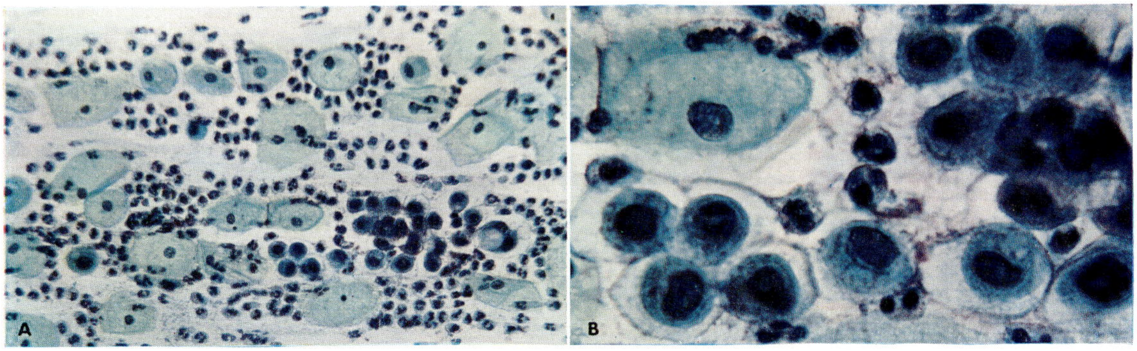

FIG. 5

Benign Basal (Inner Layer Basal) Squamous Cells in a Cervical Smear

A. Superficial, intermediate and parabasal squamous cells are interspersed with polymorphonuclear leucocytes. A large irregular group of small dark cells is present in the right lower quadrant. At low magnification their nuclei appear large and hyperchromatic. Vacuolation of cytoplasm is seen in a few cells. (×200.)

B. It is now apparent that these basal squamous cells are uniform in size and rounded in shape. Their cytoplasm is densely opaque and basophilic, forming a rim, narrower than the nuclear diameter, around the nuclei. This small amount of cytoplasm suggests that the nuclei are large whereas, in fact, they closely resemble the vesicular nucleus of the intermediate cell in the left upper quadrant. They are central in position and uniform in appearance. Only one cell shows cytoplasmic vacuolation, which is due to degeneration. (×800.)

These small benign basal squamous cells are rarely seen in smears from women in the child-bearing years. Their presence is a result of trauma, or, as in this smear, inflammation. Initially, their appearance may arouse suspicion because of their high nucleocytoplasmic ratio and dark nuclei. The uniformity, however, of cellular and nuclear size, shape and staining establishes their benign nature.

ENDOCERVICAL CELLS

The recognition of these tall columnar cells is made easier by their frequent arrangement in characteristic groups, either lying in parallel rows forming a 'picket fence' or 'palisade', or seen in cross-section forming a 'honeycomb' with clearly defined cellular borders and adequate cytoplasm separating the regularly arranged nuclei. The individual columnar cells are basophilic and have basal nuclei, beyond which may project a tail of cytoplasm. At the luminal edge of these cells eosinophilic cilia and terminal plates can often be identified. The presence of cilia or a terminal plate is a useful guide to the benign nature of a columnar cell, as these features are absent in malignant cells. The nuclei of endocervical cells are vesicular with a chromatin pattern coarser than that of the intermediate cell; they are uniformly round or oval in shape but may show considerable variation in size. Small nucleoli can occasionally be identified. Some endocervical cells show marked secretory activity with cytoplasmic vacuoles and consequent flattening of one nuclear border. At mid-cycle the nucleus may have a nipple-like projection on its superficial aspect which produces an apparent irregularity of nuclear shape. Endocervical cells are infrequently seen in smears prepared from the posterior fornix pool, but should be present in cervical scrape smears.

Endocervical columnar cells are often accompanied by cervical mucus, which appears as blue-staining amorphous strands or streaks of varying density. Histiocytes and polymorphonuclear leucocytes may be seen enmeshed in the mucus. When endocervical cells are not seen in a cervical scrape smear, these strands of cervical mucus provide useful indirect evidence that the smear is, in fact, of cervical origin.

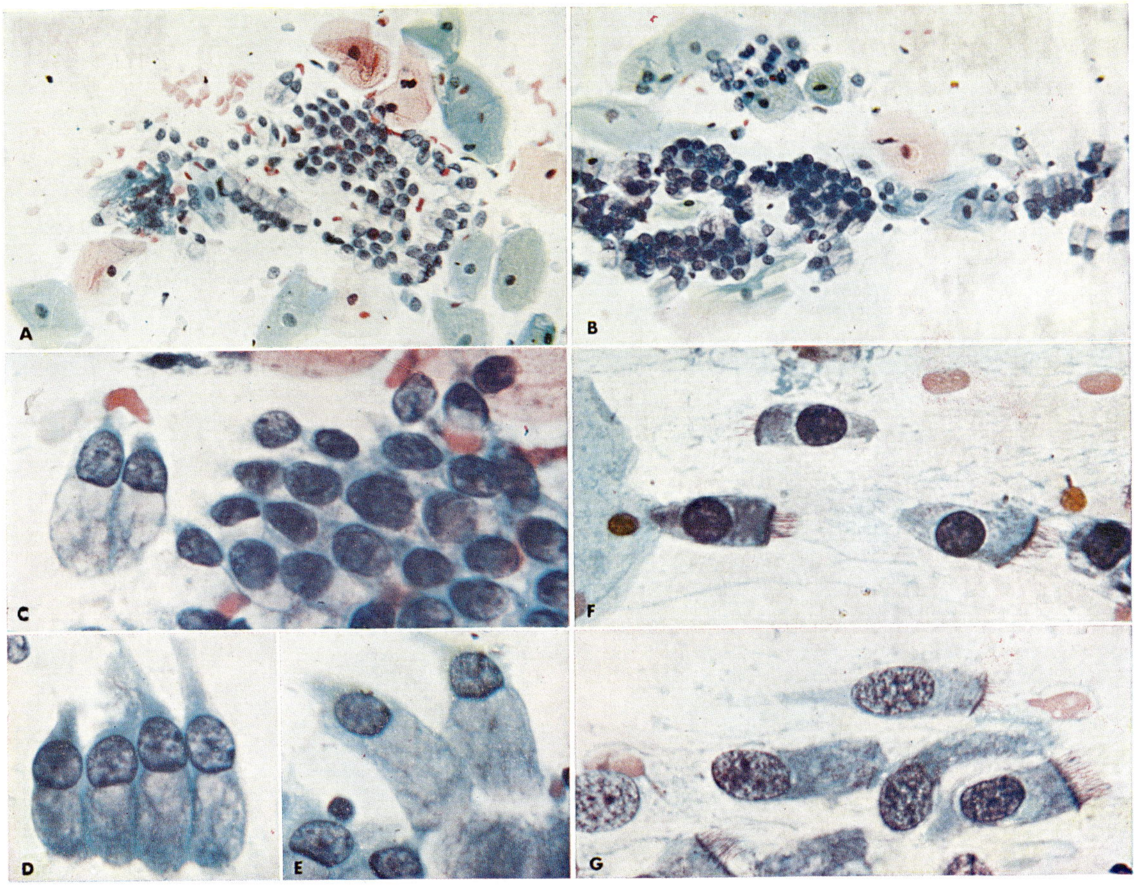

FIG. 6

Endocervical Cells

A. In the centre of the field, the columnar endocervical cells are arranged in a characteristic 'palisade' pattern. Above and to the right, cells of the same type are seen end-on in a 'honeycomb' pattern. (×200.)

B. A 'palisade' arrangement of endocervical cells is seen at the right of the field, while further 'honeycomb' groups are present on the left. (×200.)

C. The pair of endocervical cells, at the left, contain mucus which has produced slight flattening of the nuclear border. In the 'honeycomb' arrangement to the right, the cells lie in a regular linear pattern. Although the nuclei are variable in size, they are fairly uniform in shape. (See Figs. 8 and 9 where endometrial cells show irregular grouping with variability in nuclear shape but comparative uniformity in nuclear size.) (×800.)

D. A 'palisade' of mucus-containing endocervical cells showing basal tails of cytoplasm. Distinct nucleoli can also be resolved in three of the nuclei. (×800.)

E. Endocervical cells with vacuolated cytoplasm. Here the cells appear to be more spread out than in C and D. (×800).

F. Ciliated endocervical cells. The terminal plates are clearly illustrated. (×800.)

G. Ciliated endocervical cells. The chromatin pattern is slightly coarser than in the previous examples, the chromatin being evenly distributed throughout the nucleus and uniform for all the cells in the group. The cell at the top of the field clearly shows a basal tail of cytoplasm. (×800.)

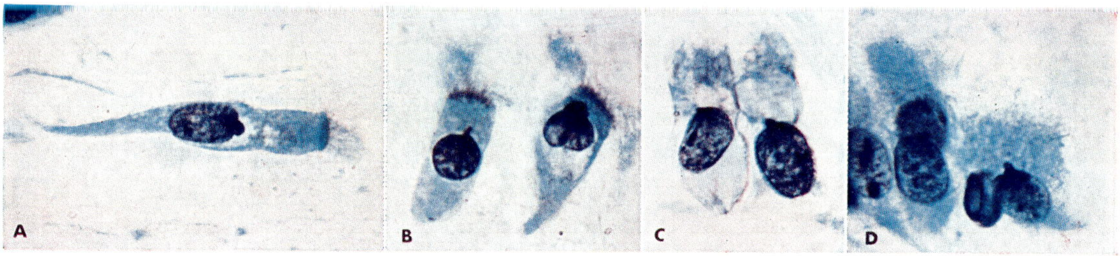

FIG. 7

Endocervical Cells with Peaked Nuclei

The oval nuclei of endocervical cells occasionally show a nipple-like protrusion or peak of nuclear material on the luminal aspect of the nucleus. This is most commonly seen at the time of maximal oestrogen activity at mid-cycle, during the child-bearing years, but also appears in menopausal patients, probably in association with unopposed oestrogen activity. This appearance may cause diagnostic difficulty by producing an irregularity in nuclear shape. In benign cells this peak is readily recognisable by its consistent relationship to the free border of the cell. These cells show normal chromatin patterns.

A. A peaked nucleus is seen in a ciliated endocervical cell. (×800.)

B. The occurence of peaking in a multinucleated cell on the right of the field produces a suspicious appearance but comparison with the change seen in the single nucleus on the left permits recognition of the typical nuclear protrusion. (×800.)
Cells shown in A and B were present at mid-cycle.

C. Peaking of nuclei is seen in mucus-secreting endocervical cells. (×800.)

D. These cuboidal endocervical cells are from a menopausal patient. A definite nuclear peak is seen in the cell to the right. There is a loss of cytoplasmic definition, which is often seen in columnar cells. (×800.)

ENDOMETRIAL CELLS

Recognition of endometrial cells is made more difficult by the fact that they are commonly seen in blood-stained smears as groups of small dark basophilic staining cells, totally irregularly arranged, with frequent overlapping of cytoplasm and nuclei. In the individual cell, the nucleus is eccentrically situated within the scanty opaque basophilic cytoplasm. The small nuclei show marked variation in shape with round, oval, crescentic or triangular forms, but are comparatively uniform in size. They show coarser chromatin clumping than endocervical nuclei, with marked basophilic chromocentres; single small eosinophilic nucleoli are occasionally seen. The nuclei are often denuded of cytoplasm.

Endometrial cells are usually present in larger numbers in vaginal smears than in cervical smears. They may, however, be seen for a day or two prior to menstruation, during the menstrual flow and for up to seven days afterwards. Their presence in smears from post-menopausal women constitutes a significant abnormality and warrants further investigation.

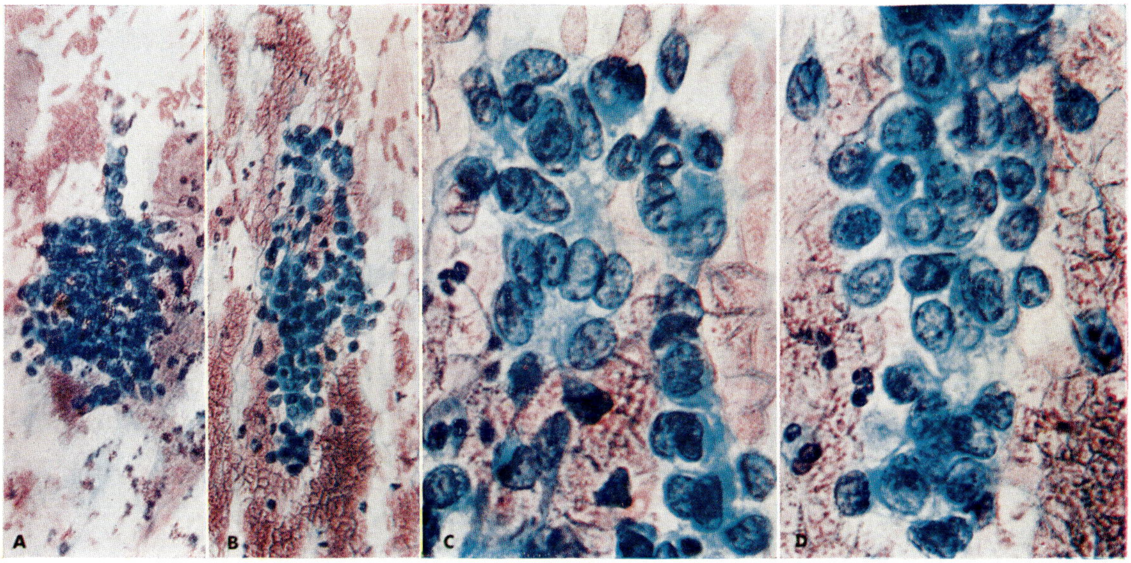

FIG. 8

Endometrial Cells

A. This group of small dark basophilic endometrial cells is seen against a background of erythrocytes. The cells are irregularly arranged and there is overlapping in the centre of the group. (×200.)

B. Although this group is less crowded than that shown in A the arrangement of the cells remains irregular and nuclear overlapping is seen. A small amount of cytoplasm is visible but the nuclei are not separated by the uniform rim of cytoplasm which is seen in the group of endocervical cells in Fig. 6 A. (×200.)

C. The nuclei of these endometrial cells show some variation in shape, with round, oval, triangular, crescentic and wedge shapes; there is little variation in nuclear size. The chromatin pattern is uniform with coarse stippling and without any large clumps; nucleoli are present in a few nuclei. When the scanty cytoplasm is identifiable the nucleus is in an eccentric position. (×800.)

D. Comparative uniformity in nuclear size and chromatin pattern is again associated with variation in nuclear shape. This overlapping, irregularly arranged group shows a typical pattern for endometrial cells. (×800.)

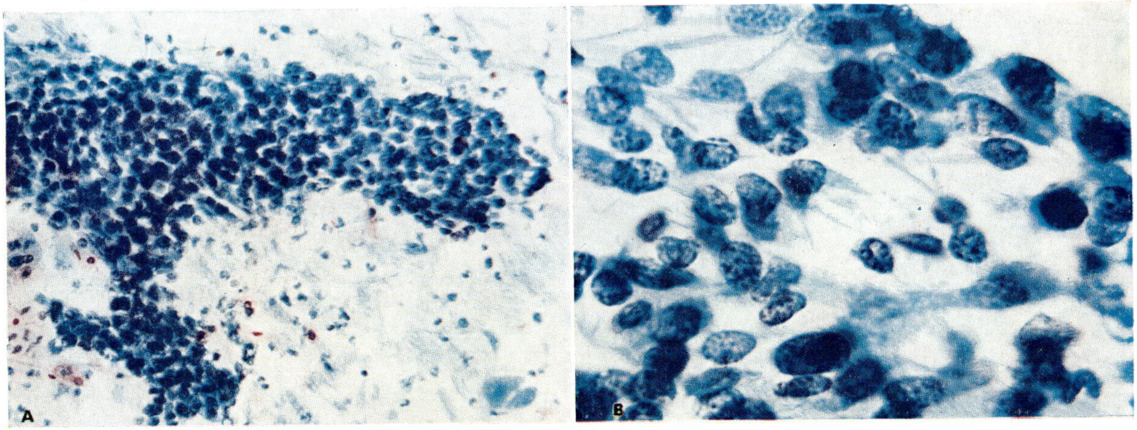

FIG. 9

Endometrial Cells

A. A large group of endometrial cells is seen; they are irregularly arranged and frequently overlap. In some areas the cells are not in sharp focus, indicating the thickness of the group. (×200.)

B. The nuclei of these endometrial cells show marked variation in shape. Oval, round, triangular, crescentic and wedge shapes are seen. The nuclei are fairly uniform in size. The chromatin is coarsely stippled but the pattern is uniform and no large clumps can be seen. Nucleoli are present but are not prominent or multiple. The cytoplasm of these cells is scanty and, where seen, it forms a narrow rim round the nucleus which is eccentric in position. (×800.)

NON-EPITHELIAL CELLS:

ERYTHROCYTES, POLYMORPHONUCLEAR LEUCOCYTES, HISTIOCYTES, LYMPHOCYTES AND PLASMA CELLS

These cells are described in standard textbooks on haematology but the description given is usually of their appearance after air drying and Romanowsky staining. Their appearance after wet fixation and Papanicolaou staining shows differences in detail, although the essential morphological appearances are unchanged.

ERYTHROCYTES

Erythrocytes usually appear as round or oval eosinophilic anucleated structures. They may, however, show basophilic blue-green staining. They appear smaller and less rounded than erythrocytes seen in air dried films.

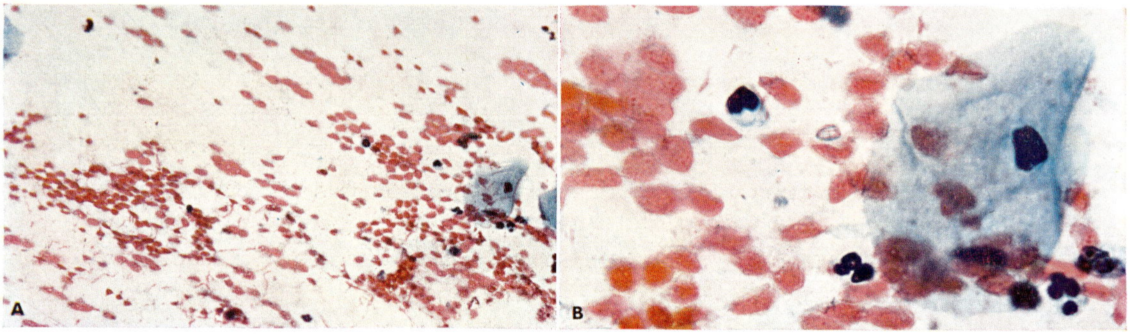

FIG. 10

Erythrocytes

A. and B. Erythrocytes lying singly and in groups are seen together with occasional polymorphonuclear leucocytes and a squamous cell. (×200.) (×800.)

Erythrocytes showing basophilic staining are illustrated in Figure 159.

POLYMORPHONUCLEAR LEUCOCYTES

The lobulated densely basophilic nuclei of these cells are their most striking feature. On careful examination the pale blue-green staining of the cytoplasm can be seen. Cytoplasmic granules are not stained and it is, therefore, impossible to subdivide these cells. In cervical smears, especially of parous patients, polymorphonuclear leucocytes are usually present in small numbers, but they are often absent from normal vaginal smears.

HISTIOCYTES

These phagocytic cells can show considerable variation and unless this is recognised histiocytes may be confused with epithelial cells and occasionally with malignant cells. Characteristically they appear as small cells, loosely arranged in groups, or lying along a strand of mucus in company with polymorphonuclear leucocytes. The borders of individual cells are ill-defined with finely vacuolated, foamy, pale staining grey-green cytoplasm. The eccentric nuclei, typically reniform but often round or oval, show coarse chromatin clumping and occasional nucleoli. There is a striking similarity, however, in the nuclear chromatin patterns of the cells in a group. Degenerate polymorphonuclear leucocytes and other ingested debris may be seen within the cytoplasm. The cell border often has an amoeboid outline.

Although histiocytes are usually smaller than, or approximately the same size as, parabasal cells, occasional forms appear which may be larger than superficial cells. In the larger forms, multinucleation occurs; usually two or three nuclei are present but the number may rise to as high as twenty-five or thirty. Histiocytes may be seen in any smear but are present in large numbers at the end of the menstrual flow and for a few days afterwards. The multinucleated giant forms are seen especially in atrophic post-menopausal smears.

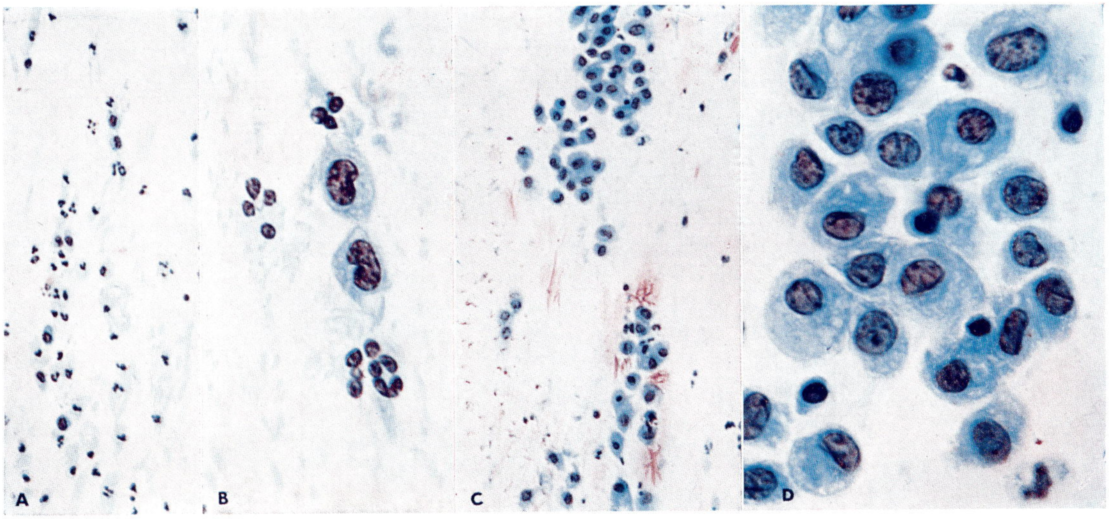

FIG. 11

Histiocytes, Polymorphonuclear Leucocytes and Lymphocytes in a Cervical Smear

A. Histiocytes are seen loosely arranged in elongated groups which are interspersed with polymorphonuclear leucocytes. (×200.)

B. These two histiocytes show the characteristic, finely vacuolated, pale basophilic cytoplasm and reniform nuclei with a coarse chromatin pattern. A few polymorphonuclear leucocytes are seen in the background. (×800.)

C. In these large groups the histiocytes are arranged in a more compact manner but still remain discrete and do not show the overlapping of cells and nuclei seen in groups of endometrial cells. Their arrangement and vacuolated basophilic cytoplasm are the features which most easily distinguish them from endometrial cells with which they could be confused. (×200.)

D. These histiocytes show the typical vacuolated cytoplasm and the wide variation in nuclear size and shape which may occur, although in a few cells the typical reniform nucleus is present. In spite of this nuclear pleomorphism there is uniformity of nuclear chromatin pattern. Occasional lymphocytes with small round deeply basophilic nuclei and a narrow rim of basophilic cytoplasm are present. (×800.)

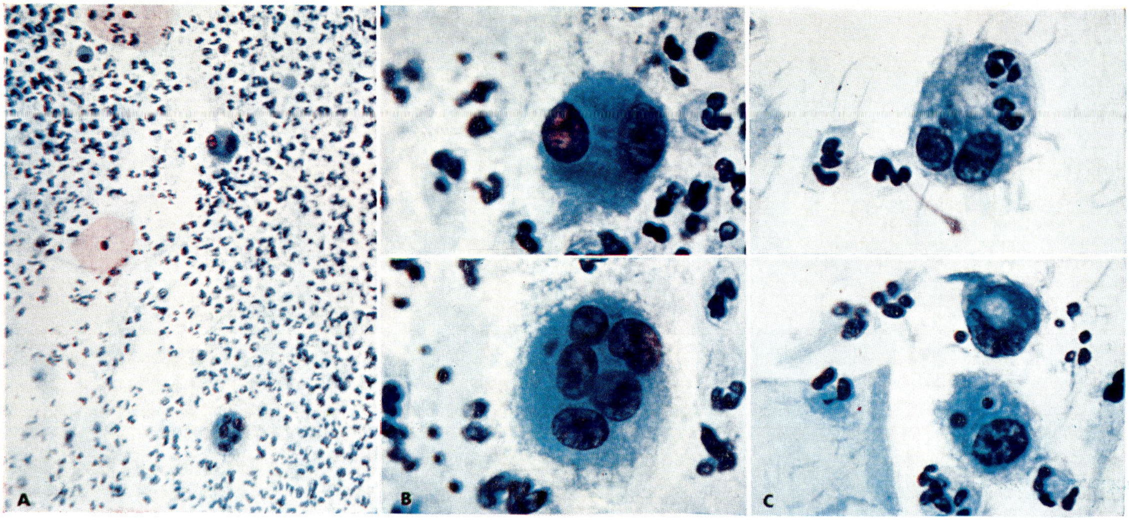

FIG. 12

Histiocytes Showing Multinucleation and Phagocytosis in a Cervical Smear

A. Two large multinucleated histiocytes are seen against a background of polymorphonuclear leucocytes. (×200.)

B. In the upper half of the field a binucleated histiocyte and in the lower half a multinucleated histiocyte are seen. In each of these cells the cytoplasm shows a lacy appearance with a poorly defined border due to cytoplasmic vacuolation. The nuclei are even in size and all show a similar rather coarse chromatin pattern. In the multinucleated cell a typical 'kidney-bean' shaped nucleus is seen (upper right). (×800.)

C. Ingested polymorphonuclear leucocytes are seen in the cytoplasm of the binucleated histiocyte. In the lower mononucleated histiocyte ingested nuclear debris is present. (×800.)

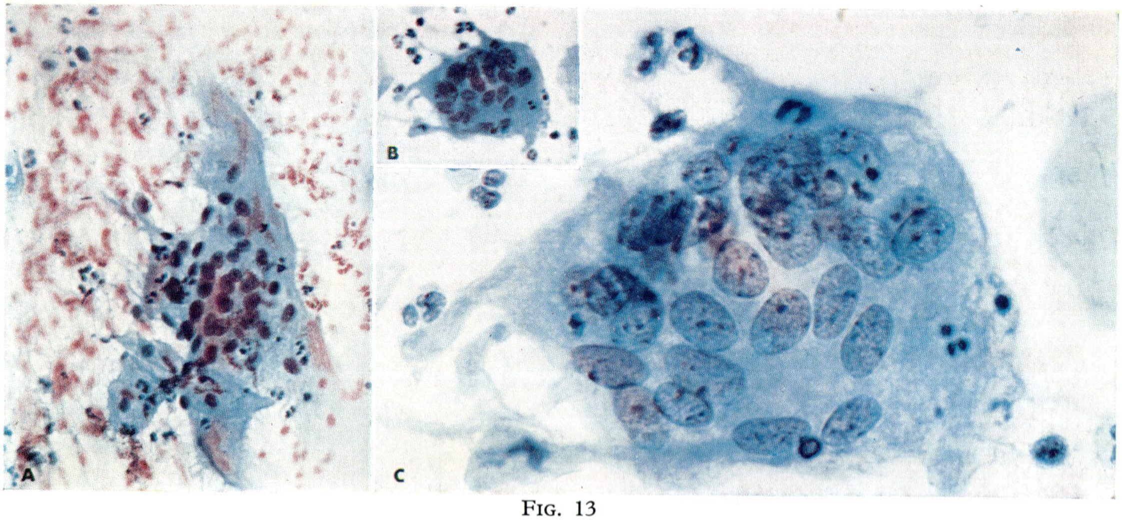

FIG. 13

Multinucleated Giant Histiocytes

These giant histiocytes are seen not infrequently in the case of the post-menopausal patient with an atrophic smear. They may also appear following vaginal surgery or radiotherapy but it is in the untreated post-menopausal patient that they are most likely to be misinterpreted.

A. The large size of this cell can be appreciated by comparing it with the intermediate squamous cells which lie below it. In spite of the bizarre size of the cell the nuclei are fairly uniform in size and shape. (×200.)

C. This high-power view of the giant cell, shown at low-power magnification (×200) in inset B, illustrates the typical features. The cytoplasm is foamy and contains ingested polymorphonuclear leucocytes and debris. Note the variation in size and shape of the nuclei including an occasional 'kidney-bean' form, the recognisable nucleoli and a rather coarsely stippled chromatin pattern. This cell demonstrates to a striking degree the amoeboid appearance of the cytoplasm. (×800.)

LYMPHOCYTES AND PLASMA CELLS

Both these cell types are rarities in vaginal smears but are occasionally found in cervical smears. Lymphocytes show the characteristic densely basophilic, often slightly wrinkled nucleus and narrow cytoplasmic border. Plasma cells, however, may cause more diagnostic difficulty; their eccentric nuclei, with punctate peripherally distributed clumps of chromatin, so called 'cart wheel' or 'clock face' nuclei, may cause confusion with small malignant cells unless plasma cells are considered in the differential diagnosis.

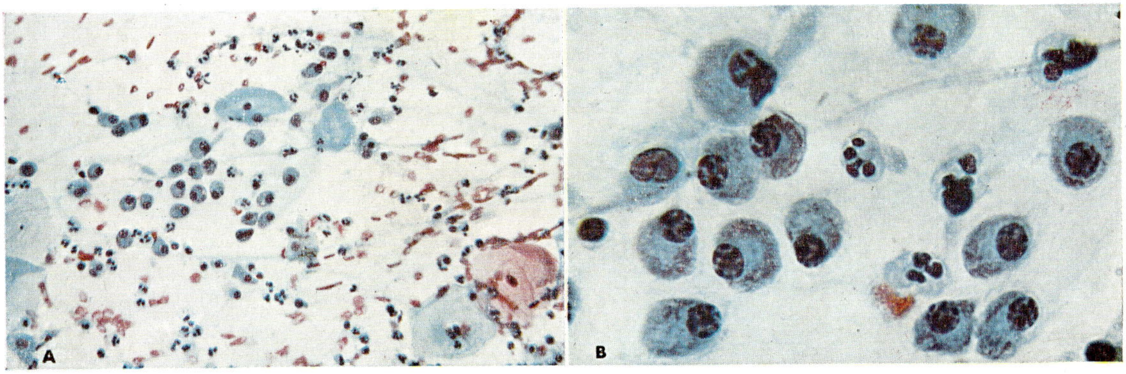

FIG. 14

Plasma Cells and Lymphocytes in a Vaginal Smear

Obtained from vault granulation tissue in a patient after total hysterectomy.

A. A group of plasma cells is seen against a background of erythrocytes, polymorphonuclear leucocytes and occasional lymphocytes. ($\times 200$.)

B. This field is composed mainly of plasma cells with uniform eccentric round nuclei and basophilic cytoplasm. The nuclei show chromatin clumping but should not be confused with those of malignant cells which would also show nuclear pleomorphism. Two lymphocytes and several polymorphonuclear leucocytes are seen. ($\times 800$.)

SPERMATOZOA

Spermatozoa are readily recognisable if well preserved, when the basophilic shaded ovoid head and body and long tail can be seen. Frequently, however, as a result of degeneration, the tail disappears and the slightly smudged outline of the head and body may cause diagnostic difficulty.

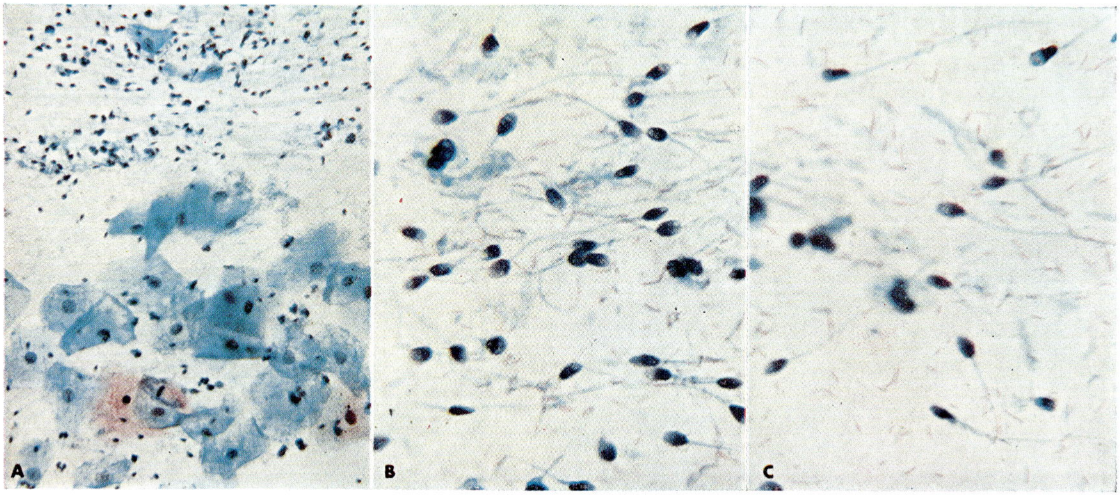

FIG. 15

Spermatozoa in a Cervical Smear

A. Large numbers of spermatozoa are present in the upper portion of the field. (×200.)

B. Well preserved spermatozoa are seen. Note the typical two-tone shaded staining of the head and body. (×800.)

C. A few degenerate spermatozoa with disappearance of the tail and enlargement of the head are present to the left of the field. (×800.)

CHAPTER 2

NON-MALIGNANT DISORDERS

CHAPTER 2

NON-MALIGNANT DISORDERS

INFLAMMATION

INFLAMMATION produces widespread changes in the appearance of the smear and of the individual cells.

CHANGES IN THE CELLULAR PATTERN

Initially, the most striking change which occurs in the presence of inflammation is an increase in the number of inflammatory cells. This increase is predominantly composed of polymorphonuclear leucocytes but, in chronic inflammation, histiocytes, lymphocytes and occasionally plasma cells may be seen. However, the squamous epithelial cellular pattern is also disturbed. In women in the child-bearing years a severe inflammatory reaction is accompanied by the appearance of a significant number of parabasal cells. In older women a different picture is seen, as an inflammatory reaction may be accompanied by an increase in the number of intermediate squamous cells and even by the appearance of appreciable numbers of superficial cells. The overall result in all age groups is a wider distribution of squamous cell types with a loss of characteristic hormonal patterns. Incidentally, this phenomenon makes inflammatory smears unsuitable for the assessment of hormonal effects.

EPITHELIAL CELL CHANGES

The epithelial cell response to inflammation has two aspects. Many epithelial cells show evidence of damage with degenerative changes and cell death. At the same time, other epithelial cells show a hyperplastic reactive response.

Degenerative changes seen in the cytoplasm include loss of definition of the cellular border, producing a curiously 'moth-eaten' appearance, alterations in the staining reaction, usually with unexpected eosinophilia but occasionally with polychromatic effects, coarse vacuolation and the appearance of phagocytic polymorphonuclear leucocytes within the cytoplasm.

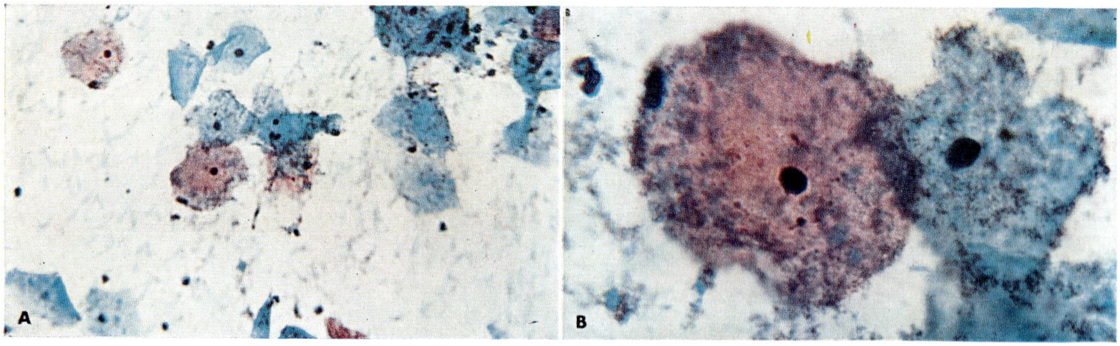

FIG. 16

Cytoplasmic Change in Squamous Epithelial Cells in Inflammation

A. The cell margins appear to be poorly defined. (×200.)

B. There is a loss of definition of the cytoplasmic border of these squamous cells producing a 'moth-eaten' appearance which represents an early degenerative change. (×800.)

Degenerative changes are also seen in the nucleus. Nuclear pyknosis may occur in intermediate and parabasal squamous cells, instead of being confined to mature superficial cells. Karyorrhexis, with coarse condensation of chromatin into large rounded densely opaque masses distributed around the margin of the nucleus, or lying free in the cytoplasm, may be seen. This change must be distinguished from the irregular angular chromatin clumping of the malignant nucleus. Karyolysis, with pale opacity of the centre of the nucleus and a relatively prominent nuclear membrane, must be differentiated from the irregular marginal condensation of chromatin found in the malignant nucleus.

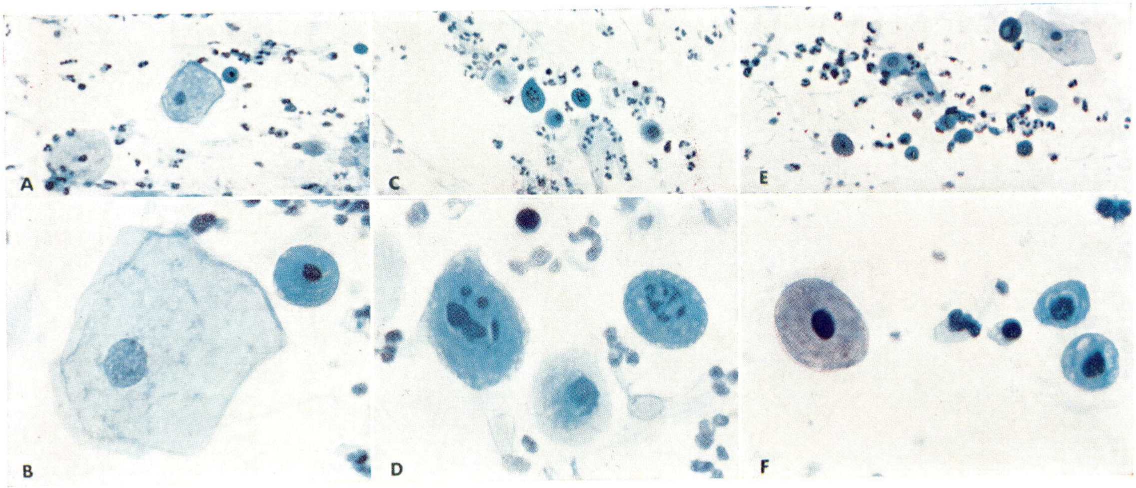

FIG. 17

Degenerative Changes in Nuclei of Squamous Epithelial Cells in Inflammation

A. Against a background of polymorphonuclear leucocytes, a superficial, an intermediate and a parabasal squamous cell are seen. The nucleus of the parabasal cell is similar to that of the superficial cell. (×200.)

B. The parabasal squamous cell shows nuclear pyknosis, the nucleus being small, hyperchromatic and homogeneous. (×800.)

C. and D. The nuclei of the parabasal squamous cells (centre) appear to be irregularly hyperchromatic. At higher magnification (D) they show karyorrhexis, the chromatin having coalesced into coarse, but rounded, hyperchromatic masses around the periphery of the nuclei. This degenerative change must be differentiated from the chromatin clumping seen in malignant nuclei. (×200.) (×800.)

E. This field shows a predominance of parabasal squamous cells with a background of polymorphonuclear leucocytes. (×200.)

F. The parabasal cells all show nuclear pyknosis. Cytoplasmic changes are also present, e.g., polychromatophilic staining and vacuolation. (×800.)

In contrast, squamous and columnar cells may show evidence of nuclear activity. Nuclei, while retaining their smoothly rounded shape, may show enlargement. Multinucleation, with striking similarity between the nuclei, can occur and, occasionally, there is an increase in the coarseness of the chromatin pattern. Small nucleoli may be seen in squamous cells or may become more prominent in columnar cells; however, despite these active nuclei, in benign conditions the cells of a group retain their uniformity.

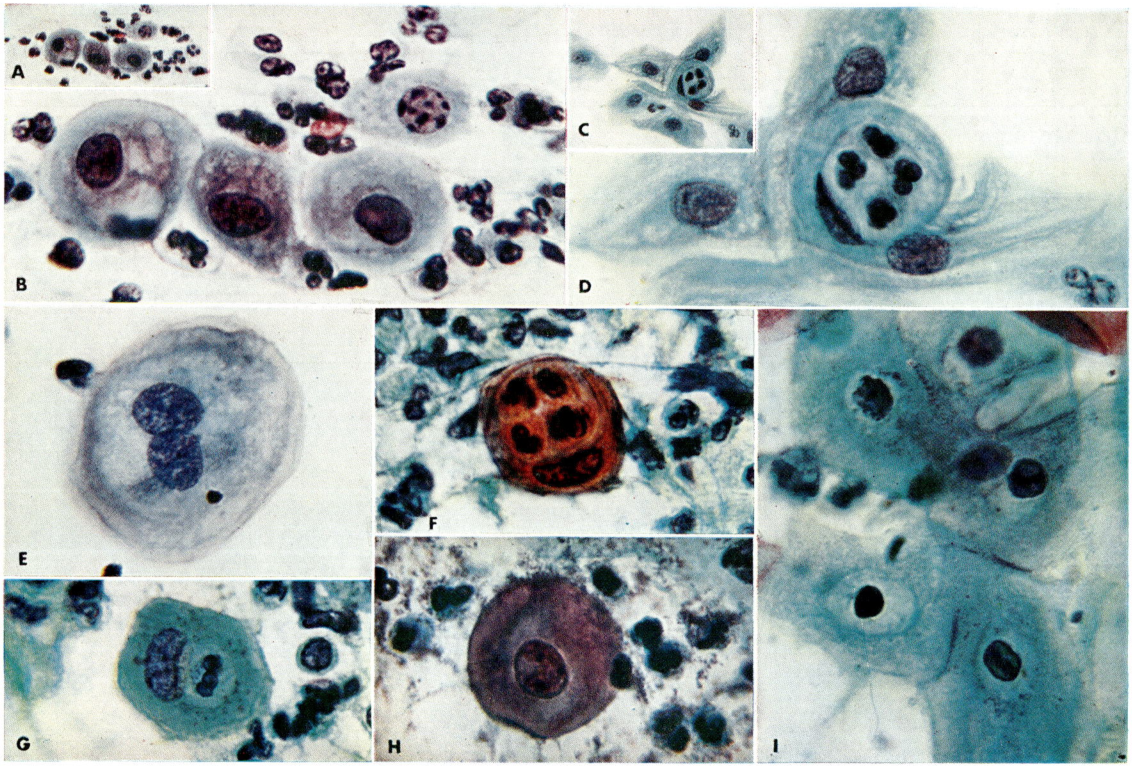

FIG. 18

Changes seen in Benign Parabasal and Intermediate Squamous Cells in Inflammation

A. and B. A group of parabasal cells against a background of polymorphonuclear leucocytes. At high magnification (B) coarse vacuolation of the cytoplasm is obvious. A single large plasma cell is present in the upper right quadrant. (×200.) (×800.)

C. and D. A large vacuole lies within the cytoplasm of the parabasal cell which at high power magnification is seen to contain four polymorphonuclear leucocytes. The cell nucleus is flattened and eccentric in position. (×200.) (×800.)

E. This parabasal cell shows multinucleation, the two vesicular nuclei being uniform in size and shape. The chromatin is evenly distributed in both nuclei. (×800.)

F. Eosinophilic staining of the cytoplasm is seen in a parabasal cell. The cytoplasm is also infiltrated by polymorphonuclear leucocytes. The nucleus is flattened and eccentric in position. (×800.)

G. A single vacuole containing a polymorphonuclear leucocyte distorts the nucleus of a parabasal cell to a 'kidney-bean' shape similar to that of a histiocyte. The dense opaque cytoplasm distinguishes the parabasal cell from the histiocyte with its finely vacuolated cytoplasm (see Fig. 11). (×800.)

H. This polychromatic staining parabasal cell shows loss of definition of the cellular border with a sharply defined heavily stained nuclear border accompanied here by a loss of chromatin pattern within the nucleus, an appearance due to degeneration (see Fig. 41 where condensation of chromatin at the nuclear border is accompanied by the coarse chromatin clumping seen in malignant cells). (×800.)

I. Clear perinuclear vacuoles or 'haloes' are seen in these intermediate cells. The lower left hand cell shows nuclear pyknosis. (×800.)

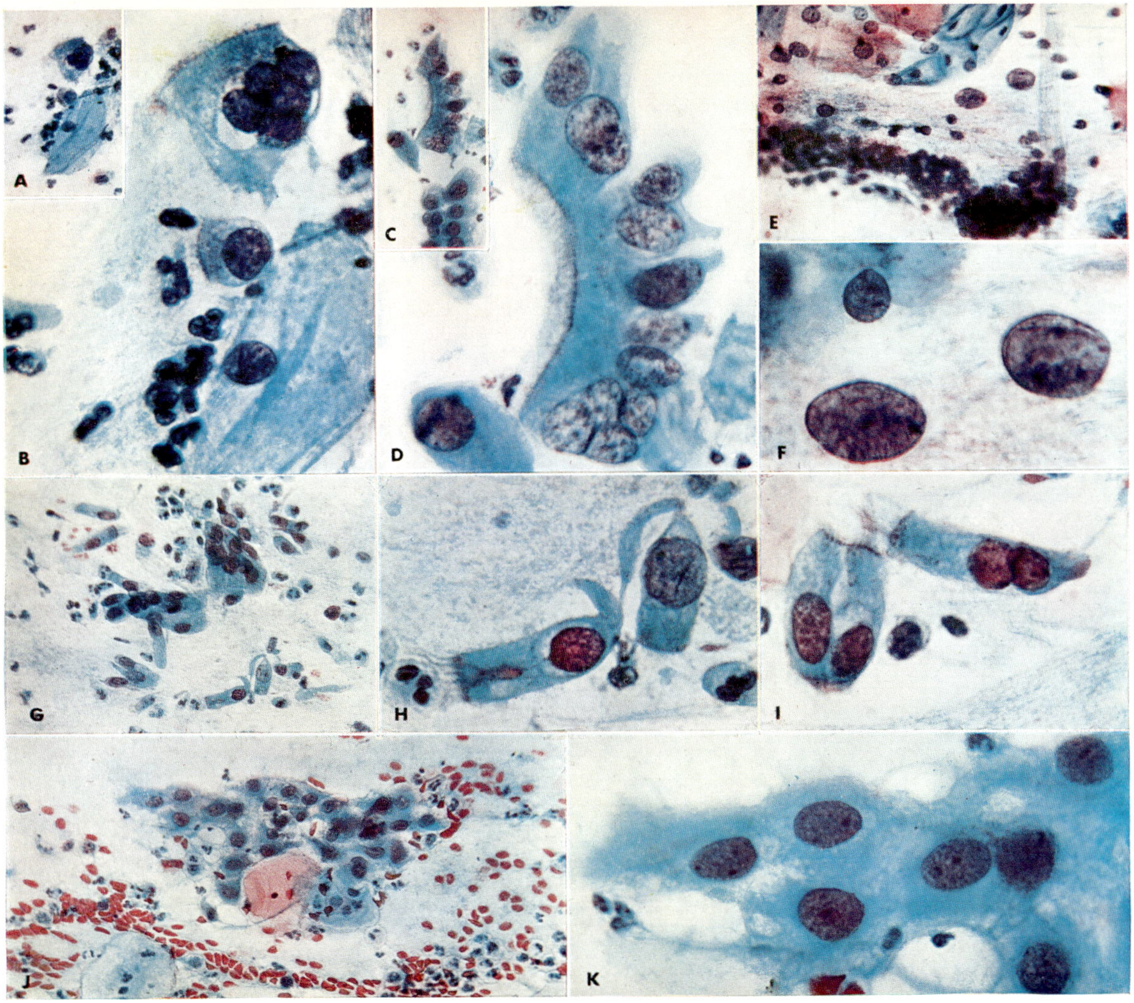

FIG. 19

Inflammatory Changes in Endocervical Cells

A. and B. The upper columnar cell appears to have a large irregular nucleus which, under higher magnification (B), is seen to be due to multinucleation. The nuclei are even in size and shape and a perinuclear vacuole is present. The terminal plate is recognisable although the cilia have almost disappeared. (×200.) (×800.)

C. and D. The lower group of cells shows the usual regularly arranged grouping of columnar cells. In the upper group the nuclei are irregularly arranged and at higher magnification (D) this is seen to be mainly due to multinucleation in the lowest cell. The nuclei show considerable variation in size but are uniformly oval except where they are distorted by close juxtaposition. Well preserved cilia and terminal plates are seen. (×200.) (×800.)

E. Both large and small nuclei of endocervical cells are seen to be denuded of cytoplasm. (×200.)

F. The three denuded nuclei show a striking variation in size but are uniform in shape and chromatin pattern. This degree of size variation is seen not infrequently in endocervical cells and is not significant unless associated with other atypical features. (×800.)

G. This group of endocervical cells shows multinucleation and perinuclear vacuoles. (×200.)

H. A vacuole forms a perinuclear halo in the lower cell. Note the well preserved cilia and terminal plate. (×800.)

I. One multinucleated cell shows a pair of uniformly matched nuclei. The cell adjacent to it contains a perinuclear vacuole. Cilia are apparent. (×800.)

J. and K. This sheet of cells shows squamous metaplasia (q.v.). A single columnar cell below and to the left of the group suggests that they originated from endocervical epithelium. The nuclei show variation in size but are uniform in shape and staining reaction. At higher magnification (K) the variation in size is more apparent and the chromatin pattern is seen to be similar to that of the endocervical cells. (×200.) (×800.)

The occurrence of any single abnormal feature is insufficient evidence on which to suspect malignant change. Careful examination of the nuclei, which remain uniform in shape and chromatin pattern in inflammation, differentiates these cells from the malignant cells of endocervical origin seen in Fig. 51. When cilia and/or a terminal plate are clearly seen, they provide additional evidence of the benign character of the cell as this degree of differentiation is not seen in malignant cells.

CYTOLYSIS

When cytolysis occurs the epithelial cells lose their cytoplasm in varying degrees while their nuclei remain clearly recognisable. This change may accompany inflammation, but it may also occur without any evidence of inflammatory change in the cells and without an inflammatory cell infiltrate. The denuded nuclei which are seen may cause difficulty in differential diagnosis from small undifferentiated malignant cells, although once this entity is recognised, the uniformity and normal chromatin pattern of denuded nuclei preclude a diagnosis of malignancy.

There are three types of epithelial cell which are especially likely to show cytolysis. The delicate cytoplasm of columnar cells is readily lost and denuded endocervical cells are often seen. They may give rise to difficulty in diagnosis by showing variation in size while remaining uniform in other respects but intact columnar cells can usually be identified in, or near, the group. In atrophy, when a smear is mainly composed of parabasal cells, denuded nuclei may be present. Occasionally, these nuclei form compact clumps, which can present a diagnostic difficulty, especially as they may show nucleoli; careful examination will reveal their similarity to the nuclei of adjacent sheets of parabasal cells. Perhaps the most striking degree of cytolysis may be shown by the intermediate cells in a pregnancy smear; lactobacilli are prominent in these preparations. Here the only difficulty in diagnosis is caused by the elongation of the vesicular nuclei of some of the folded intermediate cells, but these narrow nuclei show normal chromatin patterns.

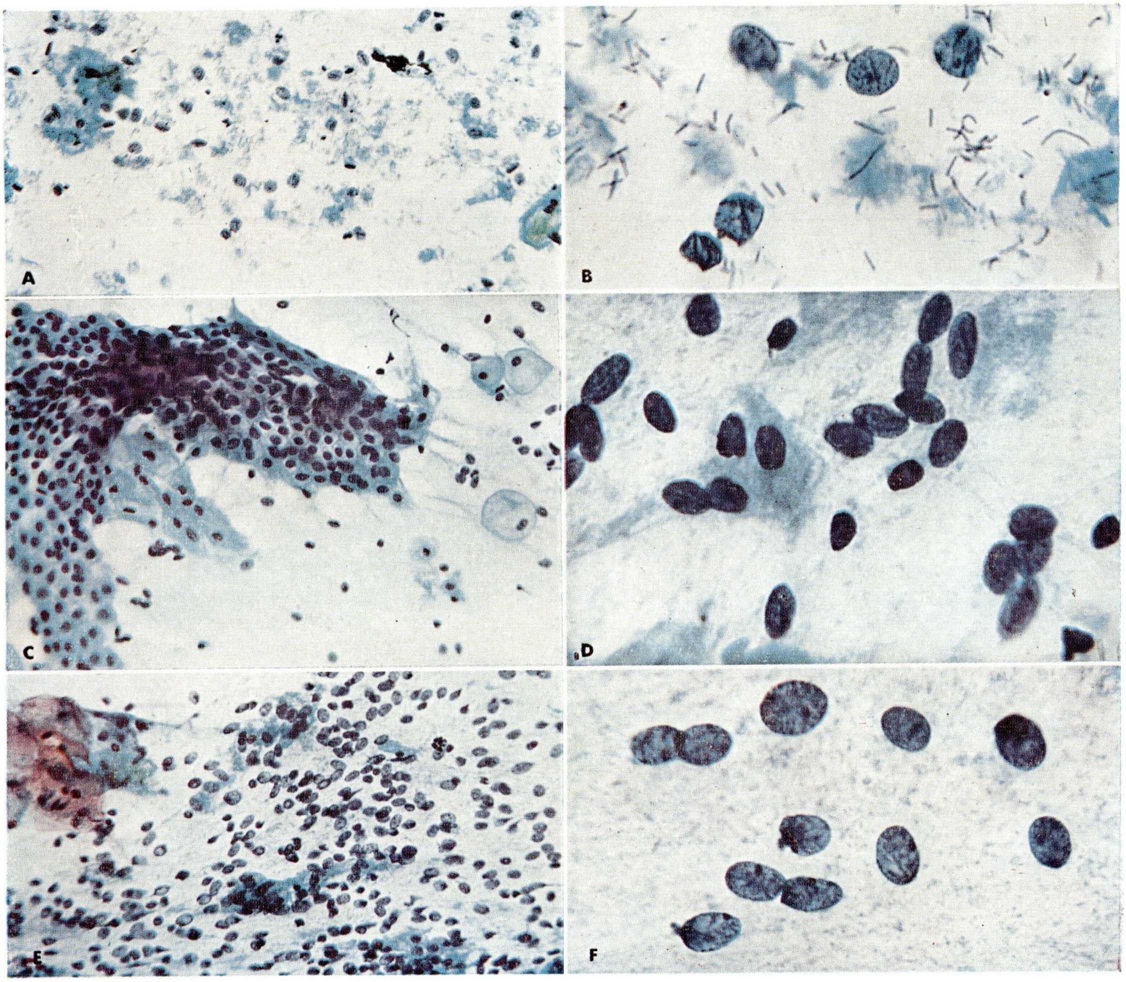

FIG. 20

Denuded Benign Nuclei in Cervical Smears

A. Initially this smear appears 'dirty'. Only a few intact intermediate squamous cells are seen (left). The rest of the smear consists of denuded nuclei and fragments of cytoplasm. Only an occasional polymorphonuclear leucocyte is present and there is no evidence of inflammation. (×200.)

B. The denuded nuclei have smooth oval outlines, are uniform in size and their chromatin is finely stippled. Fragments of cytoplasm from the intermediate cells are seen. Döderlein's bacillus (*Lactobacillus* species) is present in large numbers. Intermediate cells in the presence of Döderlein's bacillus readily undergo cytolysis. This smear was taken from a pregnant patient. (×800.)

C. The sheet of intermediate and parabasal squamous cells indicates the origin of the denuded nuclei seen scattered throughout the rest of the field. Sheets of epithelial cells in smears from the atrophic vagina and cervix frequently show cytolysis. (×200.)

D. These nuclei are uniform and smoothly oval in shape and have finely stippled chromatin. Cytoplasmic remnants are obvious. This smear was taken from a post-menopausal patient. (×800.)

E. and F. This smear consists mainly of denuded nuclei which show little evidence of their origin but close scrutiny reveals a number of intact columnar endocervical cells lying among them. Columnar cells frequently show cytolysis with a complete loss of cytoplasm. At higher magnification (F) the denuded nuclei show striking uniformity in shape and chromatin pattern . Their variation in size is compatible with their endocervical origin. (×200.) (×800.)

SQUAMOUS METAPLASIA

In the presence of inflammation epithelial cells showing the features of both squamous and columnar types may be seen. They are derived from the germinal, or reserve, cells which lie along the basement membrane of the endocervical glands. Approximately the same size as parabasal cells, they share their opaque adequate cytoplasm but are more elongated in shape and frequently show a tail of cytoplasm. The nuclei are oval and vesicular, uniform in shape but show variation in size; small nucleoli are often present. An occasional cell in the group may be frankly columnar or groups of columnar cells may be closely adjacent. If this change is not identified correctly the size variation of the nuclei, which are otherwise uniform, may be confused with a significant abnormality in squamous cells.

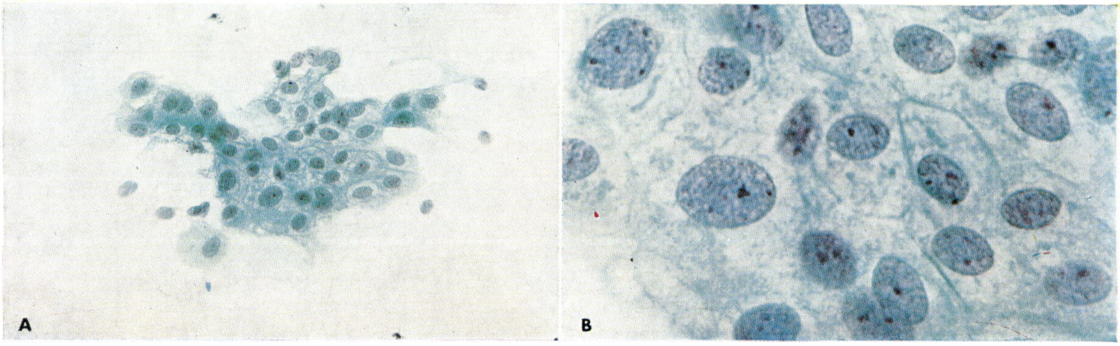

FIG. 21

'Squamous Metaplasia' in a Cervical Smear

A. This sheet of benign squamous cells is composed of angular and somewhat elongated cells with finely vacuolated cytoplasm and large vesicular nuclei. The smooth surface along the upper right border suggests that the group may have formed a part of the lining of an endocervical gland. (×200.)

B. The elongated angular cellular borders are seen in the centre of the group. All the cells show fine cytoplasmic vacuolation, the oval vesicular nuclei varying in size. Prominent nucleoli are present. These sheets of 'metaplastic' cells require careful scrutiny because of their variable nuclear size and prominent nucleoli. However their characteristic grouping, elongated cellular outline, uniform smoothly ovoid nuclei and fine chromatin network establish their benign character. (×800.)

The term 'squamous metaplasia' is a misnomer as these squamous cells are not present as a result of a transformation in mature columnar endocervical cells. They are derived from squamous maturation of the reserve, or basal, cells of the endocervical epithelium. In consequence they share with columnar endocervical cells an elongated outline, vacuolation of cytoplasm and variation in size of the uniformly oval nuclei. Their grouping, with smooth luminal surfaces, is also in accordance with an endocervical origin. Frequently these 'metaplastic' cells are found near groups of endocervical cells (see Fig. 22).

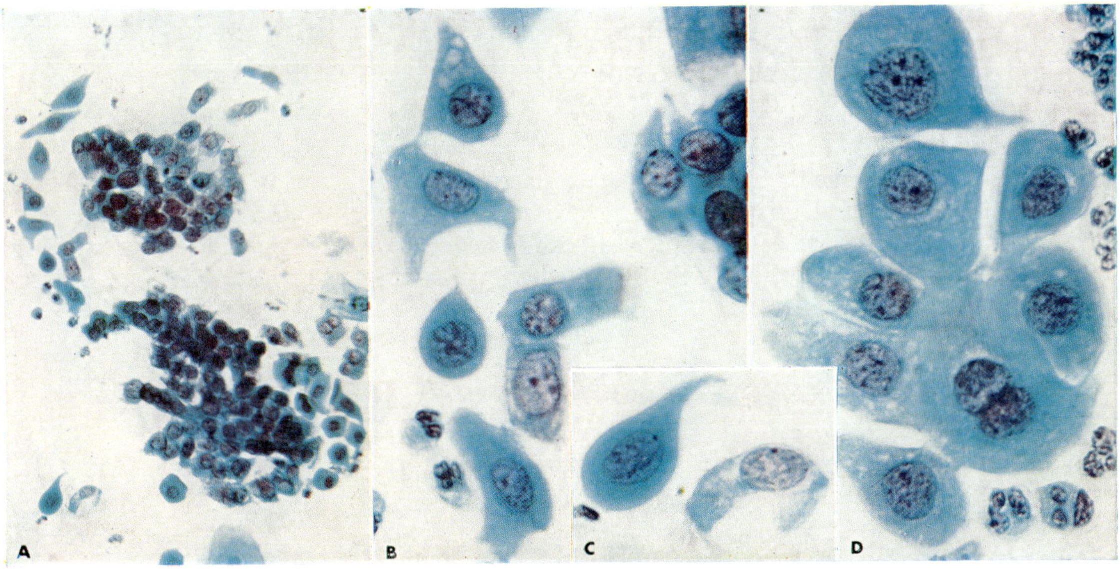

FIG. 22

Endocervical Cells Showing 'Squamous Metaplasia'

A. Most of the cells in the two large groups are recognisable as columnar endocervical cells arranged in a linear manner with uniform oval basal nuclei and elongated cytoplasm. Scattered around the two main groups are small groups and single cells of two different types. Typical columnar endocervical cells are readily identified. The other cells are elongated squamous cells with oval vesicular nuclei which contain prominent nucleoli. (×200.)

B. The squamous cells on the left vary in size and shape. Their nuclei are similar to those of the two columnar endocervical cells in the centre of the field. Nucleoli are seen in several cells. (×800.)

C. The cell on the left shows 'squamous metaplasia'; its nucleus resembles that of the ciliated columnar endocervical cell on the right. (×800.)

D. The cytoplasm of several cells shows fine vacuolation. The nuclei are variable in size but uniformly oval and contain prominent nucleoli. When compared with the 'metaplastic' cells in Fig. 21 these nuclei show slightly coarser chromatin clumping. Slight irregularity in nuclear outline is seen in the cell at the top of the group and multinucleation is present in the cell in the lower right quadrant. These features are associated with the presence of inflammation. Note the numerous polymorphonuclear leucocytes. (×800.)

Inflammation is one of the factors associated with 'squamous metaplasia'. When present it may cause mild nuclear atypia in the 'metaplastic' cells.

LEUKOPLAKIA OF THE CERVIX

The clinical appearance of a white patch on the cervix is accompanied by the presence of yellow or orange stained anucleated keratinised squames in the smear. Their significance, if they predominate in the smear, is that this keratinised layer may have prevented the collection of material from deeper cellular layers which would yield cells more suitable for diagnosis. Under these circumstances examination of a further specimen would be advisable.

CERVICAL AND ENDOMETRIAL POLYPI

It is not usually possible to make or confirm the diagnosis of cervical or endometrial polypi on cytological grounds. Occasionally, however, a papillary group of columnar cells may appear to be compatible with such a diagnosis. The surface of an endocervical polyp may show active squamous metaplasia and this may be reflected in the smear. When a polyp is large, it is advisable to take further smears after it has been removed because it may have prevented the clinician from scraping the whole of the area of the squamo-columnar junction.

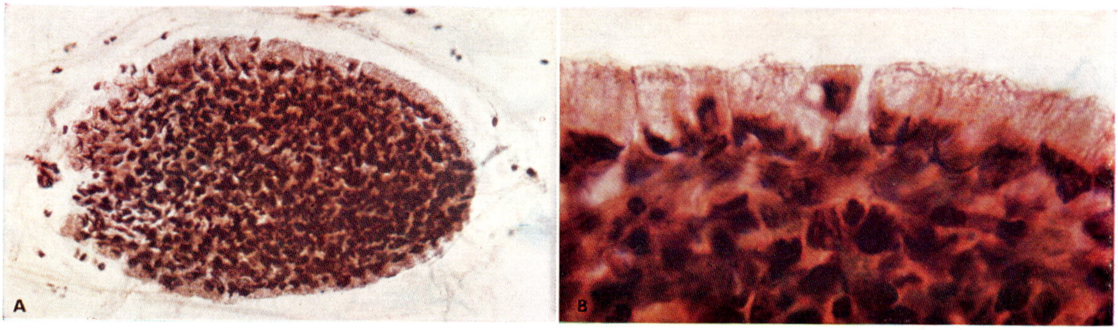

FIG. 23

Benign Endometrial Polyp

A. This group of endometrial cells shows a core of irregularly arranged endometrial stromal cells covered by a single layer of cuboidal endometrial glandular cells. (×200.)

B. The endometrial stromal cells are irregularly arranged and show some variation in nuclear configuration. The cuboidal surface cells tend to be more regularly arranged but do not achieve the strikingly uniform pattern seen in endocervical cells. Unless a polypoid fragment is present in a specimen, endometrial cells usually appear predominantly as stromal cells with only an occasional cuboidal glandular cell present on the surface of a group or lying singly in the smear. (×800.)

EROSION OF THE CERVIX

This extension of the columnar epithelium outside the endocervical canal may be reflected in the smear by the presence of many large groups of endocervical cells which frequently show squamous metaplasia. Usually, too, the smear will show evidence of inflammation in both squamous and endocervical cells and the presence of increased numbers of inflammatory cells.

SPECIFIC INFLAMMATORY CONDITIONS

TRICHOMONAS VAGINALIS

Infestation by this protozoon is common, especially in symptomatic hospital patients. Its recognition is important to the clinician for specific treatment of symptoms which may be associated with its presence. In addition, infestation may be accompanied by epithelial changes which may mask or mimic, for the inexperienced observer, the appearance of malignant cells. The epithelial cellular changes associated with trichomonal infection include features already described. Certain changes are particularly striking; the cytoplasm of squamous cells frequently shows eosinophilic or polychromatic staining and perinuclear vacuoles, forming haloes, are prominent. Karyolysis, with a marked nuclear border, may arouse suspicion until the characteristically 'empty' nucleus is identified. Alteration of the distribution pattern of the squamous cells is frequently seen and this may be marked in post-menopausal women, who may show an abnormal maturation effect. The background of the smear often shows a finely granular ('ground glass') appearance in the presence of trichomonal infection.

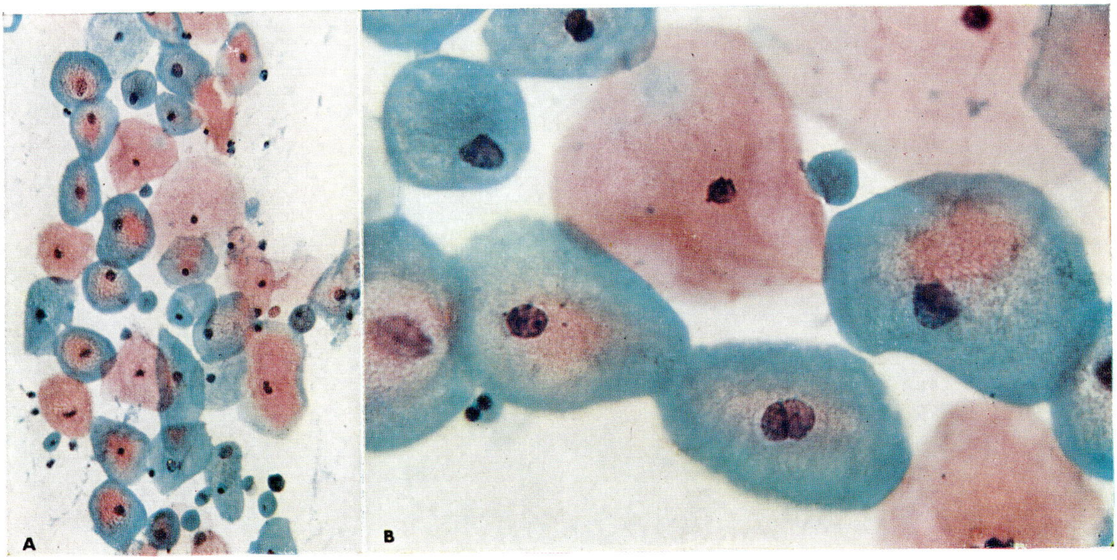

FIG. 24

Abnormal Cytoplasmic Staining in Trichomonal Infection

A. In this group of superficial and intermediate squamous cells both eosinophilic and basophilic staining of cytoplasm can be seen in the same cell. *Trichomonas vaginalis* is readily recognisable and the trichomonads are present in appreciable numbers especially in the lower half of the field. (×200.)

B. The intermediate squamous cells show areas of eosinophilic staining in a predominantly basophilic cytoplasm. A single trichomonad is present to the right of the eosinophilic superficial squamous cell. (×800.)

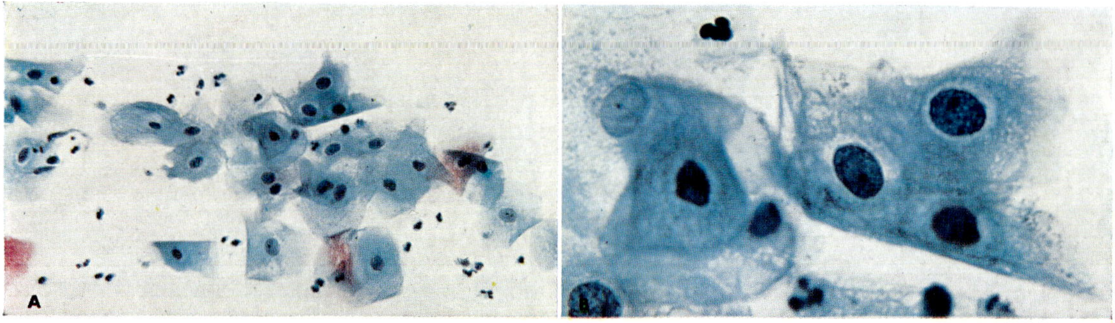

FIG. 25

Perinuclear Vacuoles in Trichomonal Infection

A. Many of these benign intermediate squamous cells show 'haloes' of clear, translucent or pale staining cytoplasm around the nuclei. A single trichomonad is present in the upper left quadrant but is not seen sufficiently clearly for definitive diagnosis. Polymorphonuclear leucocytes are numerous. (×200.)

B. Sharply defined clear or pale 'haloes' of cytoplasm are seen around the nuclei. The cytoplasm also shows a loss of definition of its peripheral border producing a 'moth-eaten' appearance. (×800.)

The perinuclear vacuole or 'halo' is seen in response to many types of infection. It is, however, especially common and well defined in trichomonal infection. When seen it should lead to a careful search for the organism. The halo may be seen around atypical nuclei, and it is then especially important to recognise the presence of *Trichomonas vaginalis* as it may be a contributory factor in nuclear atypia or, alternatively, the inflammatory changes associated with its presence may mask the diagnostic features of malignant cells.

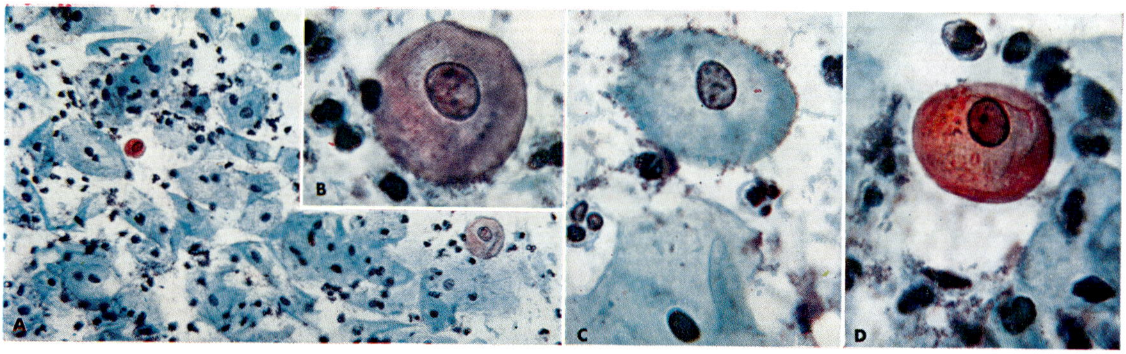

FIG. 26

Nuclear Changes seen in Trichomonal Infection

A. An eosinophilic and a polychromatophilic parabasal cell both show prominent hyperchromatic nuclear borders. (×200.)

B. Although the nuclear border is strikingly prominent the nucleocytoplasmic ratio is normal, the nucleus retains an oval shape and, within the heavy border, appears 'empty' with a loss of chromatin pattern. (×800.)

C. The parabasal cell shows a prominent nuclear border with a loss of internal nuclear structure. This cell also shows inflammatory changes in the cytoplasm with a perinuclear 'halo' and loss of definition of the cellular border. (×800.)

D. The brightly eosinophilic cytoplasm of this parabasal cell combined with its hyperchromatic nuclear border make careful examination necessary. The nucleocytoplasmic ratio is normal, the nucleus rounded and the chromatin pattern absent, leaving it typically opaque and empty. (×800.)

Although condensation of chromatin at the nuclear border occurs in malignant cells it is not evidence of malignant change when it occurs alone. It is then more likely to be evidence of degeneration. This phenomenon is commonly seen in inflammation especially when this is due to trichomonal infection.

Although certain features in a smear may suggest trichomonal infestation, the protozoon must be identified for a definitive diagnosis. Trichomonads appear as discrete round or oval bodies, varying in size from that of a small histiocyte to that of a parabasal cell. Their cytoplasm stains grey-green or grey-blue, a similar colour to the cytoplasm of intermediate squamous cells; at high magnification the cytoplasm is seen to contain tiny red granules. The tiny eccentric nucleus stains blue and sometimes appears refractile; the presence of this characteristic nucleus is essential for diagnosis. In the differential diagnosis one must consider cytoplasmic fragments, distinguished by their poorly defined outlines and variability of size and shape; denuded vesicular nuclei, differentiated by their sharply defined nuclear border and deeply basophilic staining; degenerate polymorphonuclear leucocytes, which are deeply basophilic and lack the characteristic tiny nucleus; and degenerate histiocytes and parabasal squamous cells, which stain similarly but are more rounded in outline and do not have the characteristic nucleus.

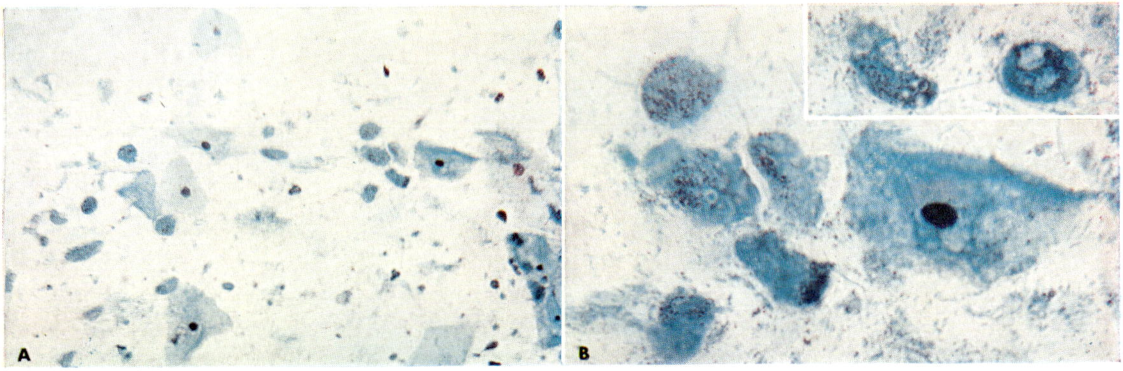

FIG. 27

Trichomonas Vaginalis

A. *Trichomonas vaginalis* is present in large numbers. They appear as multiple ovoid or pear-shaped structures their size varying from that of a small histiocyte to that of a parabasal squamous cell. The staining is lightly basophilic, (grey-blue) similar in density to that of the cytoplasm of the intermediate squamous cells. Several of the squamous cells show the clear perinuclear 'halo' frequently seen in trichomonal infection. (×200.)

B. A group of trichomonads is seen to the left of the squamous cell. They show the typical definite but not linear outline, are ovoid or pear-shaped, of variable size with lightly basophilic staining cytoplasm which is stippled with eosinophilic granules. The tiny round nucleus is clearly seen in the central trichomonad of the group and in the two in the inset. The nucleus, which is situated in an eccentric position and stains a very pale blue, may occasionally appear to be refractile. Note the typical granular ('ground glass') appearance of the background. (×800.)

Although a causal relationship has not been demonstrated between trichomonal infection and carcinoma of the cervix, there is a casual one probably due to common factors such as trauma to the epithelium and non-specific infection. In cytologic diagnosis the recognition of trichomonal infection has further significance. Its presence can produce changes in epithelial cells which may mask malignant cells by producing cellular degeneration and cytolysis. Alternatively trichomonal infection may produce cellular atypia which can mimic some of the cellular changes seen in carcinoma of the cervix.

CANDIDA ALBICANS (MONILIA)

This fungal infection, though far less common than trichomonal infestation, is found quite frequently in smears taken at ante-natal or post-natal clinics. Occasionally it is found in the post-menopausal patient who should then be checked for glycosuria. Recently fungal infection has been noted more frequently among younger non-pregnant patients. Possible explanations of this phenomenon include the reported appearance of fungal infections in patients taking progestational agents for contraceptive purposes and the increasing use of antibiotic therapy.

The characteristic brightly eosinophilic hyphae and conidia are readily recognisable but are often present only in very small areas of the smear. Smears showing marked fragmentation of the nuclei of polymorphonuclear leucocytes should be especially carefully searched for the fungus.

The presence of trichomonal infestation does not exclude the simultaneous presence of monilia, and treatment of trichomonal infection may be followed by the appearance of monilia in the vaginal exudate.

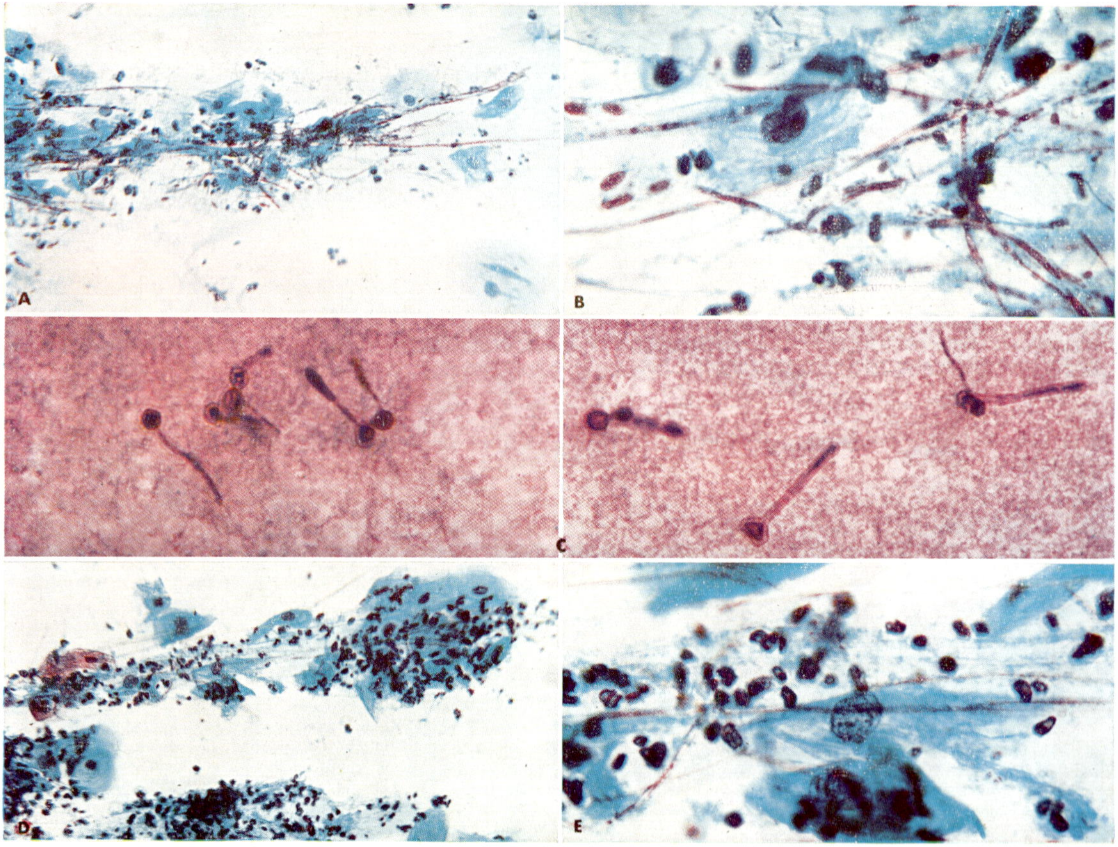

FIG. 28

Candida Albicans (Monilia) in a Cervical Smear

A. The fungus appears as long filaments of brightly eosinophilic material with adjacent small ovoid unicellular eosinophilic structures. (×200.)

B. Pseudo-hyphae and unicellular yeast bodies are now discernible at this high magnification. (×800.)

C. In this case the fungus was identified specifically as *Candida albicans* (*monilia*) by culture of the fungus in fresh human serum at 37°C. for two hours. The resulting characteristic germ tubes, producing a 'hand-mirror' appearance, are clearly seen. (×800.)

D. and E. In the presence of a fungal infection degeneration of polymorphonuclear leucocytes is often a prominent feature. The presence of the densely basophilic degenerate nuclear material seen here led to further high power examination (E) which revealed that branching hyphae and yeast bodies were, in fact, present among the debris of the polymorphonuclear leucocytes. (×200.) (×800.)

NON-SPECIFIC INFECTION

Although the presence of bacteria may be noted in many inflammatory smears, it is not possible to make a morphological bacteriological diagnosis on Papanicolaou stained smears alone. Gram stained preparations and culture are necessary for definitive bacteriological diagnosis.

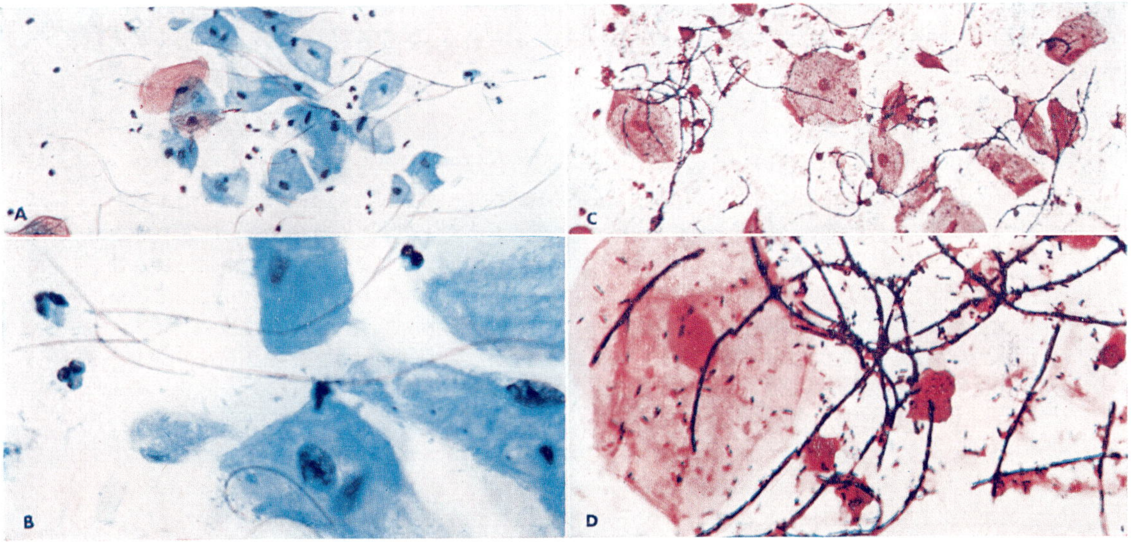

FIG. 29

Filamentous Micro-organisms

A. and B. In this Papanicolaou stained specimen eosinophilic filamentous micro-organisms which appear to show branching are seen. These might be confused with the presence of a fungus but the filaments are more slender than the hyphae of a fungus and budding is not seen. (×200.) (×800.)

C. and D. This Gram stained specimen shows filamentous and chain forms of a Gram positive bacillus. In plate D interlacing of filaments is shown. The appearance is compatible with the presence of organisms of the group *Lactobacillaceae*. (×200.) (×800.)

CHAPTER 3

THE CYTOLOGICAL DIAGNOSIS OF CANCER

In the examination of a smear, no single feature of a cell or its nucleus is diagnostic of malignant change. Success depends on careful evaluation of the appearance of individual cells and nuclei, of groups of cells and of the overall appearance of the smear.

Loss of uniformity is the most striking characteristic of malignant cells and their nuclei. This is most significant when it is the nuclei which are irregular in size, shape and chromatin pattern, but irregularity in cell size, shape and staining reaction is often the most striking feature. Overall changes in the smear are the least dependable evidence of an abnormality, but may be useful in arousing suspicion, or in supporting a diagnosis based on nuclear and cellular changes.

The Nucleus of a Malignant Cell

Nuclei of malignant cells vary in size; they are frequently enlarged, but small malignant nuclei are also encountered. There is an increase in the nucleocytoplasmic ratio of malignant cells whether the actual cell is large or small. *Malignant nuclei vary in shape* and often exhibit an angular or notched outline. *The chromatin pattern is irregular*, with clumping in angular masses or linear strands, against a comparatively clear background. Uneven condensation of chromatin at the nuclear margin produces an irregularly sharp nuclear border. Malignant nuclei frequently appear hyperchromatic, but this is not necessarily the case. *Abnormal distribution of chromatin, however, is the most significant feature of the malignant nucleus.*

Multiple or prominent enlarged eosinophilic nucleoli are seen in malignant cells but it should be borne in mind that nucleoli also appear in benign cells. In malignant cells they may be unusually large and there may be a variation in the number and size of nucleoli within cells of a group. In addition nucleoli may appear angular in malignant cells compared with the rounded outlines of their benign counterparts.

Multinucleation occurs in benign and malignant cells; the nuclear size, shape and chromatin pattern show little variation in benign multinucleated cells whereas the multiple nuclei found in malignant cells show differences in all these features.

Mitotic figures are infrequently seen in exfoliated or detached malignant cells and are as likely to occur in benign cells as in malignant ones. However, an abnormal mitotic figure may occasionally be seen in a malignant cell.

The Cytoplasm of a Malignant Cell

The cytoplasm and its relationship to the nucleus provides evidence which helps to elucidate the type of tumour which is present and the degree of differentiation. Cytoplasmic abnormalities alone should not be used for the diagnosis of malignancy, although changes in the cytoplasm can contribute to the diagnosis.

Malignant cells show variation in staining reaction, as well as in size and shape. Occasional giant cells are seen, but frequently the malignant cell is small in size. A more important factor is the variation in size which may be seen in a group of malignant cells. A few malignant cells may attain bizarre shapes, e.g., 'tadpole' and 'fibre' cells. The majority lack these gross abnormalities but a comparison of cells within a group reveals considerable differences in configuration. The staining reaction in malignant cells shows marked variation; occasionally the cytoplasm appears opaque and brightly eosinophilic, but this is also seen in inflammatory and degenerative changes. Similarly vacuolation of the cytoplasm may be a striking feature, but this also occurs in benign cells of many types. The cytoplasm of a malignant cell may contain a second cell of the same type, an appearance suggestive of phagocytosis.

Relationship of Nucleus to Cytoplasm in a Malignant Cell

Malignant cells display variations in both the quantitative and spatial relationships of nucleus to cytoplasm.

Quantitative relationship: malignant cells show a varying disturbance of the nucleocytoplasmic ratio, whereas the ratio is fairly constant in benign cells of any one type. The change in the malignant cell consists of an increase in the nucleocytoplasmic ratio; in its most striking form this may produce an undifferentiated malignant cell, in which the cytoplasm is either not apparent or appears only on a part of the nuclear margin when examined by light microscopy at dry high-power magnification. Care must be taken to distinguish between this and the complete loss of cytoplasm of the degenerate cell, in which no cytoplasm is identifiable even when examined with the oil immersion lens. In the differentiated malignant cell the nucleus is surrounded by a rim of cytoplasm, which is narrower than the greatest diameter of its nucleus and considerably narrower than would be found in benign cells of the same type. In contrast the dyskaryotic cell (p. 44) has a malignant nucleus, but retains the normal nucleocytoplasmic ratio of its cell type. In general this means that in the dyskaryotic cell the width of the rim of cytoplasm is greater than the maximum diameter of the nucleus.

The *spatial relationship* of nucleus to cytoplasm is often changed in malignant cells. In a group of benign cells the position of the nuclei within the cytoplasm is relatively constant, whether they are central as in squamous cells, or eccentric as in columnar cells; in groups of malignant cells, however, this relationship may show considerable variation.

Grouping of Malignant Cells

In addition to the readily recognisable variations in nuclear and overall cell size, shape and staining reaction, which are often found in a group of malignant cells, two other features are seen. As a result of their abnormal spatial relationship, the perimeter of a group of malignant cells may appear to consist mainly of nuclei, whereas a benign cell group would be bounded by a uniform band of cytoplasm. The second feature which may be seen in groups of malignant cells is moulding of nuclei by their neighbours. In benign cells this phenomenon is prevented by their adequate cytoplasm, except where there is distortion of the cells caused by secretory activity.

The Overall Appearance of a Malignant Smear

Many of the features present in malignant smears are also seen in benign smears. Occasionally, however, the overall pattern will lead to a further search for malignant cells. The specific abnormalities will vary according to the site from which the specimen was taken. In smears from some sites the presence of erythrocytes may arouse suspicion; in other instances cells may be seen which are themselves benign, such as squamous metaplastic cells in sputum, but which are associated with malignant change. In other specimens the actual cells seen may be normal, but their presence in a particular age group denotes abnormality. Such a case would be the presence of apparently normal endometrial cells in a post-menopausal cervical smear. Occasionally degenerate cellular material, which is not diagnostic, may be sufficiently abnormal to warrant further examination of the specimen or of additional specimens. This may be the case in squamous cell carcinoma of bronchus, where the only abnormal cells obtained initially may be degenerate.

These examples show that it is necessary to note any feature of the smear which is incompatible with the clinical symptoms and findings and then to search further for diagnostic evidence.

SUMMARY

The most reliable evidence of malignant change is nuclear abnormality, especially the irregularity of chromatin pattern. Cytoplasmic evidence serves mainly to determine tumour type and differentiation. Cell groups provide valuable material for comparative purposes. Any abnormal microscopic feature of a smear should be noted and evaluated against the clinical evidence.

CHAPTER 4

CANCER OF THE FEMALE GENITAL TRACT

At the present time the most important application of cytology is the early diagnosis of carcinoma of the cervix uteri, but it also has a useful part to play in the diagnosis of carcinoma of the corpus uteri. In comparison with the role of cytology in the diagnosis of these two lesions, its application to the diagnosis of malignant lesions of vulva, tube or ovary is relatively insignificant.

There are many similarities in the malignant cells from squamous cell carcinomata of the vulva, vagina or cervix. In the same way the malignant cells from cases of adenocarcinoma of cervix, corpus uteri, fallopian tube and ovary have many common characteristics. In fact, it is rarely possible to identify the site of a particular squamous cell carcinoma or adenocarcinoma from the appearance of its malignant cells.

A general description of malignant cells from squamous cell carcinoma and adenocarcinoma, of undifferentiated malignant cells and of the phenomenon of dyskaryosis can eliminate some repetition in the description of the cytology of individual lesions.

DIFFERENTIATED MALIGNANT CELLS OF SQUAMOUS TYPE

Differentiated malignant cells of squamous type can be divided into three groups. All the cells have malignant nuclei completely surrounded by cytoplasm with grossly abnormal nucleocytoplasmic ratios. The cells are, therefore, subdivided into groups only by their overall shape. Using this criterion the three categories are described as *tadpole cells*, *fibre cells* and *third-type malignant cells*. Although tadpole cells and fibre cells are striking because of their bizarre shapes, they are seen relatively infrequently. The third-type malignant cell is by far the commonest type of squamous malignant cell found in smears.

Tadpole Cells

The term tadpole describes the shape of this type of cell; the 'head' contains an enlarged malignant nucleus giving the cell an abnormal nucleocytoplasmic ratio; the 'tail' of cytoplasm may be keratinised and stain a bright orange colour. The nucleus is irregular, often angular in shape and hyperchromatic, with an abnormal chromatin pattern, showing uneven clumping of the chromatin to form angular masses and an irregularly thickened nuclear border. The presence of this malignant nucleus is an essential feature for the diagnosis of the differentiated malignant squamous cell of the tadpole type. The bizarre cell shape is not adequate evidence for this diagnosis as occasional benign squamous cells may show tadpole forms.

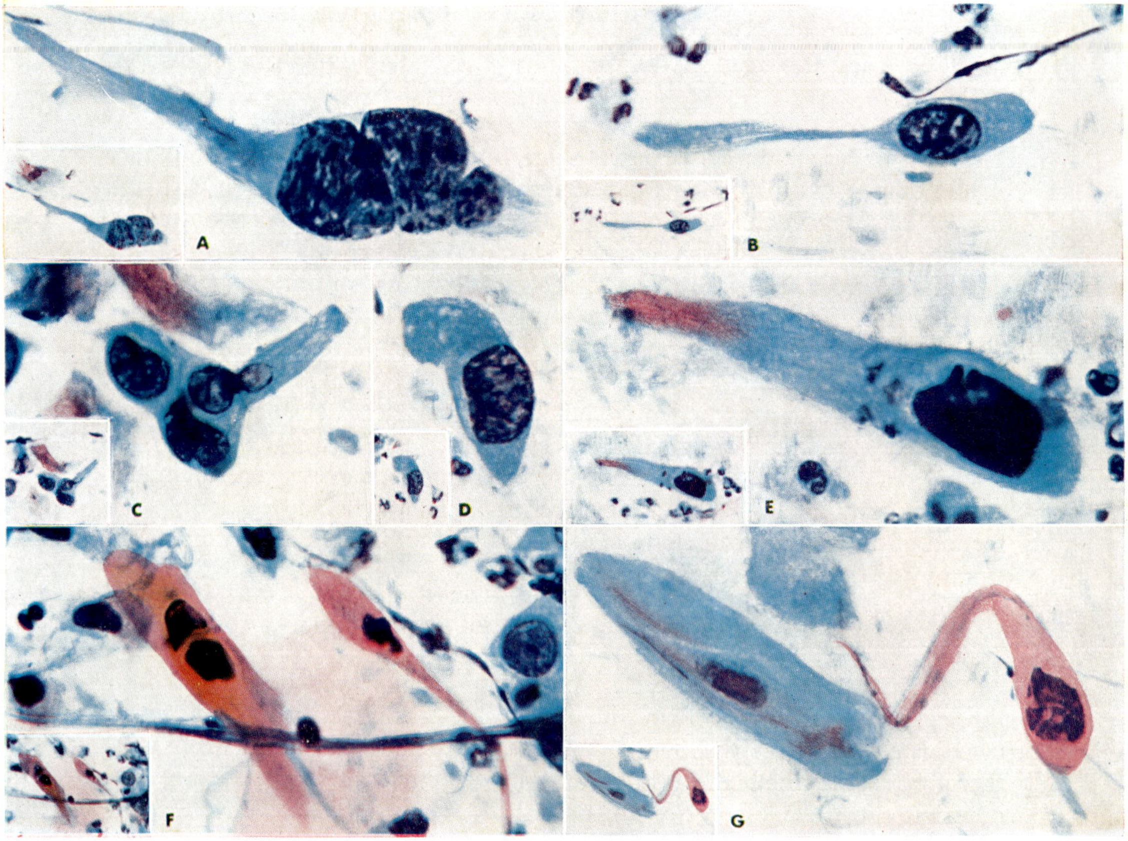

FIG. 30

Differentiated Malignant Cells of Squamous Type—Tadpole Cells—In Cervical Smears from cases of Invasive Squamous Cell Carcinoma and Carcinoma-in-situ of Cervix

For detailed description see facing page.

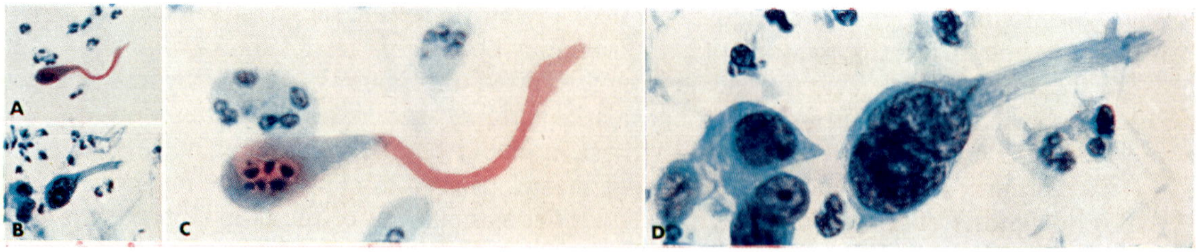

FIG. 31

Benign and Malignant Squamous Cells of Tadpole Type from Cervical Smears

For detailed description see facing page.

FIG. 30

Differentiated Malignant Cells of Squamous Type—Tadpole Cells—In Cervical Smears from cases of Invasive Squamous Cell Carcinoma and Carcinoma-in-situ of Cervix

A. B. C. and D. All these cells show the features of differentiated malignant cells of squamous type with 'tadpole' forms. All have a 'head' occupied by one or more large irregular hyperchromatic nuclei with strikingly abnormal chromatin patterns, while the relatively scanty cytoplasm forms an elongated 'tail'. In plate A, the tadpole cell is multinucleated with marked variation in the size and shape of the individual nuclei. A more common form of tadpole is seen in plate B, where the nucleus shows a strikingly abnormal chromatin pattern. A bizarre form of multinucleated tadpole is shown in plate C, while the cell in plate D shows the essential features of a malignant cell of squamous tadpole type but is not quite as elongated as the usual form. (Insets ×200.) (×800.)

E. F. and G. In these cells the overall pattern is similar to the malignant tadpole cells but degenerative changes in the nuclei make them unsuitable for definitive diagnosis. However, their bizarre shape, abnormal nucleo-cytoplasmic ratios and irregular nuclear outline should arouse suspicion and lead to a careful search for diagnostic cells similar to those illustrated in fields A to D. (Insets ×200.) (×800.)

FIG. 31

Benign and Malignant Squamous Cells of Tadpole Type from Cervical Smears

A. and B. On low-power examination, these tadpole cells appear similar in shape and nuclear hyperchromasia. (×200.)

C. Higher magnification of field A reveals this to be a degenerate squamous cell with variable cytoplasmic staining. The nucleus, which shows karyorrhexis, is small and oval. Prior to degeneration, this was probably a benign cell. No other abnormal cells were found in this or subsequent specimens taken from this patient. (×800.)

D. Higher magnification of field B reveals a bizarre differentiated malignant cell of squamous type. The two large nuclei of this malignant tadpole cell are pleomorphic and irregularly hyperchromatic. In the hysterectomy specimen from this patient an extensive carcinoma-in-situ of cervix, with an area of invasive squamous cell carcinoma, was found. (×800.)

It is not the tadpole shape which is the diagnostic characteristic of this type of differentiated malignant cell but the presence of the malignant nucleus.

Fibre Cells

These elongated narrow cells are aptly described by their name. The presence of a malignant nucleus, with an irregular shape and an abnormal chromatin pattern, is essential for the diagnosis of malignancy. A benign rolled or folded squamous cell may appear to have the same shape as a malignant fibre cell and this is seen not infrequently in pregnancy smears. The benign cell, however, can readily be distinguished from the malignant cell by the marked difference in nuclear appearance and nucleocytoplasmic ratio.

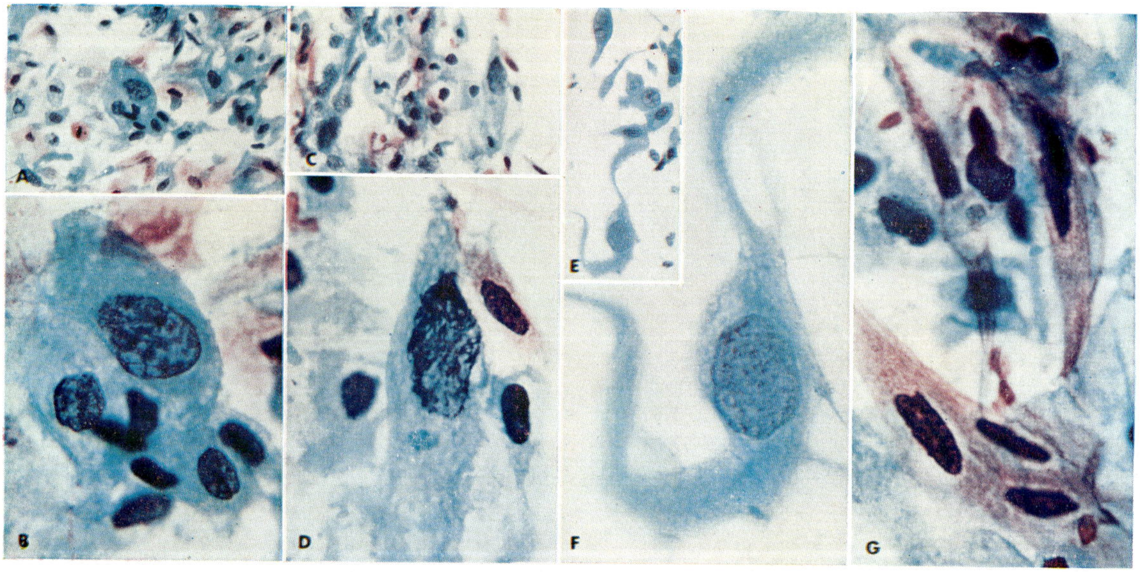

FIG. 32

Differentiated Malignant Cells of Squamous Type—Fibre Cells—in a Cervical Smear from a Carcinoma-in-situ of Cervix

A. These squamous cells show striking variation in nuclear staining, size and shape, ranging from the large pale ovoid nucleus in the centre of the field to the smaller elongated dark nuclei on the right. (×200.)

B. Marked variation in nuclear size, shape and staining is seen in the cells of this group. Although the large nucleus in the centre appeared paler than the other nuclei in field A, higher magnification shows marked clumping of chromatin, with condensation at the nuclear border. The nucleus to the left of the group shows notching of its lower border. (×800.)

C. and D. These cells show the elongated cytoplasm and nuclei of 'fibre' cells and at high-power magnification (D) the nuclei show considerable variation in size, shape and staining reaction. (×200.) (×800.)

E. and F. This group of elongated cells (E) shows marked variation in size and shape of the nuclei. A differentiated malignant cell of squamous type (F) is strikingly elongated. Its nucleus is large and irregular in shape with slight clumping of chromatin although this is less marked than in those cells seen in fields B and D. (×200.) (×800.)

G. The nuclei in this group of fibre cells are hyperchromatic and show variation in size and irregularity in shape (×800.)

Mere elongation of cytoplasm and nucleus is insufficient evidence for diagnosis of the fibre cell type of differentiated malignant cell as these features may be seen in rolled or folded benign squamous cells. Variation in nuclear size, pleomorphism and hyperchromasia are essential features for the diagnosis of the malignant fibre cell.

Third-type Malignant Cells

This classification includes all the differentiated malignant cells of squamous type, which are excluded from the two previous categories. The most common form is a round cell with a central malignant nucleus surrounded by a narrow band of cytoplasm the width of which is less than the greatest diameter of the nucleus. The nucleus shows irregularity of outline and chromatin pattern, essential features in the diagnosis of malignancy. Within this category, however, there may be a wide variation in size, shape and staining reaction of the cells including irregularly shaped and brightly eosinophilic or orangophilic keratinised cytoplasm. Third-type cells are the most common type of differentiated squamous malignant cells. They are frequently seen in groups and comparison of the cells within a group reveals marked variations both in cell size, shape and staining reaction and in the nuclear size, shape and chromatin pattern.

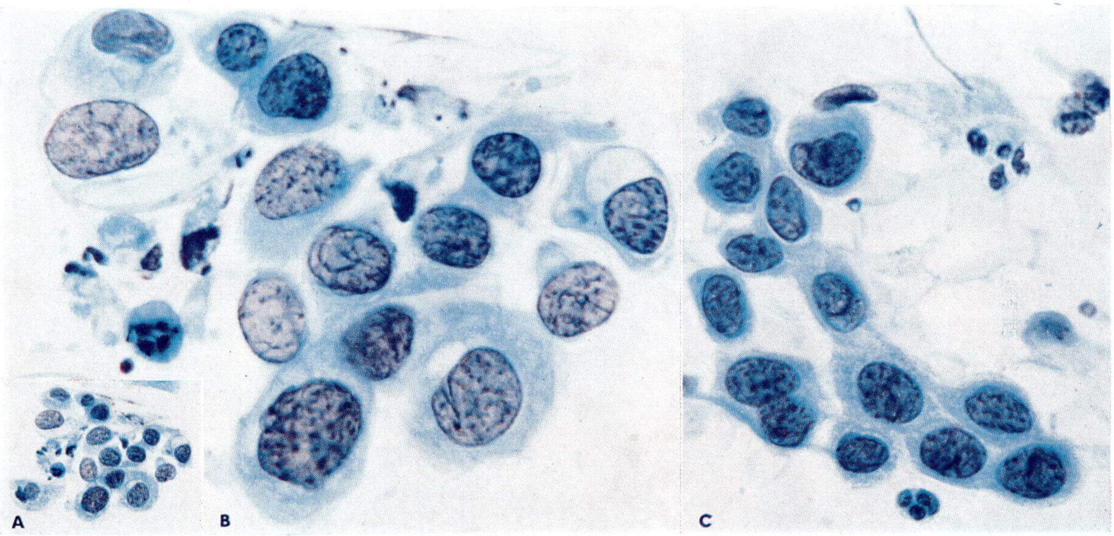

FIG. 33

Differentiated Malignant Cells of Squamous Type—Third-type—in Cervical Smears from two cases of Carcinoma-in-situ of Cervix

A. and B. Even on low-power examination, this group of 'third-type' differentiated malignant cells shows marked variation in nuclear size and abnormal nucleocytoplasmic ratios. Higher magnification reveals variation in shape of the centrally placed nuclei, irregular nuclear borders with angular and notched outlines, marginal condensation of chromatin and irregularly clumped abnormal chromatin patterns. Compared with many malignant cells, these nuclei at first examination do not appear to be hyperchromatic, but careful scrutiny shows that the chromatin pattern is abnormal. Occasional undifferentiated cells are present, in most of which, especially to the left of the field, there is some loss of nuclear structure, indicative of degenerative change. (×200.) (×800.)

C. This group of malignant cells of squamous type shows less marked nuclear enlargement, but still reveals abnormal nucleocytoplasmic ratios. The nuclei show striking pleomorphism and hyperchromasia. (×800.)

The abnormal nuclear characteristics and nucleocytoplasmic ratios constitute the diagnostic features of third-type malignant cells.

DIFFERENTIATED MALIGNANT CELLS OF ADENOCARCINOMATOUS TYPE

Although these cells have nuclei completely surrounded by cytoplasm, with an abnormal nucleo-cytoplasmic ratio, their nuclei are eccentric in position, contrasting with the centrally placed nuclei of malignant cells of squamous type. (An analogy is readily drawn with the different positions of the nuclei in benign squamous and columnar cells.) The fundamental point in the diagnosis is, again, the typical enlarged malignant nucleus with its irregularity of form and chromatin pattern. Multiple, prominent, enlarged and often angular eosinophilic nucleoli may be a striking feature but these nucleoli may exhibit what appears to be a basophilic staining reaction if they are obscured by overlying chromatin and cytoplasm. The cytoplasm of these cells is frequently vacuolated but this feature is not sufficient evidence for a diagnosis of an origin from an adenocarcinoma since vacuolation also occurs in the cytoplasm of squamous type malignant cells. Not all differentiated malignant cells of adenocarcinomatous type show the eccentric position of the nucleus. If the cell is seen in the 'end-on' position the nucleus will appear central. (A recollection of the two views of endocervical cells producing 'palisade' and 'honeycomb' patterns will elucidate this point.) When the cell is seen in this way it is indistinguishable from the third-type squamous malignant cell.

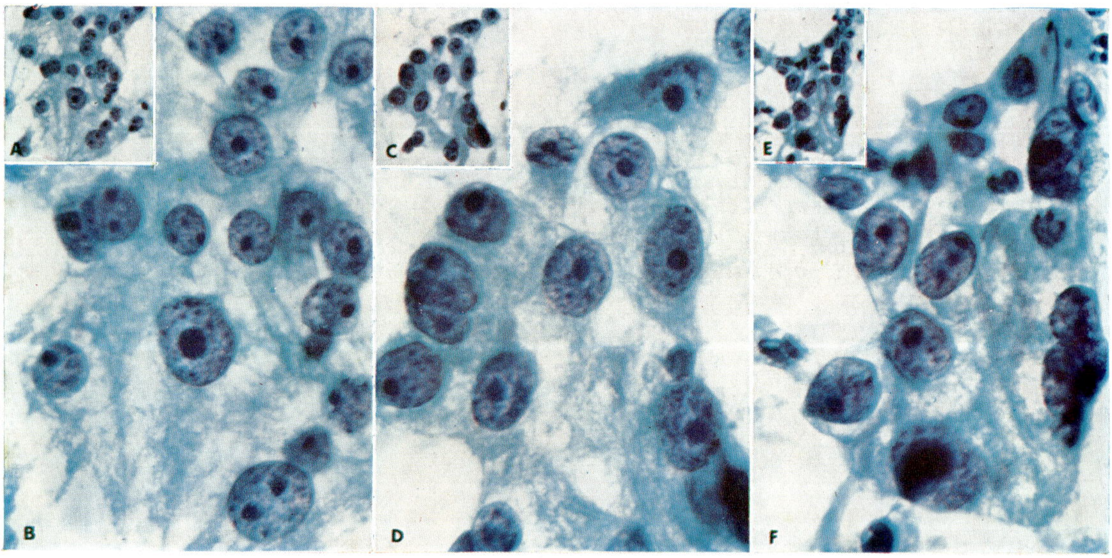

FIG. 34

Differentiated and Undifferentiated Malignant Cells in a Cervical Smear from an Adenocarcinoma of Cervix

A. Prominent or multiple nucleoli are easily resolved even at this low-power magnification. (×200.)

B. Although the prominent nucleoli are the most striking feature of these undifferentiated malignant cells, they also show abnormal chromatin patterns and marked variation in nuclear size. (×800.)

C. and D. While most of the cells in this group, shown at low- and high-power magnifications, are undifferentiated malignant cells similar to those seen in fields A and B, the cells at the right margin and upper corner of the fields show an eccentric nuclear position compatible with an origin from an adenocarcinoma. (×200.) (×800.)

E. and F. In the low-power illustration, the cells show cytoplasmic vacuolation and eccentric nuclear position, suggestive of an adenocarcinomatous origin. The high-power illustration (F) shows the variation in size and configuration, abnormal chromatin patterns and prominent angular nucleoli of the malignant nuclei. (×200.) (×800.)

In these plates the eosinophilic appearance of the nucleoli is obscured by basophilic chromatin.

UNDIFFERENTIATED MALIGNANT CELLS

These cells may arise from a squamous cell carcinoma or from an adenocarcinoma, when either the whole or a part of the tumour is poorly differentiated. The diagnosis of malignancy depends on the appearance of the nuclei, which show the usual abnormal features of variation in size, irregularity or angularity of shape and abnormal chromatin patterns, together with abnormal nucleoli. The nuclei of these undifferentiated cells, however, do not appear to be completely surrounded with cytoplasm when examined with the dry high-power objective of the light microscope; in some instances a part of the nuclear border may still show a margin of cytoplasmic covering. Care must be taken not to confuse these undifferentiated malignant cells with degenerate nuclei; this possibility can be excluded by examination at higher magnification, using an oil-immersion objective, when a very narrow rim of cytoplasm is identifiable in well preserved undifferentiated malignant cells.

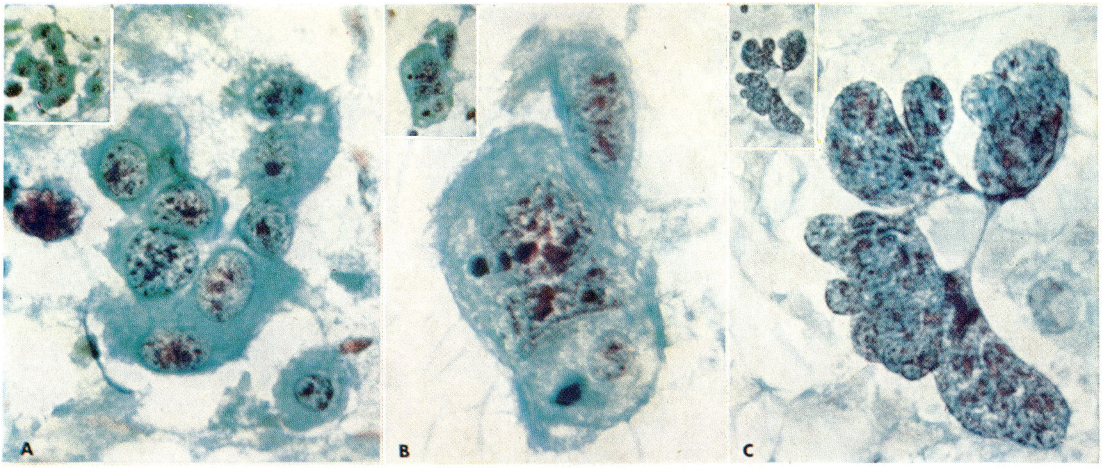

FIG. 35

Differentiated Malignant Cells of Squamous Type and Undifferentiated Malignant Cells in a Cervical Smear from a Poorly Differentiated Invasive Squamous Cell Carcinoma of Cervix

A. This group consists of differentiated malignant cells of squamous type. The nuclei are strikingly abnormal with irregular chromatin patterns and prominent nucleoli. (Inset ×200.) (×800.)

B. The multinucleated malignant cell (*centre*) shows marked variation in nuclear size, together with abnormal chromatin patterns and multiple, prominent angular nucleoli. (Inset ×200.) (×800.)

C. This bizarre aggregation of undifferentiated giant malignant cell nuclei displays an unusual polypoid nuclear pattern with an abnormal distribution of chromatin and multiple eosinophilic angular nucleoli. Bizarre undifferentiated malignant cells of this type were present in considerable numbers in the smear. This is an unusual appearance, but pathological examination of a cervical biopsy revealed many cells of a similar type. Further cells from smears taken from this patient are illustrated in Figs. 44 and 160B. (Inset ×200.) (×800.)

DYSKARYOSIS

The term *dyskaryosis* describes a cell which has a nucleus showing the features characteristic of malignancy, while retaining relatively normal cytoplasm; this phenomenon is seen in squamous and columnar cells. Dyskaryotic cells are frequently found in smears which also contain malignant cells; in some smears, however, the only abnormality may be the presence of cells exhibiting dyskaryosis. In squamous cells the appearance of this phenomenon is most important when seen in parabasal cells and least significant when seen in superficial cells. In the more mature intermediate and superficial cells dyskaryosis is easily recognised by the presence of the malignant nucleus or nuclei in a cell which retains a large amount of translucent cytoplasm and a comparatively normal cell shape. In parabasal dyskaryosis, however, it may be difficult to differentiate the dyskaryotic cell from the third-type malignant cell. In this case a useful rule-of-thumb distinction can be made by comparing the width of the surrounding cytoplasm with the greatest diameter of the nucleus. When the cytoplasmic width exceeds the diameter of the malignant nucleus in a parabasal cell, this can be considered as dyskaryosis. A nuclear diameter in excess of the width of the cytoplasmic border would place the cell in the category of a differentiated malignant cell of squamous type. This distinction, in fact, is largely theoretical since the occurrence of parabasal dyskaryosis itself warrants further investigation (see Figs. 47 and 48).

Dyskaryosis of the columnar cell type is seen in endocervical cells, but this is far less common than in the squamous cell type. In endocervical cells the normal size and shape are retained and an adequate amount of cytoplasm is present but the nucleus shows the features of malignant change.

The changes seen in dyskaryotic cells should not be confused with degenerative changes in which the nucleus shows karyorrhexis or karyolysis, nor with the nuclear enlargement and multinucleation which may be seen in inflammation. These changes may give rise to diagnostic difficulty when occurring in the same smear as dyskaryotic cells. Careful examination, however, will differentiate between the degenerative or reactive benign nuclei and the malignant nuclei.

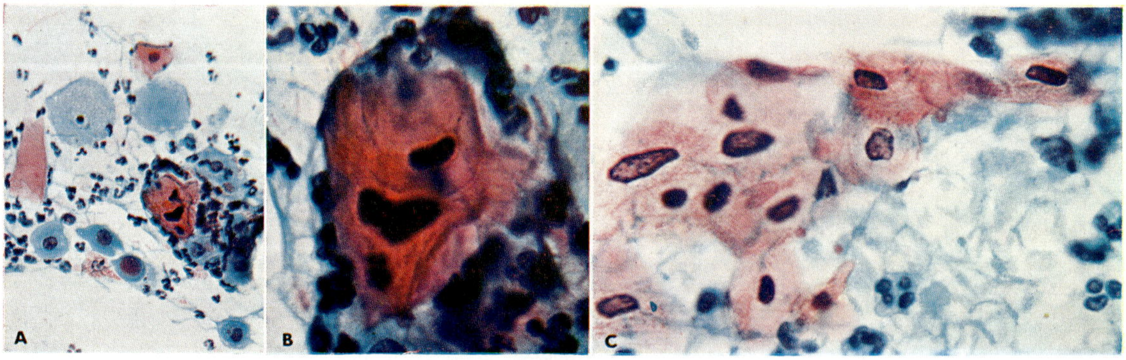

FIG. 36

Superficial Dyskaryosis in Cervical Smears from cases of Trichomonal Infection

A. and B. Squamous cells showing inflammatory changes are seen against a background of polymorphonuclear leucocytes. In the lower right quadrant, two superficial cells show marked nuclear hyperchromasia and pleomorphism. This is shown more clearly at higher magnification (B) where the abnormal appearance of these nuclei is obvious. It is apparent that these cells retain an adequate amount of cytoplasm. (×200.) (×800.)

C. These superficial cells have irregular enlarged hyperchromatic nuclei with abnormal chromatin patterns, but the amount of normal cytoplasm present exceeds the nuclear diameter, thus distinguishing these cells from differentiated malignant cells of squamous type. (×800.)

In both these cases the presence of superficial dyskaryosis is associated with trichomoniasis. This appearance should indicate that a search for the more significant dyskaryosis of intermediate or parabasal type is required. It may, however, occur alone especially in the presence of trichomonal infection.

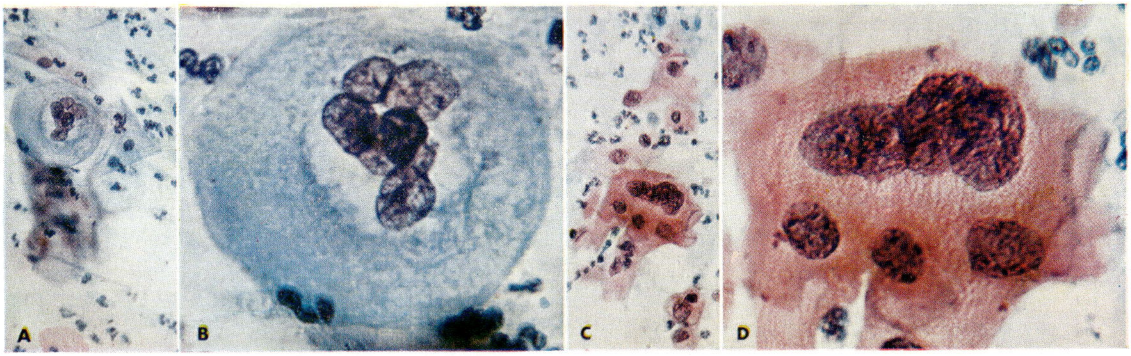

FIG. 37

Intermediate Squamous Cells showing Dyskaryosis in a Cervical Smear from an In-situ Squamous Cell Carcinoma of Cervix

A. and B. In these illustrations a large multinucleated intermediate squamous cell is seen to the left of a benign intermediate squamous cell. The nuclei of the multinucleated cell are considerably larger than that of the benign intermediate cell. At high-power magnification (B) it is seen that, in addition to their size, the nuclei show other abnormal features. There is considerable variation in size and shape, with irregularity of outline and the chromatin shows marginal condensation and clumping. The nuclei, therefore, show similar features to those seen in malignant cells. In this cell, however, they are surrounded by a large amount of clear cytoplasm. Since the width of the rim of cytoplasm is greater than the maximum nuclear diameter, this is, therefore, a squamous cell showing dyskaryosis. (×200.) (×800.)

C. The large multinucleated intermediate squamous cell and the three cells immediately below it all show marked nuclear enlargement. Several other single cells in the field also have large nuclei. (×200.)

D. The multinucleated cell and the three cells immediately below it have enlarged nuclei which vary in size and shape. Their borders are sharply defined and clumping of chromatin is obvious; these nuclei show the characteristic features of malignancy. Although these cells all retain a wide margin of cytoplasm, the nuclei are so large that their maximum diameter closely approaches that of the rim of cytoplasm. The appearance of these dyskaryotic cells is separated from that of differentiated malignant cells only by the assessment of the nucleo-cytoplasmic ratio. (×800.)

Subsequent smears from this patient showed differentiated malignant cells of squamous type. Histological examination of a cone biopsy of cervix revealed an in-situ squamous cell carcinoma of cervix and dysplasia of the squamous epithelium.

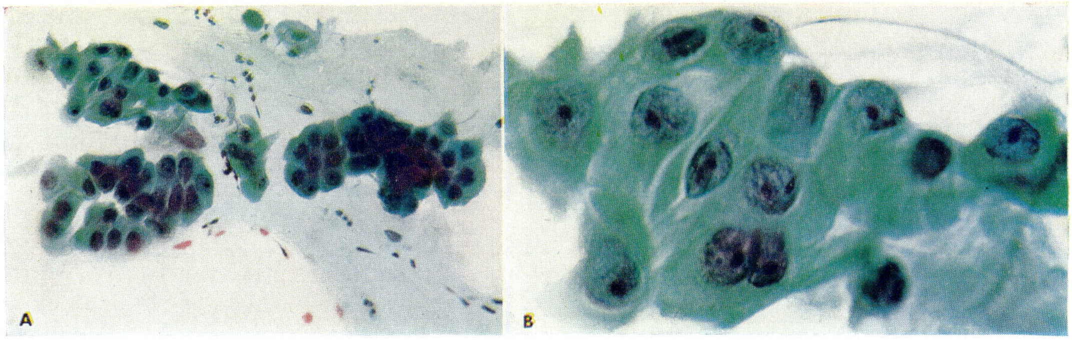

FIG. 38

Dyskaryosis in Endocervical Cells showing Squamous Metaplasia in a Smear from the Surface of an Endocervical Polyp

A. These groups of endocervical cells are fairly regularly arranged in a linear pattern. In the groups to the right and lower left the nuclei, although large, are also uniform in shape. One group (upper left) appears to be less evenly arranged and there is some variation in nuclear size and shape. (×200.)

B. High-power examination of the group (upper left, A) confirms the somewhat irregular arrangements and reveals details of the nuclei which are seen to vary in size and shape with occasional angular or notched outlines. There is considerable variation in chromatin pattern with irregular strands and masses. The nucleus at the right of the group shows irregular condensation of chromatin at its edge. Large and/or multiple, occasionally angular, nucleoli are a striking feature. In spite of their abnormal nuclei these cells retain an adequate amount of cytoplasm and should therefore be classed as dyskaryotic. (×800.)

SQUAMOUS CELL CARCINOMA OF VULVA

Cytology is only rarely of use in the diagnosis of squamous cell carcinoma of the vulva; if a suspicious lesion is present early biopsy is necessary. However, a smear taken in the out-patient clinic may, if the result is positive for malignant cells, provide confirmation of clinical suspicion prior to the patient's admission to hospital. Smears may also be useful in the post-operative follow-up of patients who have been treated surgically for carcinoma or carcinoma-in-situ of vulva and cases of clinical leukoplakia. For all these patients frequent repetition of biopsies may be impracticable.

In smears from the vulva technical difficulty due to drying may be encountered, but this risk can be minimised by the use of a moistened spatula for scraping the lesion.

In invasive squamous cell carcinoma of the vulva the malignant cells are predominantly of differentiated squamous type. As the tumour is frequently well differentiated, keratinised malignant cells and the highly differentiated tadpole and fibre cells are more common than in carcinoma of the cervix. However, in some less well differentiated tumours, undifferentiated malignant cells may also be seen.

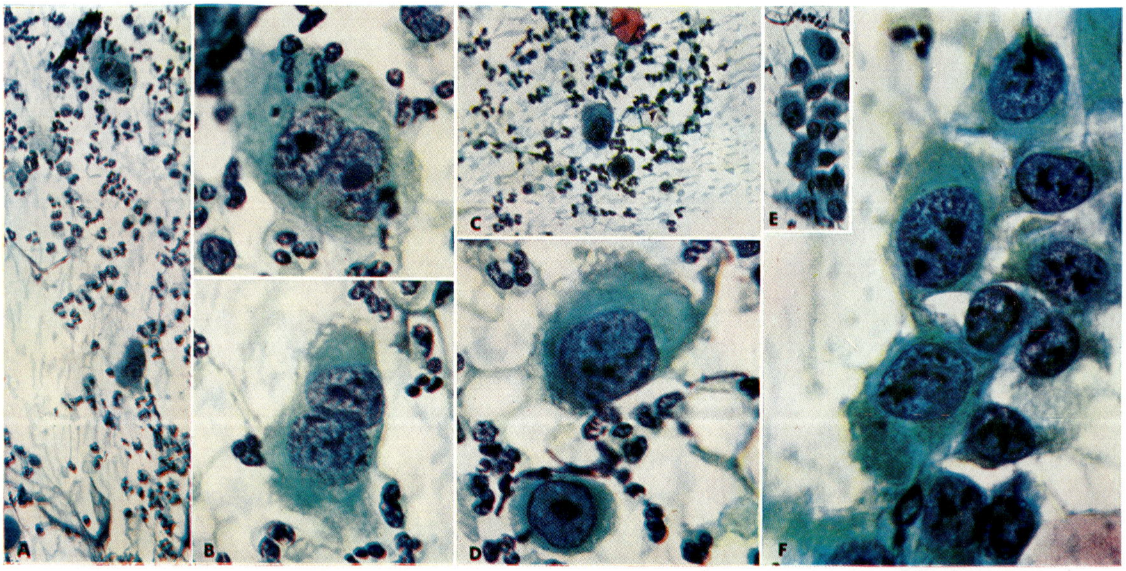

FIG. 39

Differentiated Malignant Cells of Squamous Type and Undifferentiated Malignant Cells in a Smear from an Invasive Squamous Cell Carcinoma of Vulva

A. Large multinucleated differentiated malignant cells of squamous type are seen against a background of polymorphonuclear leucocytes and histiocytes. (×200.)

B. Both these multinucleated malignant cells show loss of definition of cellular borders and abnormal nucleocytoplasmic ratios. All the nuclei exhibit abnormal chromatin clumping but this is more striking in those of the lower cell. An unusual feature in differentiated malignant cells of squamous type is enlargement of the nucleoli as shown in the upper cell. (×800.)

C. Two round differentiated malignant cells of squamous type, third-type cells, are seen while a brightly eosinophilic keratinised anucleated squame lies at the upper margin of the field. This type of keratinised cell is present in smears from the vulva and lower third of vagina, but is not usually seen in cervical smears. (×200.)

D. Loss of cellular definition, abnormal nucleocytoplasmic ratios, nuclear pleomorphism and disparity in size of nuclei are seen in these malignant cells of squamous type. Both nuclei show abnormal chromatin patterns; the lower of the two also shows marked peripheral condensation of chromatin and a prominent nucleolus. (×800.)

E. and F. This group includes both differentiated malignant cells of squamous type and undifferentiated malignant cells. The larger cells to the left of the field are well differentiated. All the features of malignant nuclei are obvious, the abnormal chromatin patterns being especially clear. (×200.) (×800.)

SQUAMOUS CELL CARCINOMA OF VAGINA

This uncommon tumour, occurring mainly in elderly women, often reaches an advanced stage before it gives rise to symptoms. At this point biopsy is indicated, although cytology may be used to confirm clinical suspicion prior to admission. In asymptomatic patients the commonly used cervical scrape smear is unlikely to detect this lesion. A vaginal smear, either from a suspicious area of vaginal wall or from the posterior fornix pool, is necessary. The possibility of detecting an occasional case of vaginal carcinoma is a minor argument for the collection of vaginal pool smears as well as cervical scrape smears from post-menopausal patients. Although cervical smears do not usually give positive results in cases of early carcinoma of the vagina, the possibility that this site is the source of otherwise inexplicable squamous malignant cells in a cervical smear should always be investigated.

The examination of vaginal smears is particularly useful in the care of those elderly patients with prolapse, who are unfit for vaginal repair operations. Pessaries may then be necessary to relieve their symptoms; if neglected, these pessaries may cause ulceration of the vaginal walls. If this occurs smears should be taken from the ulcerated areas. Even when ulceration is not present smears should be prepared from the posterior fornix pool, on each occasion when the patient returns to the clinic for the pessary to be changed, as the long-term use of these pessaries may be a predisposing factor in the aetiology of carcinoma of the vagina.

The malignant cells from a squamous cell carcinoma of vagina are not usually as highly differentiated as those from a carcinoma of the vulva. Although keratinised malignant cells are less common, differentiated malignant cells of squamous type are usually seen together with a variable proportion of undifferentiated malignant cells.

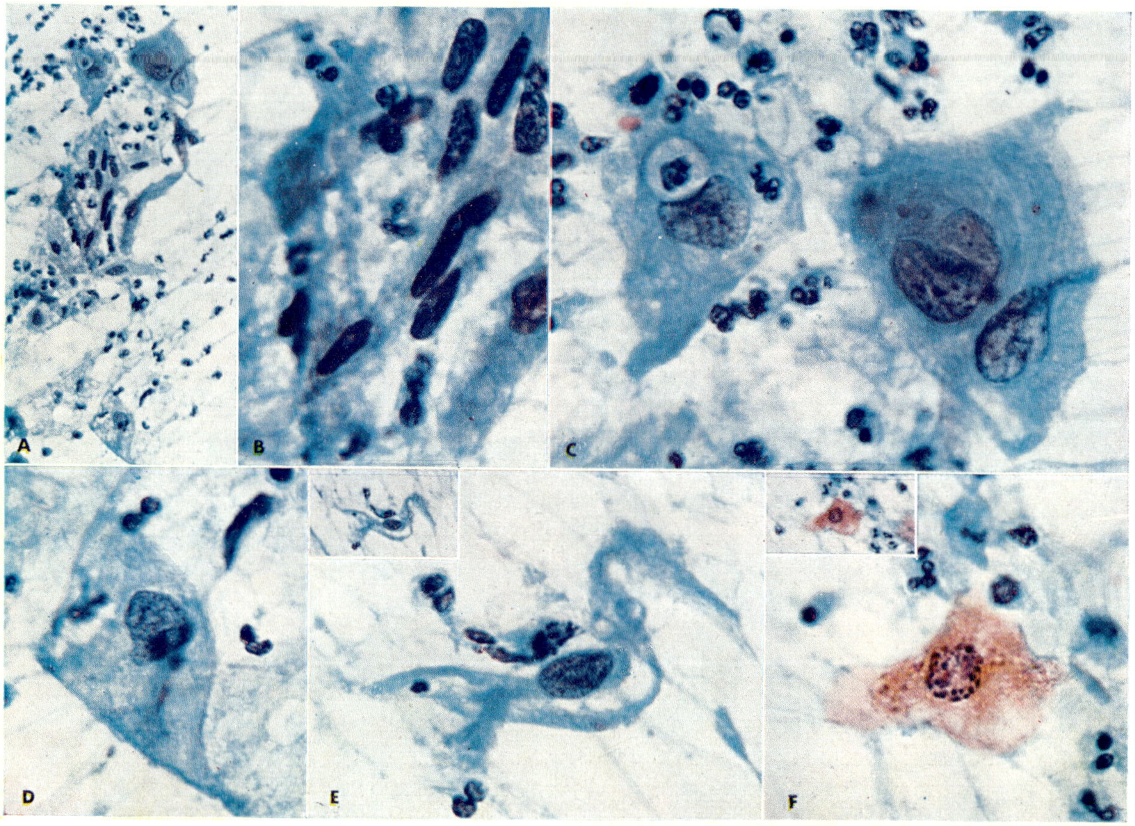

FIG. 40

Differentiated Malignant Cells of Squamous Type in a Vaginal Smear from a Squamous Cell Carcinoma of Vagina

A. This low-power illustration shows a central group of malignant fibre cells, with elongated hyperchromatic nuclei, against a background of polymorphonuclear leucocytes. At the upper margin of the field two malignant cells with large but comparatively pale-staining nuclei are seen, while at the lower margin a dyskaryotic squamous cell is present. (×200.)

B. In these fibre cells the cytoplasm is poorly defined but the varying size, irregular shape and hyperchromasia of the elongated nuclei are obvious. (×800.)

C. The multinucleated malignant cell on the right shows variation in nuclear size, with peripheral condensation and irregular clumping of chromatin to form angular wedge-shaped masses. The nucleus of the cell on the left is distorted by a vacuole containing a polymorphonuclear leucocyte, but its large size and abnormal chromatin pattern are still easily seen. (×800.)

D. The nucleus of this dyskaryotic squamous cell shows the irregularity of shape and clumping of chromatin that would be seen in a malignant nucleus, but the cell has an adequate amount of cytoplasm which distinguishes it from the differentiated malignant cell. (×800.)

E. In the centre of the field is a malignant tadpole cell with a large hyperchromatic nucleus. In the low-power view (inset) the cell immediately above it also appears to show nuclear hyperchromasia but at high magnification it becomes obvious that this appearance is due to karyorrhexis, although the impression remains that the nucleus originally had a bizarre shape. This type of degenerate cell cannot be used as a basis for diagnosis. (Inset ×200.) (×800.)

F. Again, a nucleus, on low-power examination (inset) appears to show chromatin clumping. At high magnification, however, it is clear that this is due to karyorrhexis and is not of diagnostic value. (Inset ×200.) (×800.)

SQUAMOUS CELL CARCINOMA AND CARCINOMA-IN-SITU OF CERVIX

Early invasive carcinoma and carcinoma-in-situ of the cervix are the lesions most amenable to early diagnosis by cytology. Easy visualisation of the cervix makes it possible to collect a specimen directly from the squamo-columnar junction, which is the commonest site of origin.

Although the final diagnosis of invasion in a carcinoma of cervix is essentially a pathological diagnosis, smears from invasive and in-situ lesions often show characteristic patterns. These can only be used as an approximate guide, as the invasive lesion may be accompanied by areas of carcinoma-in-situ and a predominantly in-situ carcinoma may be found to have a small area in which invasion is seen.

Invasive carcinoma often produces smears which are blood-stained and contain many polymorphonuclear leucocytes. The malignant cells present may include a considerable number of the more bizarre differentiated malignant squamous cells or undifferentiated malignant cells, while dyskaryotic cells are infrequently seen. Degenerative changes often occur in the malignant cells and make them unsuitable for definitive diagnosis; this can combine with the inflammatory cell infiltrate to obscure the real diagnosis.

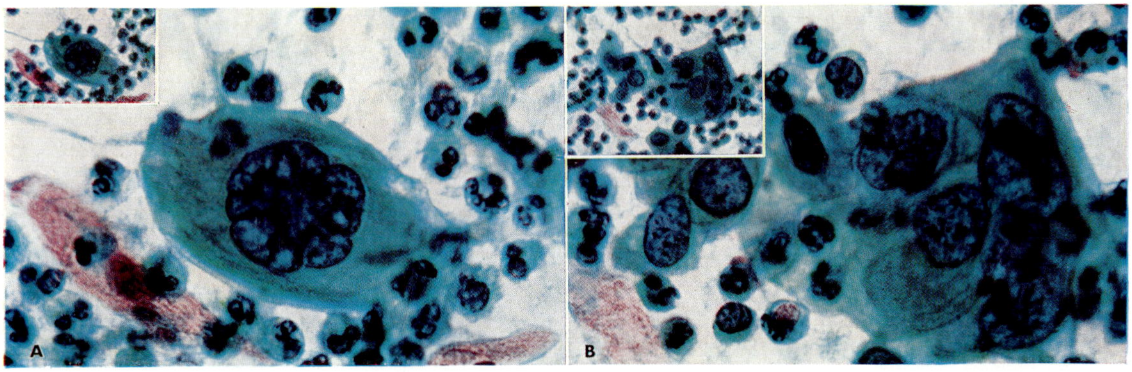

FIG. 41

Multinucleated Differentiated Malignant Cells of Squamous Type in Cervical Smears from an Invasive Squamous Cell Carcinoma of Cervix

A. In the low-power view (inset) the large cell appears to have an irregular hyperchromatic nucleus. High-power magnification, however, reveals multiple closely grouped nuclei. Individual nuclei show some variation in size and strikingly abnormal chromatin patterns. The chromatin is condensed at the nuclear margins and is also clumped in irregular masses and strands. Note the numerous polymorphonuclear leucocytes in the background. (Inset ×200.) (×800.)

B. The low-power view (inset) shows several small differentiated malignant cells of squamous type and a giant multinucleated malignant cell. The high-power magnification reveals that the multiple nuclei vary in shape and have abnormal chromatin patterns with irregular marginal condensation and clumping. (Inset ×200.) (×800.)

E

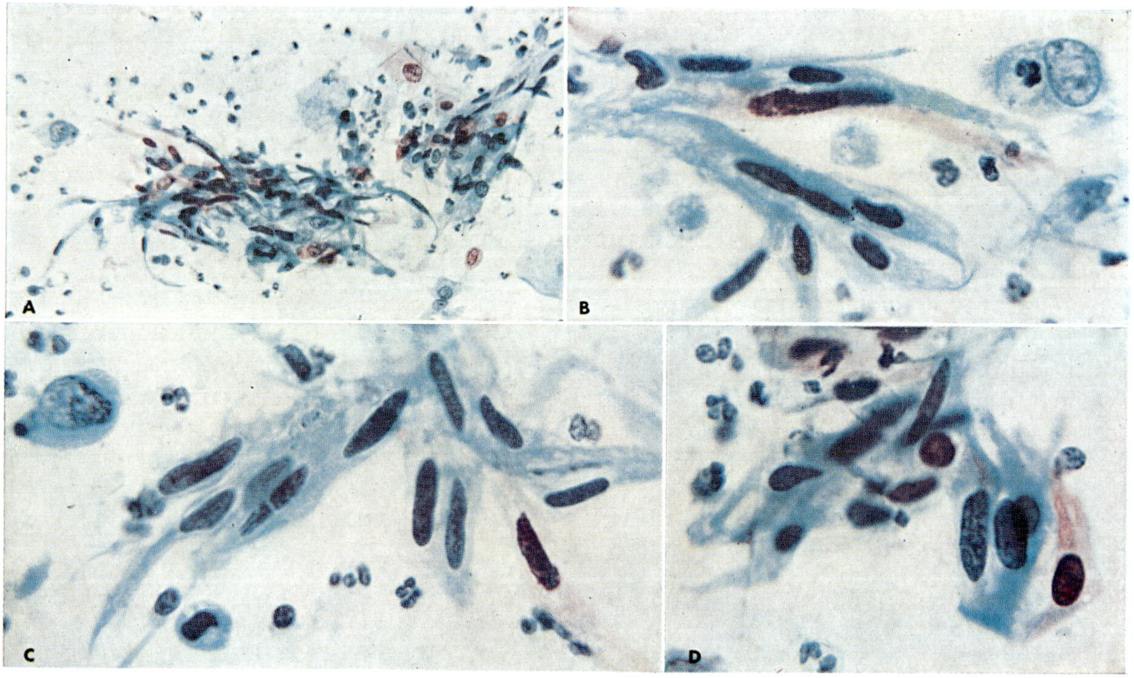

FIG. 42

Undifferentiated Malignant Cells and Differentiated Malignant Cells of Squamous Type—Fibre Cells—from an Invasive Squamous Cell Carcinoma of Cervix

A. The two large groups of cells are composed predominantly of differentiated malignant cells of squamous type with elongated whip-like cytoplasm, and deeply stained cigar-shaped nuclei. Intermingled with these fibre cells are undifferentiated malignant cells with large irregular but rather paler nuclei and scanty cytoplasm. Somewhat larger undifferentiated malignant cells are scattered singly to the periphery of the groups. (×200.)

B. The nuclei of the fibre cells, although generally cigar-shaped and hyperchromatic, show irregularity of outline and marked variation in size. The large undifferentiated malignant cell in the right upper quadrant has a relatively pale nucleus but still shows an abnormal chromatin pattern with marginal condensation and clumping. A polymorphonuclear leucocyte lies within the cytoplasm of this malignant cell. (×800).

C. This group of fibre cells shows the typical elongated cytoplasm while their nuclei show variation in size, hyperchromasia and irregularly clumped chromatin. The large undifferentiated malignant cell on the left has a pear-shaped nucleus with a heavy border and clumping of chromatin. (×800.)

D. In this field the cellular outlines are less clearly defined than in B and C, but the elongated nuclei of the malignant cells are essentially similar to those of the fibre cells. Variation in size and shape, irregularity of outline, hyperchromasia with abnormal chromatin patterns, are all present in these nuclei. (×800)

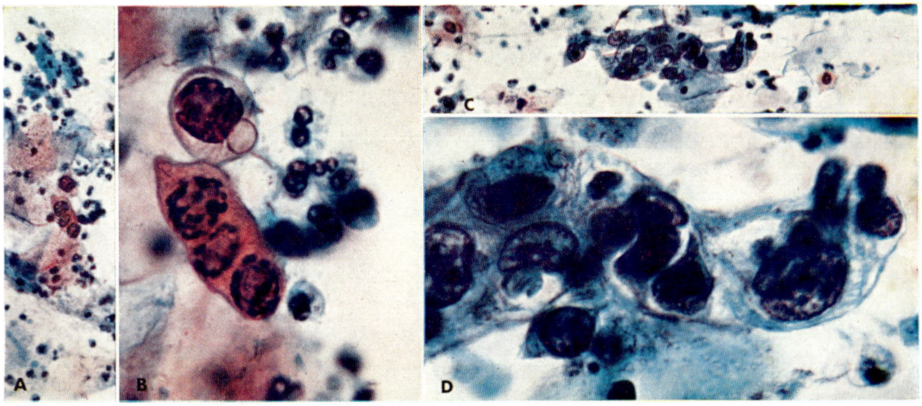

FIG. 43

Differentiated, Undifferentiated and Degenerate Malignant Squamous Cells in a Cervical Smear from an Invasive Squamous Cell Carcinoma of Cervix

A. and C. Each of these low-power fields contains a group of cells which show nuclear enlargement, hyperchromasia and pleomorphism. (×200.)

B. High-power examination of field A reveals that the three lower cells all show karyorrhexis and are, therefore, unsuitable for diagnosis. The upper single cell shows an enlarged nucleus with a wrinkled border and rounded masses of poorly defined chromatin against an opaque background; this also represents degenerative change. Although the abnormal nucleocytoplasmic ratio of this cell may arouse suspicion it is not suitable for definitive diagnosis. (×800.)

D. In contrast high-power examination of group C reveals the grossly abnormal structure of differentiated malignant cells of squamous type and of undifferentiated malignant cells. There is marked variation in nuclear size and shape. Irregular marginal condensation and angular clumping of chromatin are readily seen against an abnormally clear background and constitute diagnostic features. (×800.)

In this specimen the degenerate cells predominated and only a few groups of the diagnostic cells illustrated in plate D were seen. The smears showed severe inflammatory changes and *Trichomonas vaginalis* was present in large numbers. In this case it was thought that the presence of the protozoal infection had masked the characteristic features of many of the groups of malignant cells, but degenerative changes may also mimic malignant cells. The cytological diagnosis was confirmed by histological examination of a cone biopsy of the cervix which showed an invasive squamous cell carcinoma.

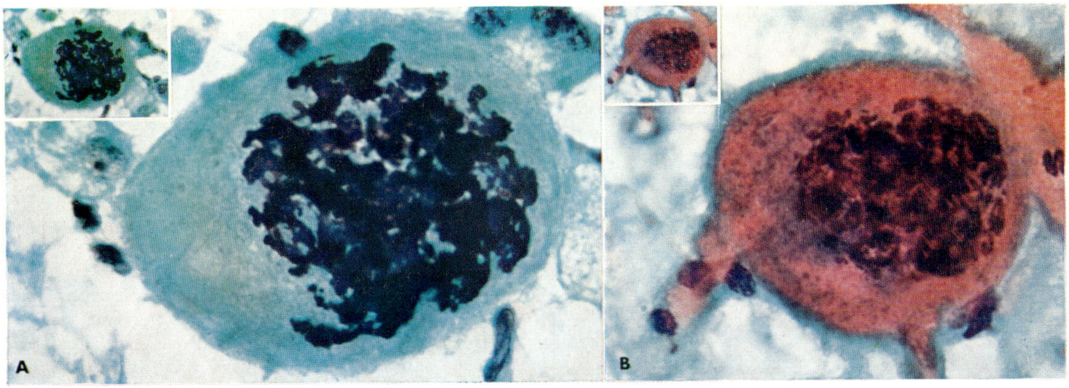

FIG. 44

Abnormal Degenerate Squamous Cells in a Cervical Smear from a poorly Differentiated Invasive Squamous Cell Carcinoma of Cervix

A. and B. At low-power magnification (insets) these giant squamous cells appear to have hyperchromatic nuclei but higher magnification reveals fragmentation of the nuclei (karyorrhexis). (Inset ×200.) (×800.)

The large size of these cells and their grossly abnormal nucleocytoplasmic ratios are highly suspicious, but the degenerative changes which have occurred make them unsuitable for definitive diagnosis. The observation of cells of this type should lead to careful examination of the smear for diagnostic malignant cells. Other areas of the smear from this patient are illustrated in Figs. 35 and 160B.

Smears from carcinoma-in-situ are often clean with only occasional polymorphonuclear leucocytes. The third-type differentiated malignant cell usually predominates and bizarre malignant cells are uncommon. It is possible to recognise in the smear the pattern of the poorly differentiated small cell carcinoma and, less commonly, the highly differentiated keratinising carcinoma. Dyskaryotic cells, especially those of parabasal type, are usually present in con-id erable numbers. A feature which is frequently seen is the presence of narrow strips of pithelium composed of differentiated malignant cells and/or dyskaryotic cells; this occurs secause the abnormal epithelium is readily detached from the underlying basement membrane.

FIG. 45

Differentiated Malignant Cells of Squamous Type in a Cervical Smear from a case of Carcinoma-in-situ of Cervix

A. The low-power illustration shows the large amount and wide variety of abnormal cellular material which may be present in a case of carcinoma-in-situ of the cervix. Striking variations in cell size and shape, including giant cells and an occasional tadpole or fibre shape, are seen. Nuclear size and shape varies considerably and hyperchromatic nuclei are a prominent feature. *Trichomonas vaginalis* is present in considerable numbers. (×200.)

In fields B to K differentiated malignant cells of squamous type are present. These cells show variation in size and shape and have abnormal nucleocytoplasmic ratios. Variation in size and shape of the nuclei with irregular outlines and abnormal chromatin patterns, including clumps and strands, are also demonstrated.

B. Two differentiated malignant cells are seen accompanied by two histiocytes and a few polymorphonuclear leucocytes. (×800.)

C. The large cell with two similar nuclei and finely vacuolated cytoplasm is a histiocyte while the smaller cells lying in close proximity are of malignant squamous type. (×800.)

D. Although these are squamous malignant cells they show coarse cytoplasmic vacuolation. (×800.)

E. The cells with eosinophilic cytoplasm, one of which is binucleated, have degenerate nuclei and are therefore unsuitable for diagnosis although, initially, they are more striking than the accompanying differentiated malignant cells with basophilic cytoplasm and large hyperchromatic nuclei. (×800.)

F. and G. Multinucleated differentiated malignant cells are the striking feature of these illustrations. Note the variation in size, shape and intensity of staining of their nuclei. At centre right are two small cells with degenerate nuclei, one having a basophilic and the other an eosinophilic cytoplasm. (×800.)

H. and J. Malignant cells showing phagocytosis are seen in these fields. The resulting 'bird's-eye' appearance is shown in field J. A benign parabasal cell in field H has an unusual appearance as a result of abnormal eosinophilic staining of the nucleus. In spite of this, the normal chromatin pattern is easily seen. (×800.)

I. The three differentiated malignant cells on the right show moulding of nuclei by closely adjacent nuclei. In this field, as in fields H, J and K, the squamous cells with eosinophilic cytoplasm are degenerate and unsuitable for diagnosis. (×800.)

K. In the centre of this field a large malignant cell with vacuolated basophilic cytoplasm and a small one with pale partly eosinophilic cytoplasm are present. The nuclei of these malignant cells contrast with the degenerate nuclei of the surrounding cells. (×800.)

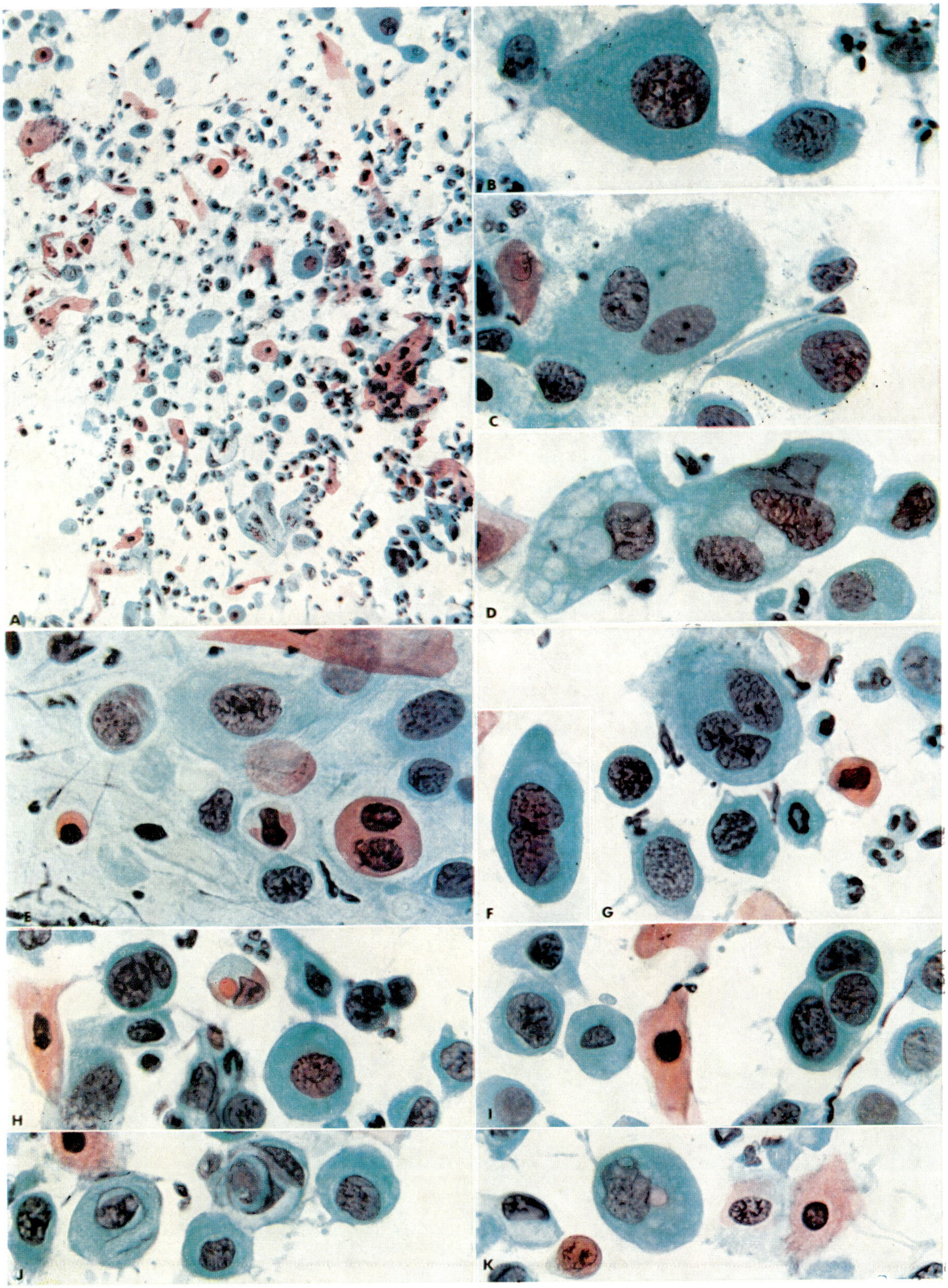

FIG. 45
For detailed description see facing page.

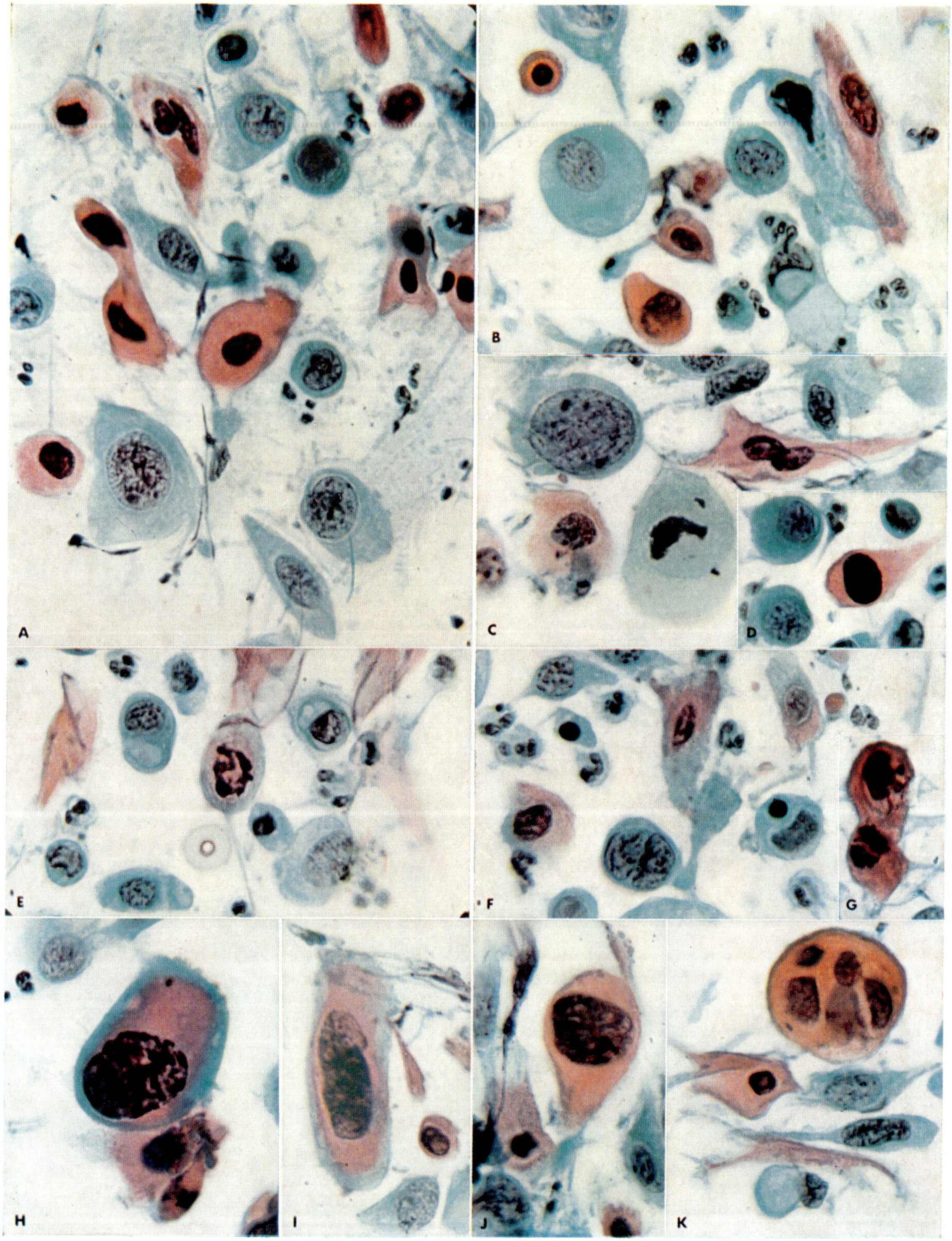

FIG. 46
For detailed description see facing page.

FIG. 46

Differentiated Malignant Cells of Squamous Type and Degenerate Squamous Cells from a case of Carcinoma-in-situ of Cervix

Initially, the cells in these illustrations appear more strikingly abnormal than those from the same specimen illustrated in Fig. 45. Closer examination, however, reveals that many of the most striking cells show degenerative changes making them unsuitable for final diagnosis; their presence indicates that a careful search for diagnostic cells is necessary.

A. This field shows a considerable number of differentiated malignant cells of squamous type but the cells with eosinophilic cytoplasm are all degenerate. (×800.)

B. and C. Differentiated malignant cells are present, one of which has a strikingly large nucleus (C, upper left). In addition to the many degenerate cells with eosinophilic cytoplasm, both fields show degenerate cells with basophilic cytoplasm and densely hyperchromatic nuclei. (×800.)

D. to K. Many bizarre degenerate cells, with eosinophilic cytoplasm and single or multiple hyperchromatic nuclei, are present in these fields. The abnormal nucleocytoplasmic ratio, irregularity of shape and variation in size of these squamous cells would arouse suspicion but a diagnosis should not be based only on such cells. In several of these fields, however, cells exhibiting the typical features of malignancy and, therefore, suitable for diagnosis, are also present. Note the phagocytosis in field F, lower right quadrant, and also the malignant tadpole cell in field K. (×800.)

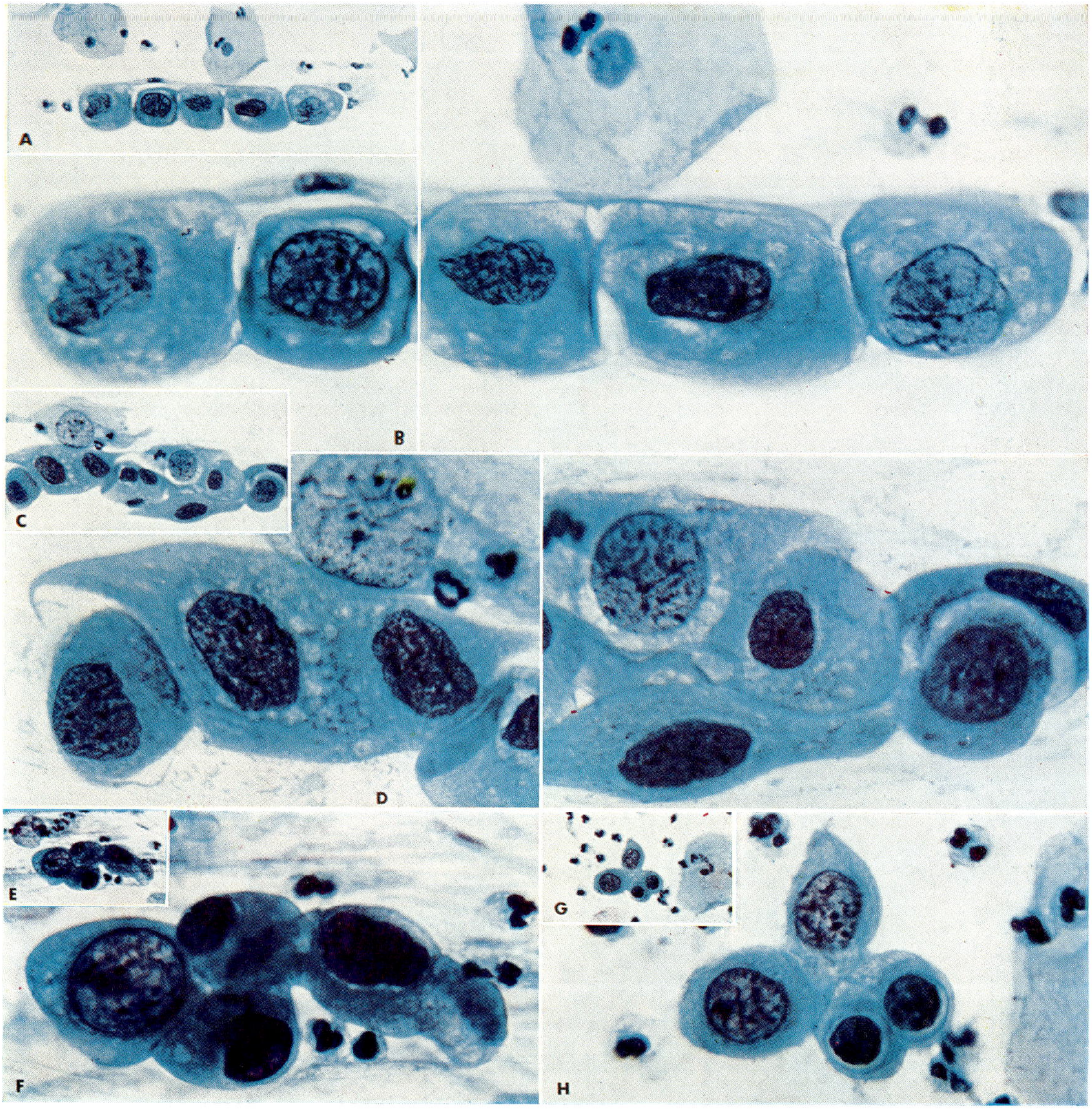

Fig. 47

Differentiated Malignant Cells of Squamous Type in Cervical Smears from cases of Carcinoma-in-situ of Cervix

For detailed description see facing page.

FIG. 47

Differentiated Malignant Cells of Squamous Type in Cervical Smears from cases of Carcinoma-in-situ of Cervix

A. C. E. and G. These low-power fields illustrate elongated strips and groups of malignant squamous cells. These patterns are often seen in cases of carcinoma-in-situ, where the abnormal epithelium is easily detached from the underlying tissue. Groups of cells such as these are especially valuable for diagnosis since comparison can be made between individual cells of a group. (×200.)

B. This strip of five cells shows marked variation in nuclear size, shape and chromatin pattern. The nuclear outlines are irregular, angular and notched. The chromatin patterns are abnormal, with marginal condensation and the formation of striking strands and clumps. Variation in nucleocytoplasmic ratio in these cells demonstrates the difficulty in drawing a distinction between parabasal dyskaryotic cells and differentiated malignant squamous cells. The four outer cells have the abnormal nucleocytoplasmic ratios of differentiated malignant cells of squamous type while the centre cell has a relatively smaller, though abnormal, nucleus and would be classified as a dyskaryotic parabasal cell. (×800.)

D. These cells show similar nuclear features to those described in plate B. Note also the comparatively pale-staining nuclei of the upper two cells in the group in spite of which the nuclei still show the characteristic features of malignancy. The cell at the extreme right of the group shows the typical 'bird's-eye' appearance produced when phagocytosis occurs in malignant cells of squamous type. (×800.)

F. The cell to the left of the field shows the abnormal nucleocytoplasmic ratio, marginal condensation of chromatin and the presence of irregular clumps and strands of chromatin which identify it as a differentiated malignant cell of squamous type. The three cells to the right of this exhibit degenerative changes and although abnormal in size and shape are unsuitable for a final diagnosis of malignancy. (×800.)

H. The two cells to the right of this group show degenerative changes while the two on the left have the characteristic features of differentiated malignant cells of squamous type. (×800.)

Not infrequently nuclei which appear to be strikingly hyperchromatic on low-power examination prove to be degenerate when examined with high-power objectives; this makes them unsuitable for diagnosis. If, however, they show abnormalities in size or shape a further search for cells diagnostic of malignancy is indicated.

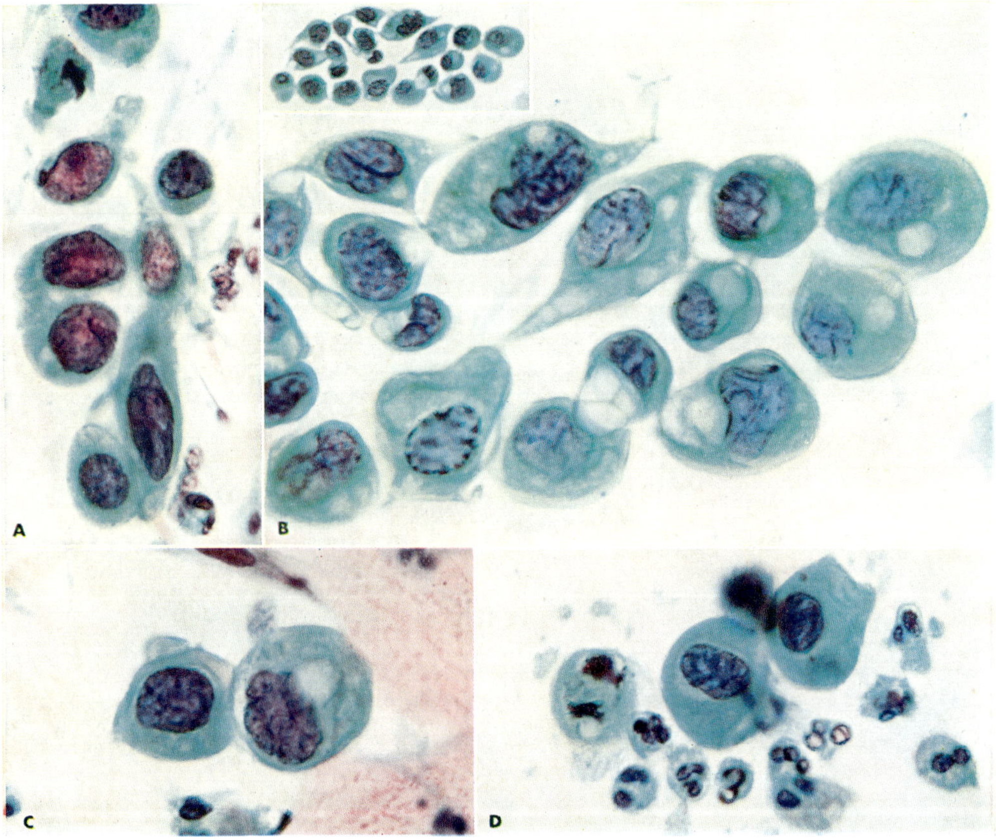

FIG. 48

Differentiated Malignant Cells of Squamous Type and Parabasal Dyskaryotic Cells in a Cervical Smear from a Carcinoma-in-situ of Cervix

A. B. and C. The nuclei of all the differentiated cells in these illustrations show varying degrees of enlargement, irregularity of outline and abnormal chromatin patterns. Many of the cells, especially in field B, show considerable cytoplasmic vacuolation. This change is not infrequently seen in malignant squamous cells and does not necessarily indicate that the cells originate in an adenocarcinoma. Some of the nuclei in field A show eosinophilic nuclear staining, possibly associated with degeneration as the two upper eosinophilic nuclei also show a loss of continuity of the nuclear border. The cell on the right in field C shows marked notching of the nuclear border giving a lobulated appearance. (Inset for B $\times 200$.) ($\times 800$.)

D. The two parabasal cells present have nuclei which show features similar to those of malignant cells but retain a normal nucleocytoplasmic ratio and are, therefore, displaying dyskaryosis. The cell to the left has undergone degenerative change and the fragmented nuclear material initially gives a false impression of a mitotic figure. ($\times 800$.)

The pattern of a smear from a carcinoma-in-situ of cervix may be considerably distorted by the presence of infection, especially of trichomonal origin, which can increase the number of inflammatory cells and produce degenerative changes in benign, malignant and dyskaryotic cells. Consequently the pattern may resemble that in a smear from a case of invasive squamous carcinoma of cervix. Pregnancy can also cause considerable changes in malignant cells. Cytolysis, occurring in both benign and malignant cells, produces denuded nuclei and under these circumstances it may be impossible to determine whether the malignant cells were basically differentiated or undifferentiated.

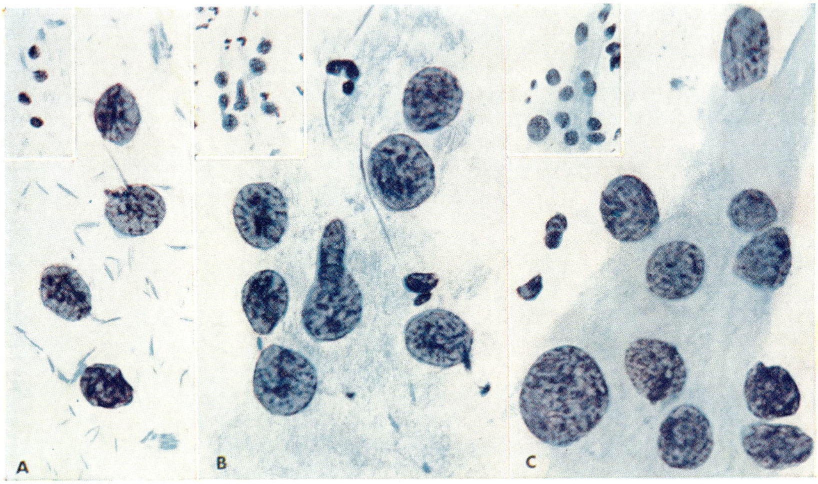

FIG. 49

Denuded Nuclei of Benign and Malignant Squamous Cells in Cervical Smears in Pregnancy

A. These four benign denuded nuclei, seen in a normal pregnancy smear, are uniform in size and oval in shape except for the folding of the lowest nucleus in the field. Their chromatin pattern is normal. (Inset ×200.) (×800.)

B. and C. These denuded malignant nuclei are seen in cervical smears from a pregnant patient who was subsequently found to have a squamous cell carcinoma-in-situ of cervix. The nuclei show considerable variation in size and shape. The chromatin patterns are abnormal with irregular marginal condensation and clumping to form small angular masses. Cytoplasm is seen merely as a diffuse background. (Insets ×200.) (×800.)

ADENOCARCINOMA OF CERVIX

This tumour arising in the endocervical canal is not so readily accessible for cytological examination as the squamous lesion of the exo cervix and squamo-columnar junction but it may extend sufficiently far down the canal for it to be accessible by scraping or it may exfoliate cells which are then available for collection at the external cervical os. The tumour is often indistinguishable on cytological grounds from the adenocarcinoma of the endometrium.

The malignant cells seen usually consist of differentiated malignant cells of adenocarcinomatous type and undifferentiated malignant cells. Occasionally, however, they still resemble tall columnar endocervical cells, an appearance suggestive of their site of origin.

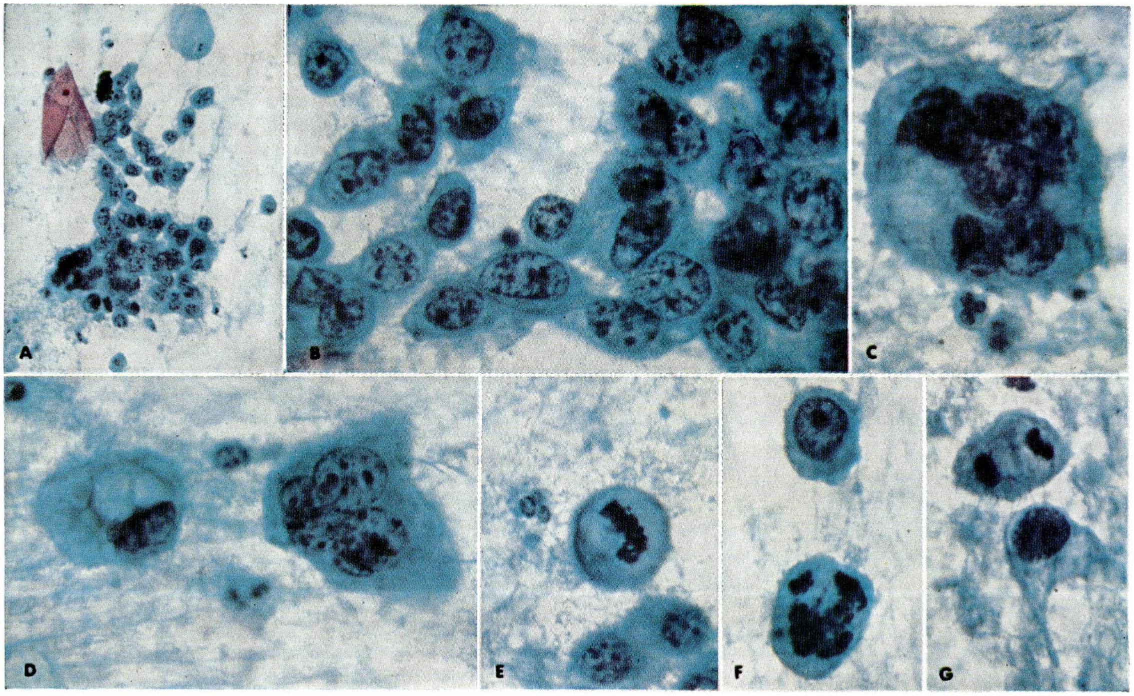

FIG. 50

Differentiated and Undifferentiated Malignant Cells from an Adenocarcinoma of Cervix

A. A large sheet of malignant cells is present and even at this low magnification, marked clumping of chromatin is readily visible. (×200.)

B. The cytoplasm of this group of cells forms a diffuse background and only a few individual cellular outlines can be identified. The nuclei show variation in size and shape and the chromatin patterns are grossly abnormal, with clumping into irregular angular masses which are separated by clear areas. Marginal condensation of chromatin is also visible. (×800.)

C. This giant multinucleated differentiated malignant cell shows a similar pattern of nuclear abnormality to that shown in field B. Cytoplasmic vacuolation is apparent. (×800.)

D. The multinucleated malignant cell shows variation in nuclear size and an abnormal distribution of chromatin. The smaller malignant cell has an eccentric nucleus and vacuolated cytoplasm. (×800.)

E. An apparently normal mitotic figure is seen in this cell, which is probably a tumour cell as it is closely associated with the two differentiated malignant cells lying below it. (×800.)

F. The upper cell is malignant and shows marked uneven peripheral chromatin condensation within the nucleus which also contains a large nucleolus. The lower cell is unsuitable for diagnosis due to the marked degeneration of its nucleus. (×800.)

G. The lower differentiated malignant cell shows the typical eccentric nucleus of a cell from an adenocarcinoma. The upper cells may have recently divided, but the nuclear detail is not sufficiently clearly seen for diagnosis. (×800.)

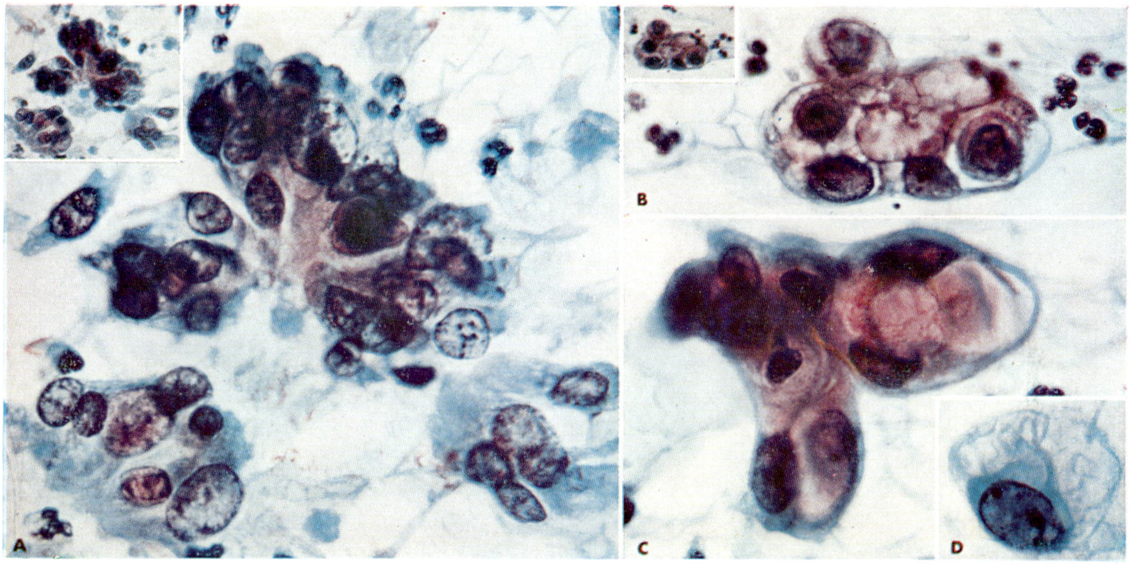

FIG. 51

Differentiated and Undifferentiated Malignant Cells from an Adenocarcinoma arising in the Endocervix

A. Many of these malignant cells show almost complete loss of cytoplasm and are undifferentiated. The two lower groups and the single cell in the upper left corner retain their elongated cytoplasm and have basal nuclei. The overall appearance and the palisaded grouping seen at low power (inset) are suggestive of their origin from endocervical epithelium. The nuclei of all the cells show marked variation in size and shape, with irregular borders. The chromatin is unevenly condensed at the periphery and in clumps within the nucleus. As a result, clear spaces are apparent within several of these nuclei. (Inset ×200.) (×800.)

B. This group of differentiated malignant cells shows the eccentric peripheral location of the nuclei and the coarse cytoplasmic vacuolation characteristic of malignant cells of adenocarcinomatous type. When seen at a lower magnification (inset) this produces a pseudo-acinar effect. The individual nuclei are hyperchromatic with marginal condensation of chromatin outlining their irregular shapes. Prominent nucleoli are present. (Inset ×200.) (×800.)

C. The arrangement of this group of differentiated malignant cells gives the impression of a cluster of cells fanning out from a focal point; this is suggestive of an origin from an adenocarcinoma, and the eccentric nuclei and cytoplasmic vacuolation support this diagnosis. Individual nuclei show marked hyperchromasia and pleomorphism with prominent nucleoli. (×800.)

D. This giant differentiated cell shows the characteristic features of a malignant cell originating in an adenocarcinoma; the nucleus is eccentric and the cytoplasm vacuolated. The large hyperchromatic nucleus shows peripheral condensation and clumping of chromatin. A prominent nucleolus is apparent. (×800.)

It is frequently impossible to differentiate between the malignant cells originating from an adenocarcinoma of the cervix and those from an adenocarcinoma of the endometrium. In the case illustrated, the obviously columnar appearance of some of the malignant cells and their arrangement in a palisade pattern was suggestive of an endocervical origin. This diagnosis was subsequently confirmed on pathological examination of the biopsy and hysterectomy specimens.

ADENOCARCINOMA OF THE ENDOMETRIUM

The cytological diagnosis of adenocarcinoma of the endometrium (corpus uteri) is not as reliable as the diagnosis of malignant lesions of the cervix. The site of the lesion is only directly accessible by a special technique, namely the endometrial aspiration smear. The cytologist is usually dependent, therefore, on exfoliated cells. Their collection is most likely to be successful if smears are prepared from the cellular material which has accumulated in the posterior fornix. A cervical scrape smear will only give positive results if bleeding is occurring or has recently occurred from the endometrium so that exfoliated endometrial cells are present in or near the external os. For this reason, in the menopausal and post-menopausal patient, it is advisable to collect a specimen from the posterior fornix pool as well as by scraping the cervix.

In the post-menopausal patient a history of bleeding or abnormal features of the smear, such as blood-staining, groups of normal endometrial cells, or many superficial squamous cells (indicating unopposed oestrogen activity), necessitate a careful search at high dry magnification for malignant cells.

The malignant cells may be differentiated, of adenocarcinomatous type, or undifferentiated and are often found lying singly or in small groups of two or three. At low-power magnification they may be difficult to identify against their background of erythrocytes, polymorphonuclear leucocytes and histiocytes and, even at high-power magnification, careful observation of the abnormal chromatin patterns and nucleoli of differentiated malignant cells of adenocarcinomatous type is often necessary in order to distinguish them from histiocytes.

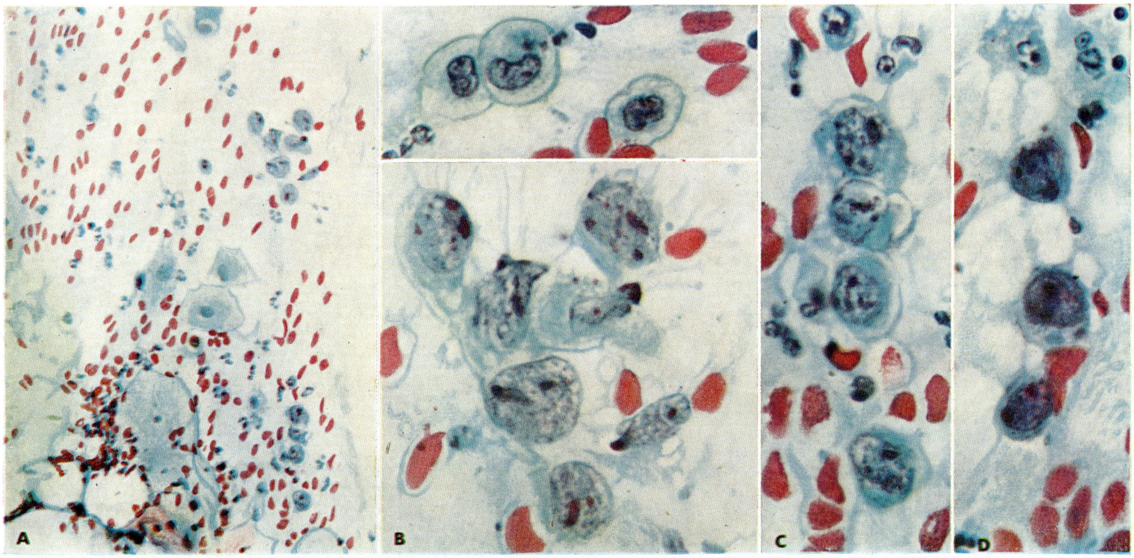

FIG. 52

Differentiated and Undifferentiated Malignant Cells from an Adenocarcinoma of the Endometrium

A. Against a background of erythrocytes and polymorphonuclear leucocytes, two groups of malignant cells are seen in the right half of the field. Note their small size when compared with the superficial and intermediate squamous cells. (×200.)

B. A group of differentiated and undifferentiated malignant cells is illustrated (inset above is a group of histiocytes for comparison), the upper cells showing the eccentric position of the nuclei characteristically seen in adenocarcinoma. In the lower cells of the group, most of the cytoplasm is lost and the appearance is that of undifferentiated malignant cells. They all show variation in nuclear shape and size, multiple or prominent angular eosinophilic nucleoli and chromatin clumping. (×800.)

C. In this group of differentiated malignant cells, the eccentric nuclei are smaller and more uniform in size than those shown in field B but prominent nucleoli and chromatin clumping are still obvious. (×800.)

D. This illustration shows three discrete but poorly differentiated malignant cells with nuclear hyperchromasia and pleomorphism, and also prominent nucleoli. This is the type of small, dark and single malignant cell which may only be recognised by careful scanning at high-power magnification. (×800.)

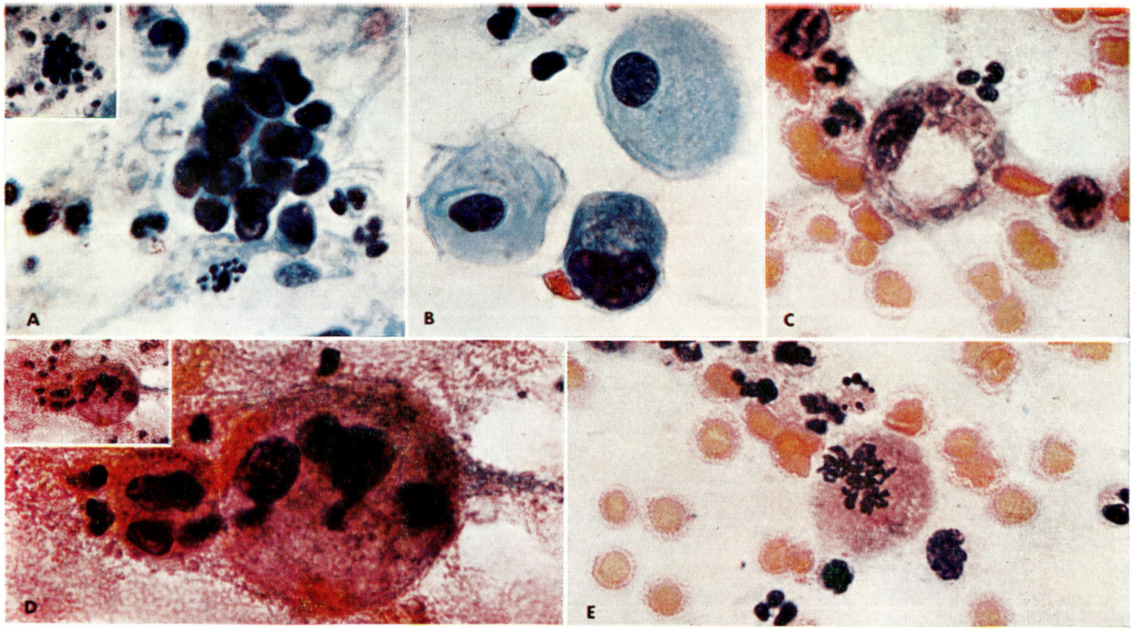

FIG. 53

Benign Endometrial Cells and Differentiated Malignant Cells from an Adenocarcinoma of the Endometrium

A. In the first smear from this post-menopausal patient, the only abnormal finding was this group of endometrial cells. The irregular arrangement and overlapping is characteristic of endometrial cells, with their scanty cytoplasm and irregularly shaped nuclei of uniform size. The group shows no evidence of malignancy, but the presence of these cells led to the examination of a second smear, in which the following cells (B to E) were seen. (Inset ×200.) (×800.)

B. A single differentiated malignant cell is present. The enlarged crescentic nucleus shows a coarsely clumped chromatin pattern; its eccentric position is compatible with an origin from an adenocarcinoma. (×800)

C. This differentiated malignant cell shows both the eccentric nuclear position and the cytoplasmic vacuolation of a cell originating in an adenocarcinoma. Note the blood-stained background of this field and of those shown in fields D and E. (×800.)

D. This giant multinucleated malignant cell shows marked variation in the size of its hyperchromatic nuclei, which also show abnormal chromatin patterns. (Inset ×200.) (×800.)

E. Mitoses are not seen frequently in exfoliated malignant cells and when mitotic figures are seen, they are as likely to be present in histiocytes as in epithelial cells. However, the general configuration of this cell is similar to the one in field C and the abnormal tripolar mitosis suggests that this is a tumour cell. Abnormal mitoses are not confined solely to tumour cells, but in the vast majority of cases their presence is associated with malignant cells. (×800.)

Curettage was performed on this patient and subsequent histological examination of the endometrium showed a predominance of benign proliferative endometrium with a relatively small area of well-differentiated adenocarcinoma. The recognition of endometrial cells in smears from post-menopausal patients should lead to an especially careful high-power examination of the smear for differentiated malignant cells of the type originating in an adenocarcinoma and also for undifferentiated malignant cells, as these cells often occur singly or in very small groups, which may not be noted on low-power screening of the smear.

ADENOCARCINOMA OF FALLOPIAN TUBE

Cells arising from this rare tumour cannot be distinguished from those originating in an adenocarcinoma of the endometrium. It should be born in mind, therefore, that this is a possible, although uncommon, source of otherwise unexplained malignant cells of adenocarcinomatous type in a cervical or vaginal smear.

OVARIAN TUMOURS

Malignant cells from carcinomata of ovary are, in the early stage of the disease, seen only occasionally in vaginal or cervical smears. The most common malignant cells in this condition are those originating from an *adenocarcinoma*, especially of *papilliferous serous type*. These cells are often indistinguishable from malignant cells of endometrial origin, although they may be larger and show marked vacuolation of the cytoplasm.

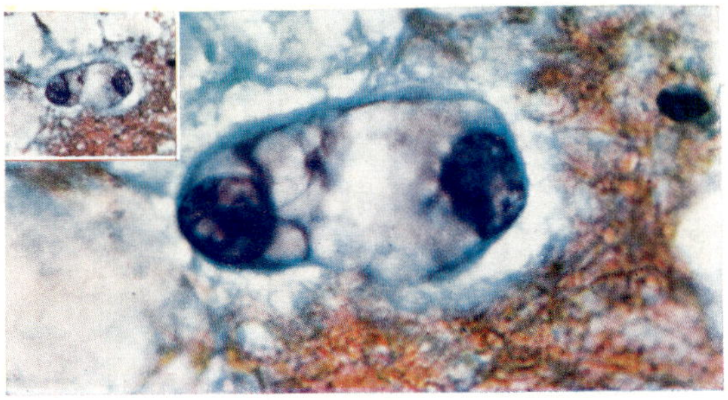

FIG. 54

Differentiated Malignant Cells in a Vaginal Smear from a case of Papilliferous Serous Cystadenocarcinoma of Ovary

A blood-stained vaginal smear from a patient complaining of post-menopausal bleeding showed groups of benign endometrial cells and these two differentiated malignant cells of adenocarcinomatous type. Both cells have eccentric hyperchromatic nuclei, which show abnormal clumping of chromatin, and marked vacuolation of cytoplasm. (Inset ×200.) (×800.)

Endometrial curettings did not show any evidence of malignancy. Subsequent laparotomy revealed bilateral papilliferous serous cystadenocarcinomata.

The other type of malignant ovarian tumour, which may be recognised by changes in the vaginal smear, is the *granulosa-cell carcinoma*. This may give evidence of its presence in two ways: i.e., poorly differentiated or undifferentiated malignant cells may be seen or, when the tumour is endocrinologically active the results of oestrogenic activity are seen, with an increase in the number of superficial squamous cells. This is most striking in the child before the menarche and in the post-menopausal woman.

Other malignant ovarian tumours may exfoliate malignant cells which appear in vaginal smears but they do not have characteristic cell types or smear patterns. In practice, the most important point is to consider the ovary as a possible source of malignant cells in cervical and vaginal smears, if investigation of the cervix and corpus uteri gives negative results.

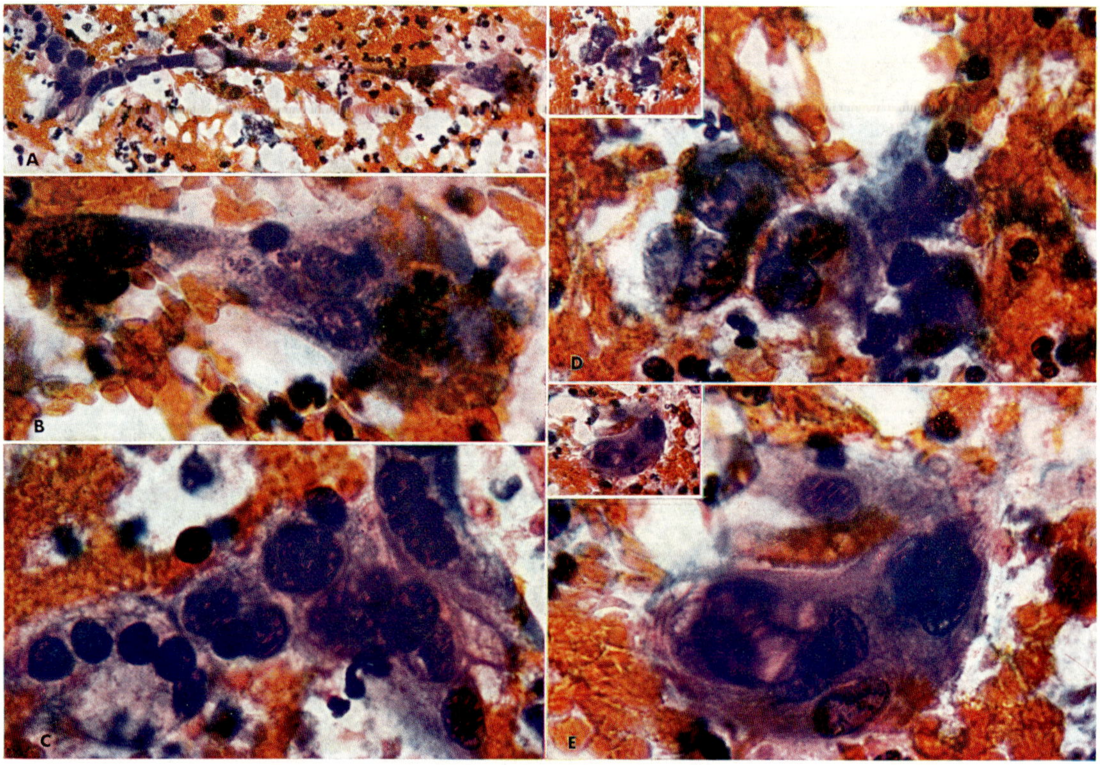

FIG. 55

Vaginal Smear showing Cells arising from a Granulosa-Cell Carcinoma of Ovary

A. In this low-power illustration an elongated group of malignant cells and a giant cell (right margin) are seen against a background of erythrocytes and polymorphonuclear leucocytes. (×200.)

B. At high-power magnification the malignant multinucleated giant cell exhibits marked variation in nuclear size; two of the nuclei show abnormal patterns, with clumping and marginal condensation, of chromatin. (×800.)

C. In this group of malignant cells hyperchromasia and variation in nuclear size are striking features. In some cells crowding of nuclei is associated with moulding of one nucleus by its neighbour. In the lower left quadrant the arrangement of the cells is suggestive of part of a rosette or so-called 'Call-Exner body'. (×800.)

D. These malignant cells show abnormal chromatin patterns with marked peripheral condensation. (Inset ×200.) (×800.)

E. This malignant multinucleated giant cell shows striking variation in nuclear size, shape and hyperchromasia. This type of multinucleated malignant cell was an unusual feature of the histological appearance of the ovarian tumour. (Inset ×200.) (×800.)

METASTATIC TUMOURS

When a metastasis is found in the female genital tract the commonest site for the primary tumour is either elsewhere in the genital tract, in the cervix, endometrium or ovary, or less frequently in other pelvic organs. A previous history of a malignant tumour should not lead to the conclusion that malignant cells seen in cervical or vaginal smears are necessarily of metastatic origin. The possibility of a second primary lesion should always be considered.

The actual appearance of the cells from a metastasis may closely resemble that of malignant cells from primary lesions of the area. Occasionally, however, unusual features may suggest a search for a primary lesion outside the genital tract.

BORDERLINE LESIONS

Borderline lesions occur which are intermediate between benign hyperplasia and carcinoma. The lesions show many of the features of carcinoma, but do not show all the characteristics necessary to establish a diagnosis of malignancy. The exact definition of these lesions, the criteria by which they are distinguished from carcinoma, even of in-situ type, and the question of their relationship to carcinoma, are problems on which there are considerable differences of opinion. They are of practical importance for two reasons. Firstly, such lesions often accompany the presence in adjacent areas of a frank carcinoma. Secondly, it is important to detect these early changes and to follow up carefully patients in whom they are seen, in view of the possibility that some, at least, of these changes may precede the development of malignant disease.

Dysplasia of the squamous epithelium of the cervix and atypical hyperplasia of the endometrium both fall into the category of borderline lesions. Even on pathological examination there may be differences of opinion on the exact nature of the lesion in some cases. It is hardly surprising, therefore, that on cytological evidence it may be impossible to distinguish such conditions from carcinoma-in-situ or even from carcinoma. Occasionally the overall distribution pattern of the abnormal cells, or cellular features, may suggest that the lesion which is present is of borderline type. In practice it is extremely inadvisable to try to make this distinction on purely cytological grounds. The main purpose of the cytological diagnosis is to detect the presence of an abnormality which requires further investigation, by curettage or cone biopsy, to provide a suitable specimen for pathological diagnosis.

In histological sections of a cervix showing dysplasia, the deeper layers of the epithelium show changes essentially similar to those seen in carcinoma-in-situ and in carcinoma of the cervix. Dysplasia is distinguished pathologically from carcinoma-in-situ by the polarity and normal maturation of the more superficial layers of epithelium and from invasive carcinoma by the integrity of the basement membrane. As a result cellular material seen in cytological smears prepared from cervical scrape specimens, which abrade the deeper epithelial layers, may include considerable numbers of cells which may be indistinguishable from those obtained from a carcinoma-in-situ or carcinoma of the cervix. Thus many of the cells illustrated in Figs. 33, 45, 47 and 48 might equally well have been seen in cervical scrape smears obtained from patients with dysplasia of the cervical epithelium.

However, the distribution of abnormal cells is usually somewhat different in dysplasia from the patterns seen in carcinoma-in-situ and in carcinoma. In dysplasia the commonest abnormal cells are dyskaryotic cells of superficial, intermediate or parabasal type as illustrated in Figs. 36, 37 and 47, together with smaller numbers of differentiated malignant cells, usually of third-type. Bizarre differentiated malignant cells of squamous type and undifferentiated malignant cells are rarely seen. This pattern differs only slightly from the typical pattern of carcinoma-in-situ with its predominance of third-type squamous malignant cells and rather smaller numbers of dyskaryotic cells. There is, however, a marked difference from the pattern of invasive carcinoma, which consists mainly of malignant cells, including bizarre differentiated forms and undifferentiated cells, and shows few, if any, dyskaryotic cells.

In vaginal smears from cases of dysplasia and carcinoma-in-situ in which only exfoliated cells are present, dyskaryotic cells may predominate or even be the only abnormal cells seen, as malignant cells from the deeper layers may be absent or present only in very small numbers. This makes it essential to obtain cervical smears from patients whose vaginal smears exhibit any abnormality.

In atypical hyperplasia of the endometrium both the cellular abnormalities and the crowding of glands may be similar to the changes seen in a well-differentiated adenocarcinoma but this condition is distinguished by the absence of either stromal or muscular invasion. However, the exfoliated cells from an atypical hyperplasia may show a considerable resemblance to those from an adenocarcinoma and, occasionally, the cells may be indistinguishable. In some cases endometrial cells are seen which show variation in overall or nuclear size, pleomorphism or hyperchromasia, but do not show all the characteristic features of malignant cells. In particular they do not show the abnormal distribution of chromatin or the large angular eosinophilic nucleoli of malignant cells. Nevertheless the presence of these atypical cells should always indicate further investigation by curettage.

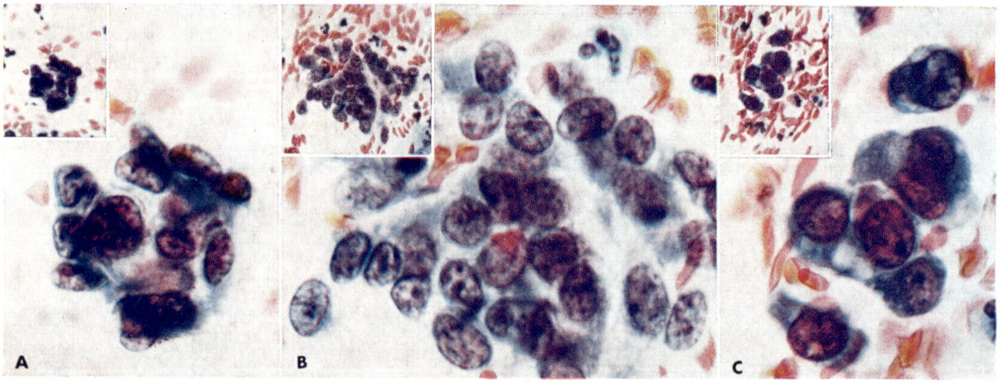

FIG. 56

Atypical Endometrial Cells in a Cervical Smear

These fields all show endometrial cells seen in a cervical smear from a thirty-one-year-old nulliparous patient with a history of a single prolonged episode of bleeding, lasting for four weeks; the specimen was collected at the end of the third week. Curettage revealed areas of atypical hyperplasia of the endometrium. This patient is being carefully followed-up.

A. The endometrial cells show marked variation in the size of their nuclei, two of which are unusually large. They also appear hyperchromatic when compared with those of normal endometrial cells seen in field B. However, no gross abnormalities of chromatin pattern are seen. (Inset ×200.) (×800.)

B. The endometrial cells seen here are comparatively uniform in size and chromatin pattern and show no significant abnormality. (Inset ×200.) (×800.)

C. These endometrial cells show both overall cellular and nuclear enlargement. The nuclei appear to be hyperchromatic but do not reveal an abnormal distribution of chromatin. Occasional single enlarged but round nucleoli can be seen. Although these cells do not show the abnormalities of chromatin pattern seen in malignant cells, they are otherwise similar in several respects to the malignant cells from an adenocarcinoma of the endometrium illustrated in Fig. 52 C and D. (Inset ×200.) (×800.)

CHAPTER 5

HORMONAL EFFECTS ON THE VAGINAL SMEAR

GENERAL CONSIDERATIONS

THE study of vaginal cytology can provide information which contributes to the assessment of gynaecological endocrine function and this information may be useful in a variety of ways. Normal physiological changes occurring in the progression from childhood, through the child-bearing years to the post-menopausal era are reflected in vaginal smears. It is essential for the cytologist to be familiar with these normal patterns as they constitute the background against which the diagnosis of malignant cells must be made. These normal patterns are also the basis for the comparisons which are necessary when vaginal smears are used in the investigation of disturbances of endocrine function. Finally observations on vaginal smears may be a convenient method of assessing the effects of hormone therapy both for endocrine disorders and for non-endocrine diseases.

The advantages of vaginal cytology are readily recognised. The investigation can be performed with minimal patient discomfort. This, combined with the comparatively low cost of the investigation, makes it possible to carry out serial studies over considerable periods of time. The vaginal epithelium is usually more sensitive to endocrine changes than other tissues such as the endometrium and, therefore, failure of end-organ response is very rarely a problem. The vaginal smear mirrors changes in endocrine levels more rapidly than other tissues and the results of a smear can be quickly available. This short time factor can be of practical value in clinical diagnosis and treatment.

In spite of these advantages considerable difficulties are often encountered both in practical application of the method and in studying reported results in the literature. There often appear to be discrepancies between the results of cytological methods of investigation and the results of other methods, such as basal body temperature records, endometrial biopsies and biological assays. In addition, there are often considerable variations in the observations of different workers reporting cytological results. Some of the difficulties are due to technical differences, but others arise from fundamental problems. Differences in technique can give rise to discrepancies in results in several ways, including variations in the site and method of specimen collection, in smear preparation, in fixation and staining methods, and in the use of a variety of subjective and objective criteria for qualitative and/or quantitative assessment of the smears. In practice most of these problems can be resolved by the use of standardised methods, while an awareness of the changes produced by modifications of technique will often elucidate discrepancies between the results of different workers.

There are, however, certain fundamental problems which are less readily soluble. The vaginal smear frequently indicates the end-result of the interplay of several hormones and it may be difficult or impossible to determine whether changes seen in smears are the result of alterations in one or several hormonal factors. This problem is exacerbated by the fact that certain changes in the smear may be produced either by oestrogens or progesterone or androgens, while these hormones may appear to act as antagonists to each other in producing other effects in the smears. In addition non-hormonal factors, such as trauma and inflammation, can produce changes in the smear which are independent of the hormonal status of the patient. Time alone may be responsible for some of the differences that are seen in the results of various assessments of hormone function. Although the vaginal epithelium is relatively rapidly responsive to hormonal changes it inevitably lags behind assays on specimens of blood, while changes in the vaginal smear occur more rapidly than those in the endometrium. Serial studies should be carried out whenever it is possible to do so, as single observations may be misleading. Analogies with animal experiments have also led to some confusion by introducing the terms cornification and mucification into the discussion, when neither of these changes normally occur in the human vaginal epithelium.

It would be impossible to discuss in detail the whole range of clinical problems in which vaginal cytology may be a useful diagnostic method, but a brief description of the technique and practical problems combined with an account of the effect of the various hormones and of the physiological changes seen in vaginal smears during the child-bearing years and the post-menopausal era may provide a basis for the study of abnormal smears due either to malignant change or endocrine disturbance.

COLLECTION METHODS

Specimens should be collected from the upper third of the vagina, as cervical smears do not show the same response to hormonal stimuli and smears from the lower vagina show the presence of anucleated squamous cells. To avoid contamination of the specimen with lower vaginal or cervical cells, the specimen should be collected under direct vision after an unlubricated speculum has been passed. The specimen can then be collected either by scraping the lateral vaginal wall in the upper third with a wooden spatula or by aspirating with a pipette the cellular material collected in the pool in the posterior fornix. Although some workers consider that it is rather more difficult to collect the scrape specimen uniformly on successive occasions, the posterior fornix pool contains cellular material exfoliated from the cervix as well as from the vagina and the material will also include cells exfoliated over a period of time which may show degenerative changes. Scraping of the lateral vaginal wall in its upper third is, therefore, the method of choice. After collection the material should be gently but rapidly smeared on a glass slide and immediately fixed while the cellular material is still moist. Forcible spreading of the smear can break up clumps of cells which may have a diagnostic significance. Drying of the smear before fixation can produce considerable distortion of the cells with the appearance of nuclear enlargement and eosinophilic staining (see Fig. 146).

STAINING TECHNIQUES

The stain used should show good nuclear morphological detail and differential cytoplasmic staining. The standard Papanicolaou technique gives excellent nuclear detail and translucent differential cytoplasmic staining, but it has certain disadvantages for hormonal studies. The method is relatively time-consuming, the cytoplasmic colour changes are somewhat unstable and show variations with alterations in pH, with infection and in pregnancy. Even in ideal conditions there may be several different shades of cytoplasm in the eosinophilic range. The Shorr technique is a rapid staining method producing differential cytoplasmic staining, which is more stable than the Papanicolaou technique and shows a straightforward division between eosinophilic and basophilic or cyanophilic cytoplasm. There is, however, a loss of nuclear detail which makes the method unsuitable for the diagnosis of malignant cells. Modification of the Shorr technique by preliminary staining with haematoxylin improves the nuclear detail, while retaining the advantages of the Shorr cytoplasmic staining, but the nuclear detail is still not as clearly defined as with the Papanicolaou technique and there is some loss of cytoplasmic translucency. In view of the fact that the Papanicolaou technique is the one in routine use in most cytology laboratories, including that of the authors, the smears illustrated in this chapter were all stained by the Papanicolaou method.

THE ASSESSMENT OF THE VAGINAL SMEAR

The smear can only be assessed intelligently in the light of information provided by a full clinical history which must include the patient's age, parity, menstrual history with the dates of menarche, last menstrual period and/or menopause, a detailed account of any relevant symptoms or signs, a note of surgical or radiotherapeutic treatment and a record of the timing and dosage of any hormonal therapy.

The evaluation of the smear depends on a critical assessment of the epithelial cells and their background of non-epithelial cells and micro-organisms. *The vaginal epithelium responds to hormonal stimuli in three ways, by proliferation, maturation and exfoliation*, and these three changes are all reflected in the smear. It is, therefore, necessary to assess the smear not only for the types of squamous cell that are present, but also for their arrangement, whether lying flat and singly or folded and clumped.

Although the epithelial cells provide the essential evidence for the hormonal assessment, it is also important to examine the overall appearance of the smear. First, the smear may be technically unsuitable because of insufficient or excessively thick cellular material or distortion due to drying or because of evidence in the smear that it has been collected from an unsuitable site. Secondly, the smear may show evidence of inflammation, especially trichomonal infection, or cytolysis which makes it unsuitable for hormonal assessment. Finally, although the main hormonal changes occur in squamous cell patterns other epithelial cells, such as endometrial cells and non-epithelial cells, such as histiocytes and polymorphonuclear leucocytes, form a part of certain characteristic hormonal patterns.

In assessing the squamous epithelial cells the problem can be approached in two ways. An attempt can be made to assess the smear quantitatively by counting the proportions of the various cell types present or the observer can rely on a subjective assessment of the smear. The numerical indices may be useful to confirm a subjective impression and also for the inexperienced observer, but two smears which yield similar numerical ratios may in fact create quite different overall impressions. Moreover those indices which depend on cytoplasmic staining reactions are subject to the errors, which can arise from faults in staining technique or from instability of the stain. In spite of these criticisms the numerical indices can provide a useful method of assessing the effect of oestrogens, where the changes are predominantly those of proliferation and maturation, but they are less valuable in assessing progesterone, where the exfoliation pattern is a fundamental point in the diagnosis.

The Numerical Indices

The principle in counting all these indices is the same and consists of counting cells over a wide area of the smear, avoiding any areas showing drying distortion or contamination with cervical columnar or metaplastic cells. This can be achieved by counting the top two cells in each of two hundred separate fields. To obtain the eosinophilic index the percentage of cells with eosinophilic cytoplasm is counted, omitting parabasal cells from the count. This count is most easily carried out at low magnification. Wide variation in results may arise from alterations in pH, from drying and from the shade variation of the Papanicolaou stain. To obtain the karyopyknotic index the percentage of cells with pyknotic nuclei is counted, again omitting parabasal cells from the count. On the strictest definition these pyknotic nuclei should be hyperchromatic, with a complete loss of chromatin pattern and should measure less than 5 μ but, in practice, the average observer recognises a pyknotic nucleus as one measuring less than 6 μ, and this shows a good correlation with oestrogen levels. High dry magnification is necessary to count these nuclei accurately. The karyopyknotic index is more reliable than the eosinophilic index as it is based on morphological rather than colorimetric criteria. It is useful in assessing the effect of oestrogens before the menarche, in the first half of the menstrual cycle, in the post-menopausal patient and in the castrate, but it is less valuable when the effect of oestrogens is modified by other hormones. To obtain the maturation index the percentages of parabasal, intermediate and superficial squamous cells are counted under high dry magnification. In the child-bearing years, when parabasal cells are not normally present, the figure at the right of this triple index will be the same as the karyopyknotic index. However, in the post-menopausal patient and in many disturbances of endocrine function the maturation index provides a wider assessment of changes in the smear than the karyopyknotic index.

Subjective Criteria

These depend essentially on the individual observer's assessment of the proportions of parabasal, intermediate and superficial cells present and their distribution patterns, whether lying flat and singly or folded and clumped. Various schemes have been prepared to classify these impressions, but their use can be misleading by suggesting that a quantitative assessment has been made when in fact they are only numerical or alphabetical expressions of a qualitative judgement. However, the subjective opinion of a trained experienced observer, whether expressed by a formula or a verbal opinion, shows a close correlation with clinical findings and other laboratory evidence. Regardless of whether numerical indices are also used, the ultimate interpretation of a vaginal smear depends on the correlation of clinical symptoms and signs with changes seen in the smear and comparison with known normal physiological patterns and with previous smears from the patient; as a result the final assessment is necessarily subjective.

THE EFFECT OF OESTROGENS ON THE VAGINAL SMEAR

Oestrogens produce proliferation and maturation of the vaginal epithelium. As a result the smears consist mainly of superficial and intermediate squamous cells, which usually lie flat and singly. At the same time as maturation of the squamous cells occurs, there is a disappearance of inflammatory cells, producing a strikingly 'clean' smear. When oestrogens are administered to a patient whose atrophic smear consists mainly of parabasal cells, there is a progressive increase in the percentage of superficial cells, but if therapy is prolonged the percentage of superficial cells may eventually fall below the maximum level which has been attained.

Only oestrogens produce this maturation to superficial cells. In view of the specificity of this change, a count of superficial cells can provide an accurate assessment of the effect of oestrogens, whether endogenous or exogenous. For this purpose the karyopyknotic index is a more reliable method than the eosinophilic index, which is subject to wider variations from technical difficulties and non-hormonal effects.

The effects of oestrogens are modified by progestogens and androgens, so that the assessment of oestrogen activity by vaginal smears is most reliable when the patient is not simultaneously subjected to the effects of other hormones (see "Karyopyknotic index" for details).

A smear showing the changes typical of the effect of oestrogens is illustrated in Fig. 60. This smear, taken at mid-cycle, consists of single flat superficial squamous cells with eosinophilic cytoplasm and pyknotic nuclei, against a clear background. This picture represents the normal physiological peak of oestrogen activity occurring at mid-cycle in the child-bearing years. The effect of exogenous oestrogens in a post-menopausal patient is shown in Fig. 66B. Here the oestrogen therapy has produced superficial cells with eosinophilic cytoplasm and pyknotic nuclei.

THE EFFECT OF PROGESTERONE ON THE VAGINAL SMEAR

In physiological conditions the effect of progesterone alone is not seen, as it normally acts on an epithelium which is also subject to the effect of oestrogens. However, if progesterone is given to the surgically castrated or post-menopausal patient it produces proliferation and maturation of the epithelium, with the appearance in the smear of many intermediate squamous cells. Progesterone alone will not produce eosinophilic superficial squamous cells with pyknotic nuclei. More striking than the proliferative effects of progesterone are the effects on exfoliation. In contrast with the effect of oestrogens, which produce an exfoliation pattern of single, flat superficial cells, progesterone causes desquamation of large compact clumps of folded intermediate cells. Since the maturation effects of progesterone are shared with oestrogens and androgens it is this typical exfoliation pattern which should be regarded as characteristic of the effect of progesterone.

When the effect of progesterone is seen in a patient who is already under the influence of oestrogens the smear shows a decrease or even, if the dosage of progesterone is sufficiently high, an almost complete disappearance of superficial squamous cells. The smear again consists of clumps of folded intermediate cells. In this way progesterone appears to 'antagonise' the effect of oestrogens, but this appearance is due to the massive desquamation of intermediate cells, rather than to an inhibition of proliferation.

Owing to the preponderance of glycogen-rich intermediate cells the smear pattern produced by progesterone often shows marked cytolysis with denuded vesicular nuclei, only occasional intact folded intermediate cells and a background of Döderlein's bacilli.

The effect of progesterone in physiological conditions is seen in the illustration of a premenstrual smear in Fig. 63 and in the normal pregnancy smear in Fig. 70. In each case the predominance of intermediate cells, their arrangement in compact groups and their folded edges are shown. A cytolytic smear, with Döderlein's bacilli in the background is seen in Fig. 71.

THE EFFECT OF ANDROGENS ON THE VAGINAL SMEAR

Androgens produce proliferation and maturation of vaginal epithelium, but only to intermediate cell level. If androgens are administered to a surgically castrated or post-menopausal patient with an atrophic smear, there is an increase in the number of intermediate cells, which eventually predominate in the smear. Unlike progesterone, androgens do not produce a clumped folded exfoliation pattern. The cells may be flat or folded, usually lying singly or in small groups.

When androgens are administered to a patient under the influence of oestrogens there is a decrease in the number of superficial cells, and again a predominance of intermediate cells, but this differs from the effect of progesterone by the absence of dense clumps of folded cells. The precise end-result depends on dosage effects of oestrogen/androgen balance.

In view of the fact that the proliferative and maturing effects of androgens are shared by oestrogens and progesterone and that their exfoliation pattern does not have the strikingly characteristic features which enable one to recognise progesterone activity, there is considerable difference of opinion on the specificity of androgenic effects. Pundel, especially, recognises an androgenic smear pattern and describes an androgenic cell, which is an intermediate cell with hypochromatic cyanophilic, often vacuolated, cytoplasm and a pale oval vesicular nucleus. Other workers consider that androgenic smears may represent only the effects of minimal oestrogen or progesterone activity.

THE VAGINAL SMEAR IN INFANCY AND CHILDHOOD

Specimens can be collected in infancy and childhood by modifications of technique. Especial care should be taken to avoid introducing infection. In the infant, material can be collected by aspiration with a plastic pipette and rubber suction bulb. The pipette is inserted very gently with the suction bulb compressed. In this way the risk of collecting cellular material from the lower vagina is decreased. When the pipette comes to rest the pressure on the bulb is released and the specimen is then aspirated by suction. Usually only a small amount of cellular material is obtained and rapid smearing and fixation are especially important. In childhood a similar technique can be employed, but in the older child the preliminary introduction of a nasal speculum before inserting the pipette avoids contamination by vulval or cervical cells.

For the first three or four days of life the vaginal smears consist mainly of intermediate cells in groups with some superficial squamous cells. From the fifth to fifteenth day there is a gradual change in the smears with a disappearance of superficial and then of intermediate cells, which are replaced by parabasal cells. During this time erythrocytes and polymorphonuclear leucocytes appear in the smear. Occasionally, frank vaginal bleeding may occur.

From this time onwards to the menarche the smear consists predominantly of parabasal cells, with occasional intermediate cells, and polymorphonuclear leucocytes are usually present in small numbers. Clinically the thin epithelium of childhood is significant because it is vulnerable to infection and severe vaginitis can occur in this age group. Otherwise, the cytological pattern in this age group is not subject to cyclic variations and is comparatively stable. A single smear may, therefore, give useful information. For example, the effect of an oestrogen-producing ovarian tumour can readily be recognised in the vaginal smear by the appearance of significant numbers of superficial squamous cells.

THE MENSTRUAL CYCLE

Although a cyclic pattern is recognisable in the vaginal smears of most women during the child-bearing years, this pattern can show many minor variations. Clinically the cycle is not infrequently longer than twenty-eight days and this prolongation of the cycle is reflected in the smears. When the cycle is prolonged the extended phase, in an ovulatory cycle, is the one preceding ovulation, with a relatively constant interval of fourteen days from ovulation to menstruation. This prolongation of the proliferative phase is shown in the changes in the smear patterns. Again occasional anovulatory cycles occur in otherwise normal women and these too are reflected in the smear patterns. Moreover, slight changes in oestrogen/progesterone balance can produce variations in the patterns seen in the second half of the cycle. For these reasons it is advisable to take serial smears, preferably for two or three cycles, for diagnostic purposes. Daily serial smears usually pose considerable practical difficulties and are only essential around mid-cycle in order to establish the exact day of ovulation. Otherwise smears can be taken at intervals of two or even three days for the remainder of each cycle, excluding the first five days of the menstrual flow. In certain circumstances a single smear on a particular day of the cycle may provide information to answer a specific clinical question but, in general, it is unwise to base a diagnosis on a single smear. More reliable results are obtained from the study of serial smears especially if they are correlated with endometrial biopsies.

The cyclic pattern in a normal ovulatory twenty-eight-day cycle, free from inflammatory changes, can be divided into six phases: the menstrual phase, the post-menstrual phase or early proliferative phase, the late proliferative phase, the ovulatory phase, the secretory phase and the late secretory or pre-menstrual phase.

During the menstrual phase, which occupies the first five or six days of the cycle, the most striking features of the smear are the large numbers of erythrocytes and the numerous groups of endometrial cells. Histiocytes and polymorphonuclear leucocytes are also present in considerable numbers. The squamous cells are often obscured by mucus and erythrocytes, especially during the first two or three days, but they are exfoliated in large numbers, compact groups of folded intermediate cells predominate initially but some superficial cells and occasional parabasal cells are present. After the first few days the number of superficial cells increases.

During the post-menstrual or early proliferative phase, which occupies the next two or three days, the predominant feature of the smear is the presence of large numbers of histiocytes intermingled with polymorphonuclear leucocytes. Endometrial cells may also still be present but tend to appear singly or in small groups which are now more loosely arranged than during the early part of the menstrual flow. The squamous cells show a predominance of the intermediate type but the number of superficial cells gradually increases and they constitute about 20-30 per cent. of the total squamous cells by the ninth day. Some cells may still be arranged in groups and show cell folding during this period but the majority are flat and lie singly.

Smears in the proliferative phase, from the ninth to the fourteenth day, show a progressive decrease in the number of polymorphonuclear leucocytes and histiocytes. The squamous cells, which now tend to lie flat and singly, show a steady increase in the proportion of superficial cells which may constitute 50-60 per cent. of the total squamous cell population.

The so-called ovulatory smear pattern seen on the fifteenth and sixteenth days is not in itself adequate evidence that ovulation has occurred, as this type of smear may be produced by unopposed oestrogen activity arising from a variety of causes. The actual appearance of the smear, however, is very striking as it consists almost entirely of superficial squamous cells with eosinophilic translucent cytoplasm and pyknotic nuclei. These cells lie flat and singly against a clear background, which is almost completely devoid of polymorphonuclear leucocytes.

The next stage in the cycle, the secretory phase, continues from the seventeenth to the twenty-fourth day. For a diagnosis to be made, on cytological evidence, that ovulation has occurred the changes of this phase must follow the typical ovulatory smear. These classical diagnostic changes are only seen in a proportion of patients. Three changes are seen during this period: there is a gradual decrease in the number of superficial cells present, the intermediate cells show cytoplasmic wrinkling, folding and clumping, and polymorphonuclear leucocytes reappear in the smear. Of these changes the most striking is the progressive increase in the number of intermediate cells and their arrangement in thick clumps of folded cells.

During the pre-menstrual phase, from the twenty-fifth day onwards, there is an increase in the number of polymorphonuclear leucocytes present in the smear. The squamous cells now consist mainly of the intermediate type, arranged in groups of folded cells, but in some patients superficial squamous cells may again appear in increased numbers at this stage. During the last day of the cycle, prior to overt menstruation, an occasional group of well-preserved endometrial cells may be seen in the smear.

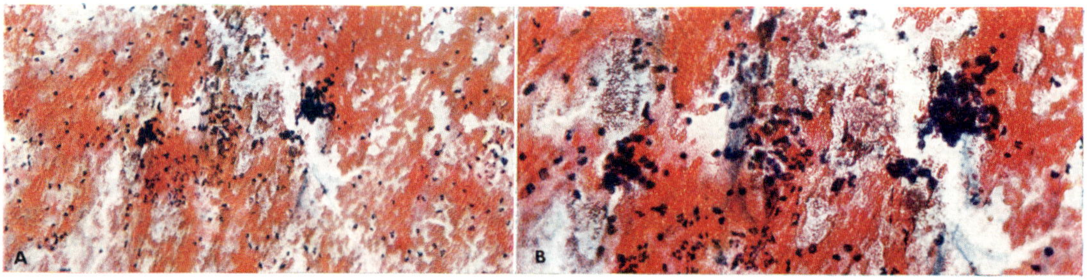

FIG. 57
The Menstrual Phase
This vaginal smear, taken on the fourth day of the menstrual flow, shows three dark compact groups of endometrial cells against a background of erythrocytes, polymorphonuclear leucocyctes and basophilic mucus. (×100.) (×200.)

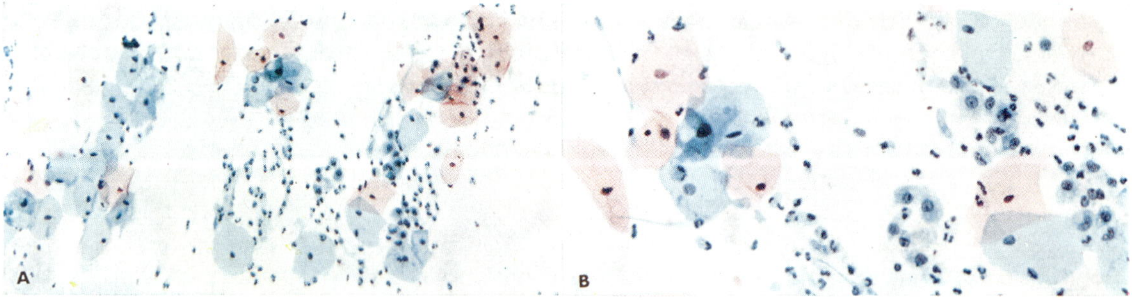

FIG. 58
The Post-menstrual Phase
These illustrations of a vaginal smear, taken on the seventh day of the cycle, show squamous cells against a background consisting of numerous histiocytes intermingled with polymorphonuclear leucocytes. The squamous cells are mainly of intermediate type, and their vesicular nuclei are well shown in the basophilic cells and in some of the eosinophilic cells in field B. In field A it is apparent that some of the cells are still arranged in groups, while others lie singly. The grouped cells still show some folding of cytoplasm, while the single cells lie flat. (×100.) (×200.)

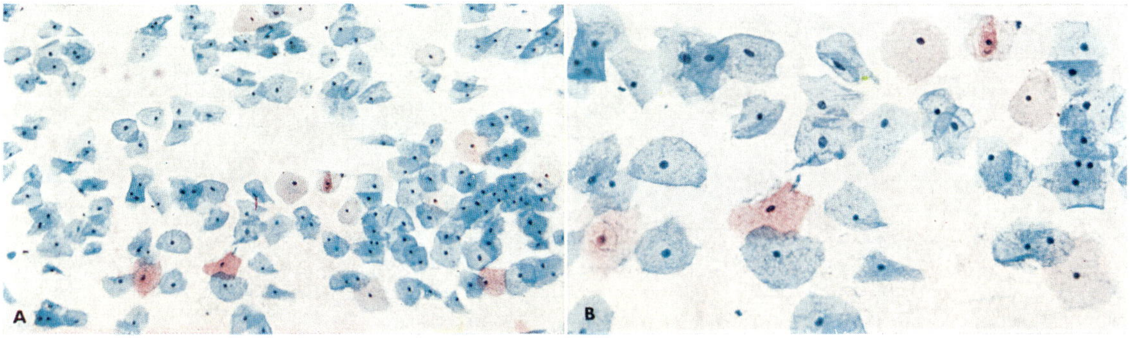

FIG. 59

The Proliferative Phase

In this vaginal smear, taken on the eleventh day of the cycle, the squamous cells are now lying flat and singly. Although the cytoplasm of many of these cells remains basophilic, a considerable proportion of them have pyknotic nuclei; this is seen better at higher magnification in field B. Under these circumstances the use of the eosinophilic index as evidence of maturation would yield misleadingly low values. (×100.) (×200.)

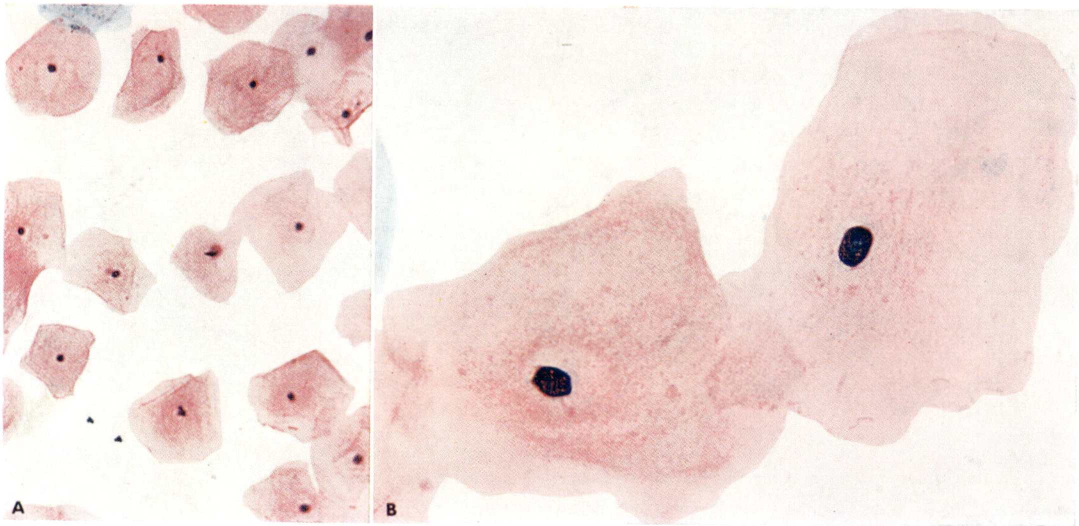

FIG. 60

The Ovulatory Phase

On the fifteenth day of the cycle this vaginal smear is seen to consist almost entirely of superficial squamous cells, with translucent eosinophilic cytoplasm and pyknotic nuclei, lying flat and singly. Note the clean background. The presence of this type of smear is not, in itself, sufficient evidence for cytological diagnosis of ovulation; its appearance must be followed by smears showing the typical clumped folded intermediate cells of the secretory phase. (×200.) (×800.)

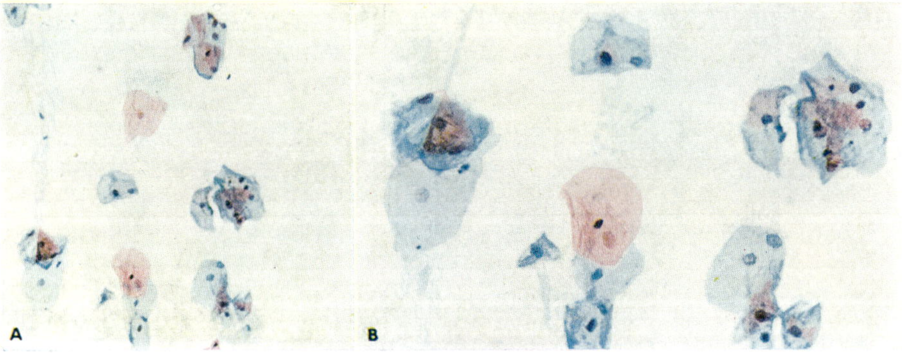

FIG. 61

The Secretory Phase—Early

This vaginal smear, taken on the eighteenth day of the cycle, shows a striking decrease in the number of superficial squamous cells. The cytoplasm of the intermediate cells appears less translucent and the edges are folded, although they are arranged singly or in small groups. (×100.) (×200.)

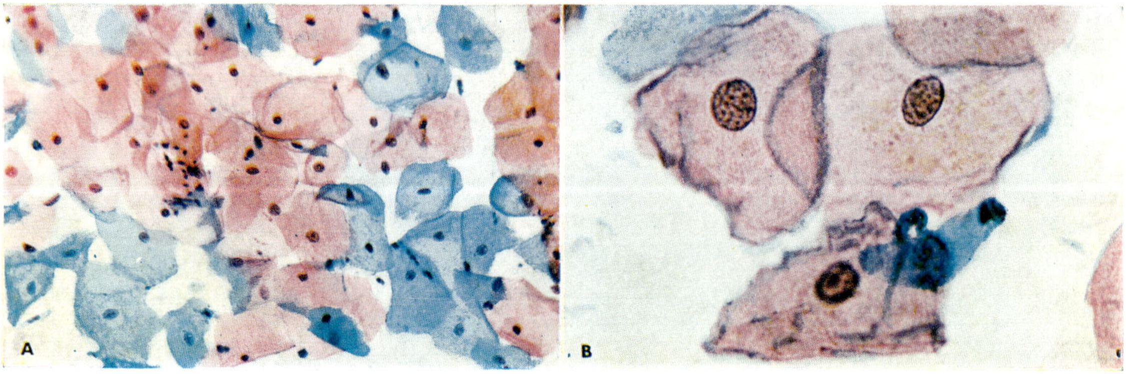

FIG. 62

The Secretory Phase—Late

This crowded sheet of squamous cells is seen in a vaginal smear on the twenty-second day of the cycle. The cells are predominantly intermediate cells with vesicular nuclei and folded wrinkled cytoplasmic borders, arranged in overlapping sheets. Despite their eosinophilic staining field B confirms that these are intermediate cells with vesicular nuclei and folded cytoplasm. This again demonstrates the fallibility of assessment by a purely colorimetric method when the Papanicolaou staining technique is employed. (×200.) (×800.)

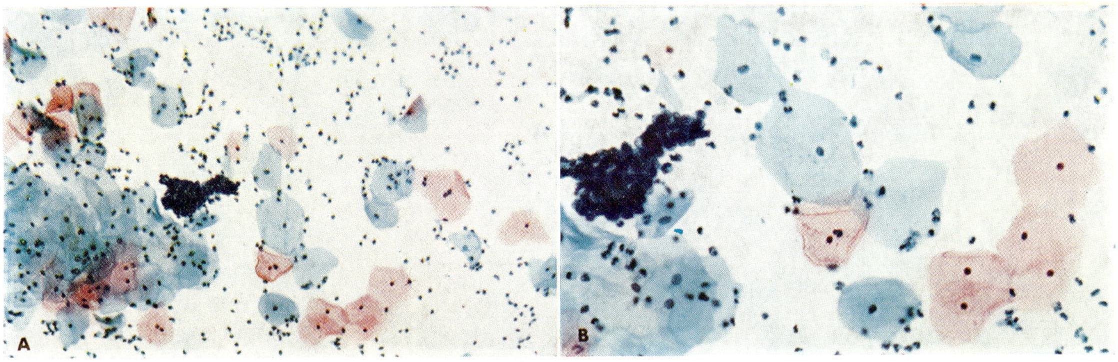

FIG. 63

The Pre-menstrual Phase

These illustrations of a vaginal smear, taken on the twenty-eighth day of the cycle, show the features of the pre-menstrual smear. A single group of small dark endometrial cells is seen in the centre of field A. To the left of it the smear consists mainly of a large group of overlapping intermediate squamous cells with rolled cytoplasmic edges. To the right a few superficial cells, which may reappear at this stage, are seen. The background of the smear contains polymorphonuclear leucocytes, which are now present in large numbers. Field B shows the various features of the smear at higher magnification. (×100.) (×200.)

DISTURBANCES OF THE MENSTRUAL CYCLE

Using the normal biphasic pattern as a basis for comparison it is possible to use serial smears as a part of the investigation of various disturbances of menstrual function. Essentially the information required is evidence of cyclical hormonal activity and evidence of ovulation. The information obtained can be used not only for diagnosis but also to control treatment, whose ultimate aim is the restoration of a biphasic cyclic pattern.

THE EFFECT OF PROGESTOGENS ON THE VAGINAL SMEAR

The use of progestogens both therapeutically and as oral contraceptives is rapidly increasing. In practice they are usually combined with oestrogens in varying dosage. They may be prescribed in a variety of ways, but these fall into three main groups: a daily dose from the fifth to the twenty-fifth day of the cycle, or daily in the second half of the cycle from the fifteenth to the twenty-fifth day, or as continuous therapy. For contraceptive purposes, in Great Britain, they are usually prescribed from the fifth to the twenty-fifth day of the cycle. A brief consideration of the changes produced by this pattern of administration is important as, only too often, these changes may be seen in smears without an accompanying history of hormone therapy. The other patterns of administration are usually a part of planned therapy and therefore less likely to be overlooked when interpreting the smears. Cyclic fifth to twenty-fifth day use of a progestational agent produces the appearance early in the cycle of those changes usually seen in the smear in the secretory phase. By the ninth or tenth day the smear may show a predominance of intermediate cells with folded cytoplasm and by the thirteenth or fourteenth day these cells are aggregated in clumps. This predominance of folded intermediate cells persists throughout the cycle, but is most striking in the first half of the cycle when normally it would not be seen. Other smear patterns also occur in patients on progestational therapy. The preponderance of intermediate cells is associated with the occurrence of cytolysis; cytolytic smears with denuded nuclei, fragments of cytoplasm, groups of intermediate cells with frayed cytoplasm and Döderlein's bacilli in the background are often seen. Not infrequently, smears showing inflammatory changes appear in patients who have had cyclic progestational therapy for some time; such smears are unsuitable for hormonal assessment.

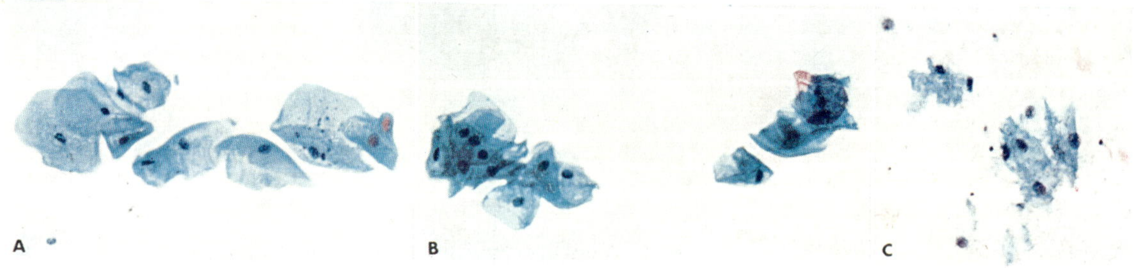

A B C

FIG. 64

The Effects of Progestogens on Vaginal Smears

A. This illustration of a vaginal smear, taken on the ninth day of the cycle (the fifth day of therapy), consists of intermediate squamous cells with folded cytoplasmic borders; their arrangement shows some degree of clumping. (×200.)

B. A vaginal smear, taken on the thirteenth day of the cycle (the ninth day of therapy), again consists of intermediate cells, but cytoplasmic folding and close clumping of the cells into groups are now marked features. (×200.)

C. In this illustration of a vaginal smear, taken on the fifteenth day of the cycle (the eleventh day of therapy), cytolysis is seen. Denuded vesicular nuclei, fragments of cytoplasm and the tattered edges of the otherwise intact intermediate cells are also seen. (×200.)

THE VAGINAL SMEAR AT THE MENOPAUSE AND IN THE POST-MENOPAUSAL ERA

Both clinically and cytologically the changes during the menopause and the post-menopausal era show enormous individual variations. Only occasionally is the menopause a definite event with a sudden disappearance of the cyclic occurrence of menstruation. More commonly there are increasing irregular intervals between periods and a decreasing duration for the menstrual flow. This may make it difficult to date the occurrence of the menopause with any certainty. In practice it is usual to consider that any patient with an interval of a year since her last period is post-menopausal. However, where the patient's problem is the recurrence of bleeding it is wiser to consider that an interval of six months since the last period makes such a recurrence significant and warrants further investigation. Even more striking than the variation in the pattern of the actual menopause is the gradation in the incidence and severity of menopausal and post-menopausal symptoms, which range from minor discomfort to considerable disability. It is, therefore, hardly surprising that the changes seen in vaginal smears during and after the menopause should include a wide variety of patterns. Several factors contribute to the complexity of the problem. The first factor is the persistence in some patients of cyclic changes in the smears; this may continue for months or even for a few years after the cessation of the menstrual flow. As a result serial smears may still be necessary for the assessment of the post-menopausal patient, especially in the early post-menopausal years. The second factor is the immense individual variation of the time interval which elapses before a recognisable post-menopausal pattern emerges. In occasional patients, even fifteen or twenty years after the menopause, an atrophic pattern may still not be seen. The third factor is the frequent incidence of infection in the post-menopausal patient with an atrophic smear. This causes clinical symptoms and signs, but it also produces changes in the vaginal smear which may create a false impression of hormonal activity. The fourth factor is the technical difficulty which may be encountered in this group of patients; often the cellular material is scanty and drying may occur before fixation, with consequent eosinophilic staining, apparent nuclear enlargement and the presence of dried mucus streaks (see Fig. 146).

Nevertheless, a consideration of some of the common patterns seen may make it possible to identify the occasional markedly abnormal pattern which arises from unusual hormonal activity of endogenous or exogenous origin. Familiarity with the common variants in post-menopausal smears is also a safeguard against incorrect identification of benign epithelial cells or non-epithelial cells as malignant cells.

At the menopause or in the early post-menopausal stage there is usually a decrease in the number of superficial cells while, at the same time, parabasal cells, previously almost completely absent, appear in significant numbers. Occasionally patients may have several anovulatory 'periods' at the menopause and in these cases the superficial cells, instead of decreasing, may predominate in the smear as a result of the unopposed oestrogen activity. The latter type of smear is illustrated in Fig. 72.

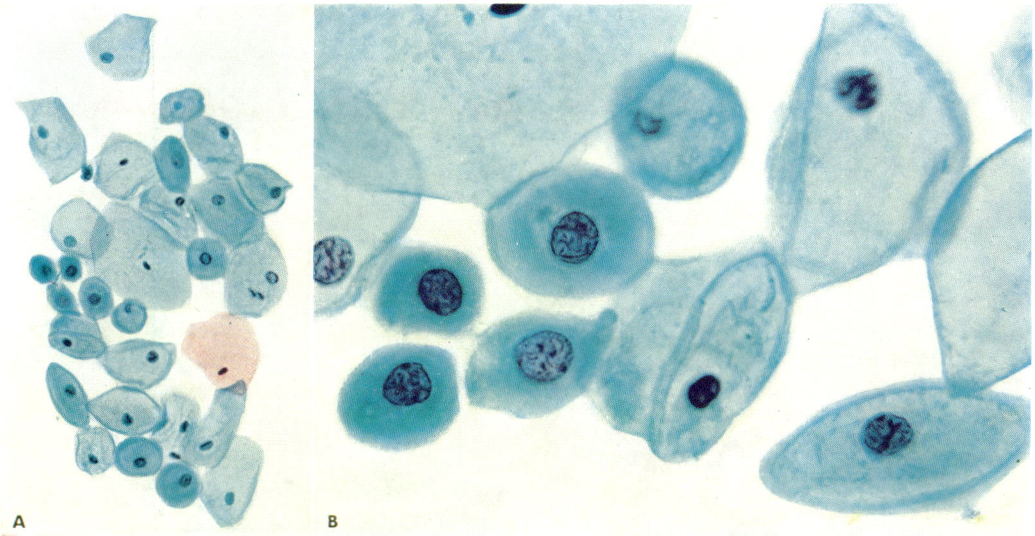

FIG. 65

The Menopausal or Early Post-menopausal Smear

The vaginal smear illustrated here was obtained from a forty-four-year-old patient whose periods had been occurring at intervals of two or three months for the previous nine months. Her last menstrual period was two months prior to this smear being taken.

A This clean smear consists of superficial, intermediate and parabasal cells with a predominance of intermediate cells. (Parabasal cells were not present in vaginal smears taken one year previously.) (×200.)

B. Parabasal cells are clearly seen on the left. The two intermediate cells in the lower right quadrant have rolled edges, which produce a navicular appearance. (×800.)

G

After the menopause the general tendency is towards a decrease and eventual disappearance of superficial cells and an increase of intermediate and/or parabasal cells. Eventually two patterns emerge with either intermediate or parabasal cells predominating, although it is apparent that in the most elderly group of patients the majority will ultimately show a single pattern with a predominance of parabasal cells.

When intermediate cells predominate, as illustrated in Fig. 66A, the smears are usually clean and the background is comparatively free of polymorphonuclear leucocytes. The arrangement of the intermediate cells may show clumping while folding of the cytoplasm may also be seen. When this is the case it may be difficult to distinguish the smear with any certainty from a pregnancy smear. This dilemma can be resolved by the administration of small doses of oestrogens, to which the menopausal or post-menopausal patient is very sensitive, so that the smear shows a striking increase in the number of superficial cells, as illustrated in Fig. 66B. In contrast the smear of the pregnant patient will show no significant response to this treatment. Two variants are seen on the intermediate cell pattern. The cells may have flat pale-staining cytoplasm and hypochromatic vesicular nuclei; this type of smear has been attributed to androgenic effects. The smear consisting of intermediate cells may show cytolysis, as seen in Fig. 20A and B; this type of smear is not suitable for definitive hormonal diagnosis as it may represent the end-result of a number of combinations of hormonal activity. In practice it serves only to exclude high unopposed oestrogen activity.

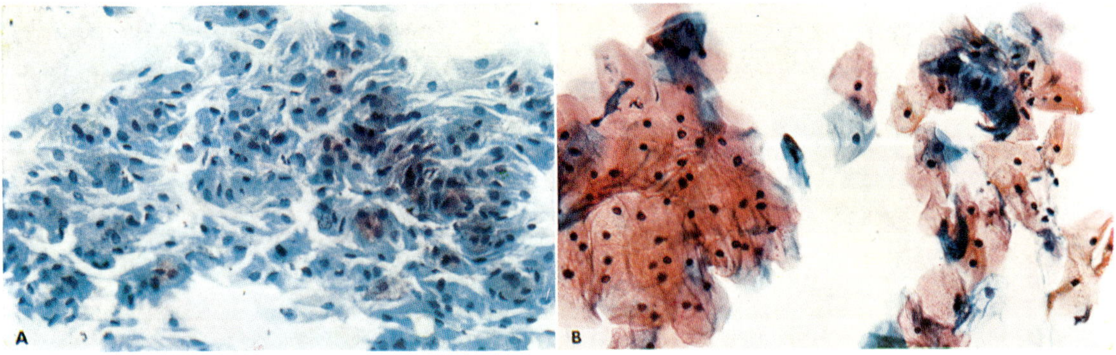

FIG. 66

The Effect of Oestrogens on a Post-menopausal Smear of Intermediate Cell Type

A. This vaginal smear was taken from a fifty-six year old woman four years after the menopause; it consists mainly of intermediate cells. Occasional polymorphonuclear leucocytes are present in the background. (×200.)

B. After two weeks treatment with dienoestrol cream to alleviate the symptoms of vaginitis, the smear now shows a predominance of superficial cells, with some intermediate cells, while the background of the smear is clean. (×200.)

When parabasal cells predominate in the atrophic type of smear the cells are of the type illustrated in Fig. 3. Frequently the smear shows sheets of larger parabasal cells and some intermediate cells. This 'crowded' menopausal pattern, illustrated in Fig. 67, shows patchy eosinophilic staining of the cytoplasm and close crowding of nuclei. These parabasal cells may also show cytolysis and small groups of denuded nuclei may be present, as seen in Fig. 20C and D. These nuclei, whether with or without their cytoplasm, appear large and hyperchromatic; this may arouse suspicion until careful examination reveals the uniformity of their chromatin patterns. When deeper parabasal cells predominate and basal cells appear, infection is often present and the smear may then show the features illustrated in Figs. 5 and 18. Even in the absence of inflammatory changes atrophic post-menopausal smears frequently contain numerous histiocytes including multinucleated giant forms, as illustrated in Figs. 13 and 68. Once recognised these present little difficulty, but some of their features might suggest the presence of differentiated malignant cells of adenocarcinomatous type to an inexperienced observer.

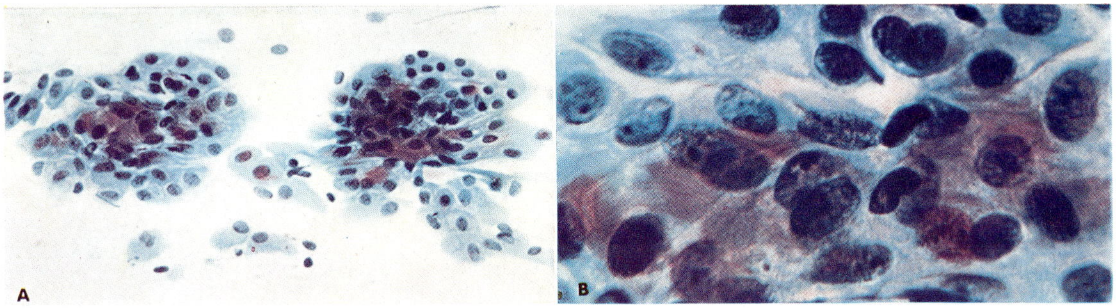

FIG. 67

The Crowded Atrophic Post-menopausal Vaginal Smear

A. This low-power illustration shows compact groups of parabasal and deep intermediate cells. The cytoplasm displays patchy eosinophilic staining and the nuclei appear to be crowded together. (×200.)

B. Higher magnification reveals that the nuclei have normal chromatin patterns, also that eosinophilic nucleoli are present in the nuclei on the left of the field. (×800.)

The nuclei of these crowded groups of cells may cause diagnostic difficulty in two ways. They may be confused with endometrial cells but endometrial cells, would have less cytoplasm, eccentric nuclei and more coarsely granular chromatin patterns. They may also be mistakenly diagnosed as malignant cells, but their benign nuclei do not show the chromatin abnormalities characteristic of malignant cells.

One type of smear which should attract attention in the established post-menopausal patient is that consisting mainly of superficial squamous cells. It may arise from exogenous oestrogen therapy, as illustrated in Fig. 66B, and the first step in the investigation should be an attempt to exclude such treatment as a cause of the changes in the smear. However, it may also arise from the effect of endogenous oestrogens as seen in Fig. 69. When this is the case it is advisable to investigate the patient further, by examination under anaesthesia and diagnostic curettage, with careful follow-up.

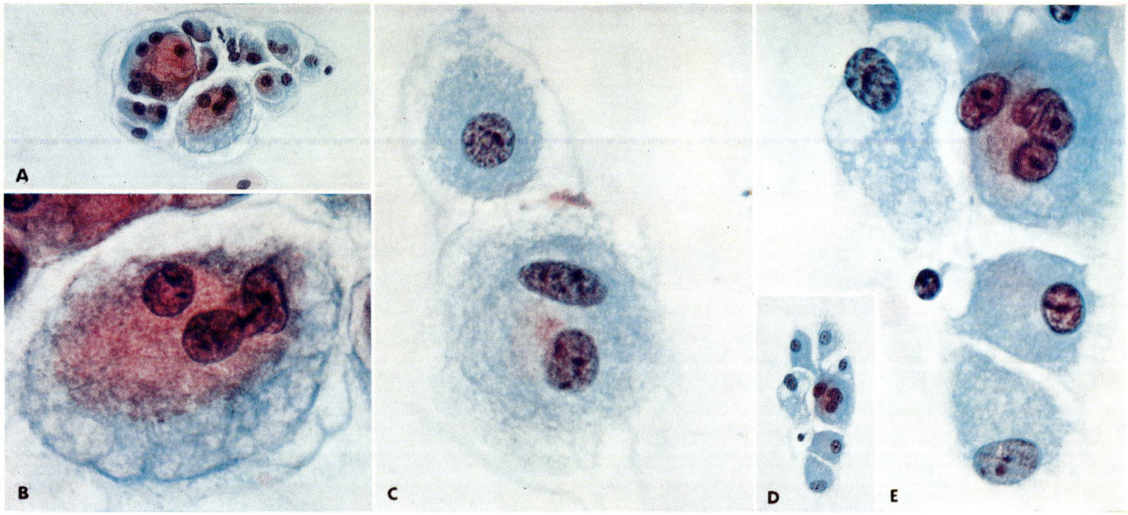

FIG. 68

Multinucleated Histiocytes in a Post-menopausal Vaginal Smear

A. A group of histiocytes including multinucleated forms seen in a vaginal smear from a patient with post-menopausal atrophy. (×200.)

B. Note the large size of this histiocyte which has foamy cytoplasm and multiple nuclei of uniform size. (×800.)

C. These histiocytes show the typical foamy cytoplasm and prominent round chromatin clumps evenly distributed within the nuclei. (×800.)

D. The arrangement of these cells with 'tails' of cytoplasm is suggestive of parabasal cells or metaplastic endocervical cells. (cf. Figs. 3 and 22.) (×200.)

E. At higher magnification, however, their foamy cytoplasm and eccentric nuclei with marked nuclear membrane and prominent chromatin patterns identify these cells as histiocytes. A comparison of this illustration with the malignant cells seen in Fig. 52 shows that eccentric nuclei and foamy cytoplasm are common to both histiocytes and those cells originating in an adenocarcinoma of the endometrium, but the malignant nuclei show variation in size, shape and chromatin distribution in contrast with the uniform nuclei of the histiocytes. (×800.)

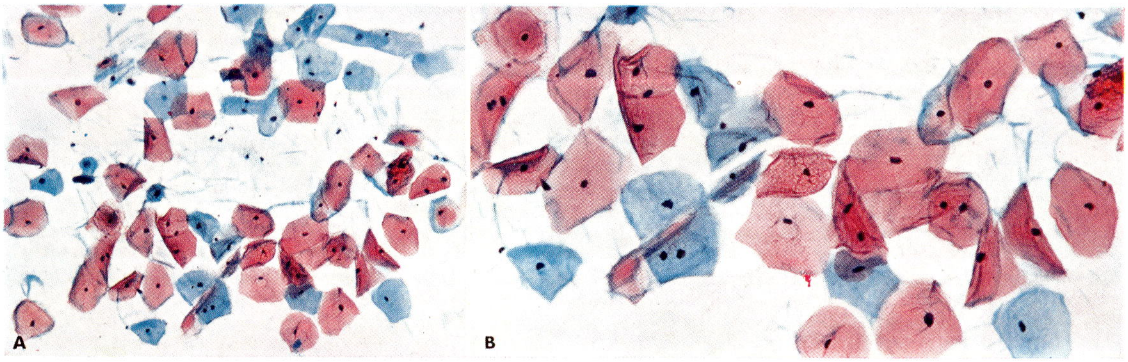

FIG. 69

An Atypical Late Post-menopausal Vaginal Smear

This vaginal smear was obtained from a sixty-six-year-old patient, twenty-four years after the menopause. She complained of vaginal bleeding and clinical examination revealed only a cervical polyp. There was no history of oestrogen therapy.

A. The smear consists mainly of superficial cells. (×100.)

B. Higher magnification reveals a striking preponderance of superficial cells with eosinophilic cytoplasm and pyknotic nuclei. (×200.)

This type of smear indicates high unopposed oestrogen activity. In the post-menopausal patient, after excluding exogenous sources of oestrogen, further investigation is advisable with curettage forming an essential part of the investigation. In this patient curettage was performed and the endometrium showed marked benign cystic hyperplasia. The polyp consisted of benign cervical tissue with dilated Nabothian follicles. The patient is being followed up and has not had any further episodes of bleeding.

THE VAGINAL SMEAR IN PREGNANCY AND THE PUERPERIUM

In pregnancy the main feature is the disappearance of the cyclic changes from the smear and the appearance of a stable pattern, which persists throughout the pregnancy. This pattern is usually established by the end of the first trimester but is often seen much earlier. Although the presence and persistence of a characteristic pattern is compatible with a normal pregnancy and its maintenance may provide useful evidence of the satisfactory progress of a known pregnancy, this pattern should not be used for the diagnosis of pregnancy, since similar changes may be seen in pre-menstrual smears, in the presence of corpus luteum cysts and at the menopause.

The vaginal smear in normal pregnancy consists predominantly of intermediate cells. These cells are arranged in dense clumps and show marked folding of their cytoplasm and rolling of their cellular borders. As a result individual cells may be described as 'navicular' cells or as 'oyster' cells, depending on their shape. The nuclei are vesicular and a significant proportion (10-25 per cent.) appear elongated and cigar-shaped. The intermediate cells are rich in glycogen, which produces areas of yellow or orange staining in the cytoplasm. This high content of glycogen makes these cells especially suitable for true cytolysis in the presence of micro-organisms of the species *Lactobacillaceae*. When this cytolysis occurs the smear then consists of denuded vesicular nuclei against a background of fragments of cytoplasm and Döderlein's bacilli. Although this type of cytolytic smear is not normally suitable for hormonal diagnosis, its presence in pregnancy is compatible with normal progress and it may persist throughout pregnancy. Cytolysis is occasionally responsible for a white vaginal discharge which responds readily to local antibiotic treatment.

Superficial cells are usually only present in small numbers in normal pregnancy smears. During the first trimester their presence is less significant than later in the pregnancy and the percentage of superficial cells may vary during this first three months, but should not exceed 20-30 per cent. of the total squamous cells. After the first trimester, few superficial squamous cells are seen and they should not exceed 10-15 per cent. of the squamous cells in the smear. In the evaluation of any individual patient attention must be paid to serial smears during pregnancy, as the numerical values given are only an approximate guide to the upper limits of normal levels. In a particular patient a persistent increase in the percentage of superficial cells may well represent a significant abnormality, even though the value obtained at any one count falls within normal limits.

Parabasal cells are rarely seen in a smear in normal pregnancy and their presence is most often due to inflammatory changes.

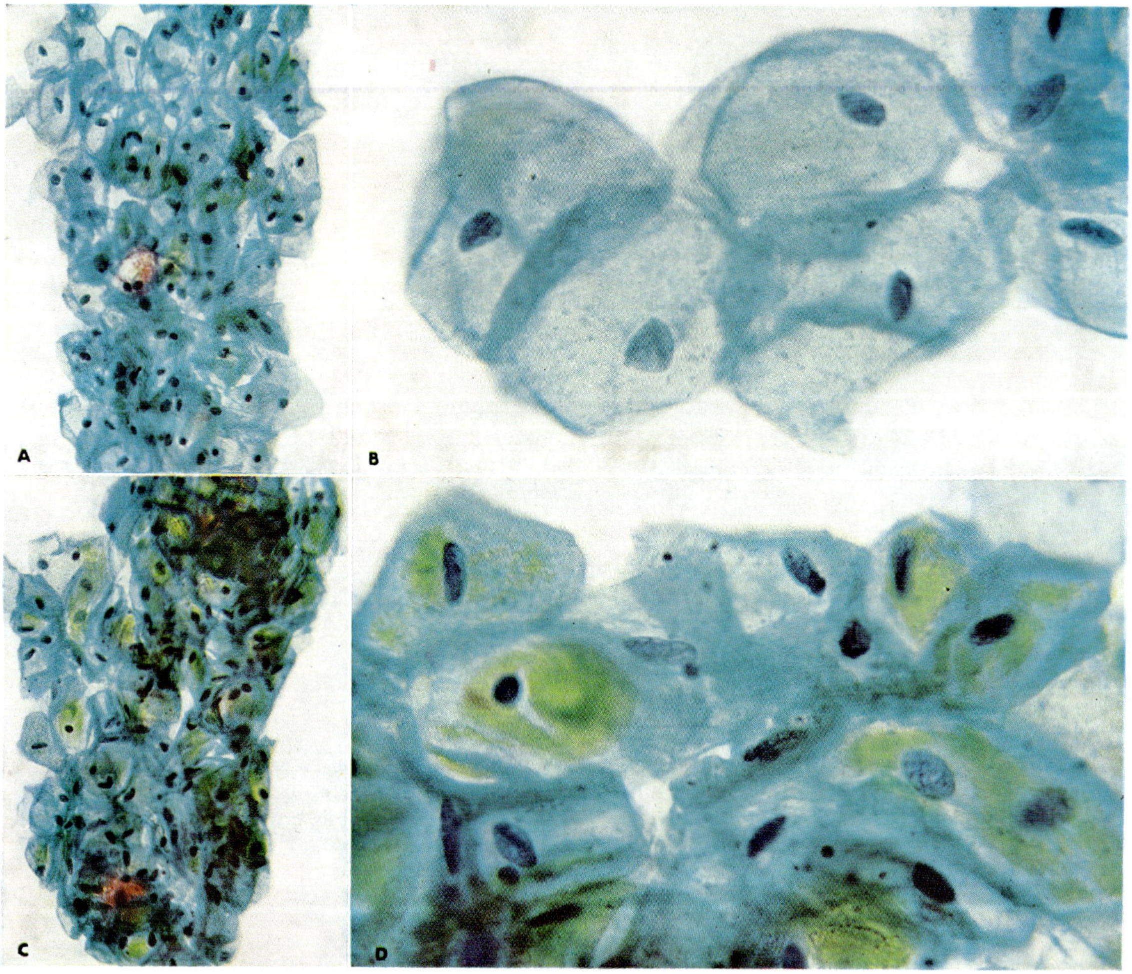

FIG. 70
The Normal Vaginal Smear in Pregnancy
For detailed description see facing page.

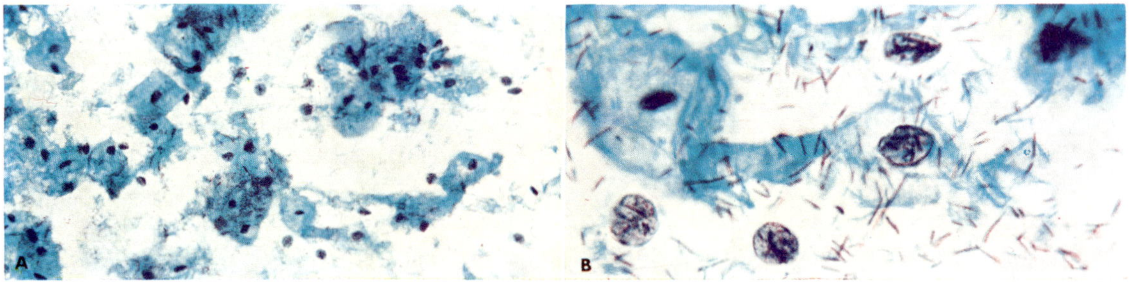

FIG. 71
The Cytolytic Vaginal Smear in Normal Pregnancy
For detailed description see facing page.

FIG. 70

The Normal Vaginal Smear in Pregnancy

This smear was taken at the thirty-sixth week of a normal pregnancy.

A. This closely clumped overlapping grouping of intermediate squamous cells is a smear pattern compatible with a normal pregnancy. (×200.)

B. The individual intermediate cells show folded cytoplasm with rolled cellular borders. The eccentric nuclei are vesicular and some show elongation to a 'cigar shape'. In consequence, there is some variation in nuclear shape in the cell group, but the nuclei are otherwise uniform in size and chromatin pattern. (×800.)

C. This compact overlapping cell group shows areas of yellow staining in the predominantly basophilic cytoplasm. (×200.)

D. The yellow staining of the cytoplasm is due to the presence of large amounts of glycogen. Several cells have 'cigar-shaped' nuclei, but note their normal nucleocytoplasmic ratios. (×800.)

FIG. 71

The Cytolytic Vaginal Smear in Normal Pregnancy

This smear was taken at the twenty-eighth week of a normal pregnancy.

A. Initially this smear appears 'dirty'. The clumps of intermediate cells have ragged edges with frayed cytoplasm. Numerous denuded nuclei are seen, with many detached fragments of cytoplasm. (×200.)

B. Higher magnification shows the denuded vesicular nuclei and their background of fragments of cytoplasm. Many Döderlein's bacilli are present and polymorphonuclear leucocytes are conspicuously absent. Note the 'cigar-shaped' nucleus and rolled cytoplasmic border of the intermediate cell on the left. (×800.)

This type of non-inflammatory cytolytic smear is compatible with a normal pregnancy.

A variety of changes in the normal pregnancy smear have been associated with the approach of term and with post-maturity. The changes associated with the approach of term include a decrease in the compact clumping of the intermediate cells, with the appearance of single intermediate cells, and a slight increase in the number of superficial cells in the smear; in the cytolytic pregnancy smear, a decrease in the cytolysis may be seen. There is often difficulty in correlating these changes with clinical events, as there may be a considerable discrepancy between the calculated maturity of a pregnancy and the onset of labour.

Abnormal Smears in Early Pregnancy

Although the normal pregnancy pattern is not necessarily diagnostic of pregnancy, certain abnormal patterns may indicate either that the patient is not pregnant or that her pregnancy is not continuing. Occasionally, the smear in a patient with amenorrhoea and uterine enlargement may show only superficial squamous cells and this is not compatible with normal pregnancy. This pattern is occasionally seen in patients at or approaching the menopause and helps to differentiate their amenorrhoea from the amenorrhoea of pregnancy. More often, however, the menopausal and pregnancy smears show too many similar features for cytology to be helpful in diagnosis.

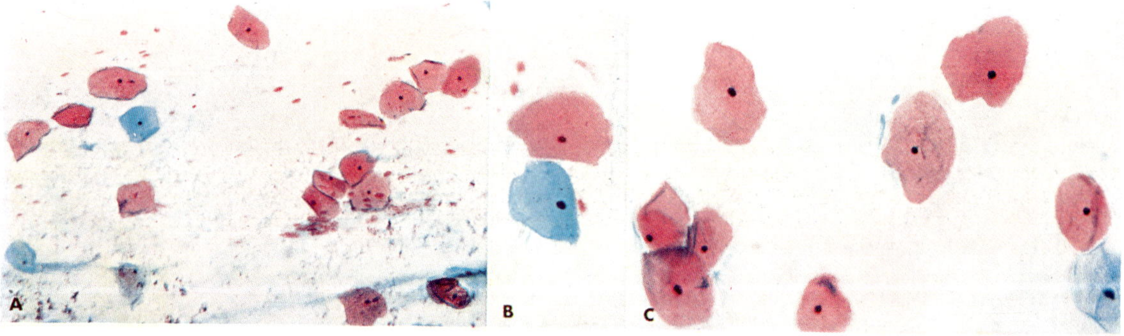

FIG. 72

A Vaginal Smear from an Amenorrhoeic Patient with a Bulky Uterus

This smear was taken from a forty-year-old patient with a history of two months amenorrhoea, who was found to have a bulky uterus.

A. B. and C. The smear consists almost entirely of single flat superficial cells. (×100.) (×200.)

This type of smear is not compatible with a normal pregnancy. Other pregnancy tests were also negative and subsequent laparotomy revealed that the uterine enlargement was due to fibroids.

In the patient who has been pregnant, but whose pregnancy is no longer continuing, the smears show a striking change. Single flat superficial squamous cells predominate against a background of polymorphonuclear leucocytes and histiocytes. This type of smear may appear before abortion actually occurs, as well as being present at the time of an abortion. Not all abortions show this picture in the smear; when abortion does not arise from a hormonal imbalance the smear may show no significant abnormality until the abortion actually occurs.

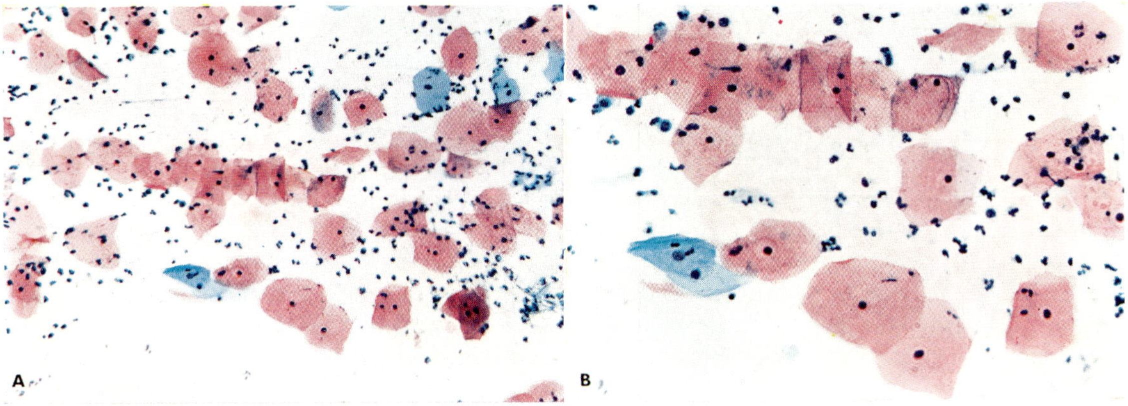

FIG. 73

The Vaginal Smear in an Abortion

This smear was taken at the twelfth week of pregnancy from a patient who had had some vaginal bleeding for the preceding week and abdominal pain for the preceding twenty-four hours. This patient had previously had six successful pregnancies and three abortions. After admission the patient had a spontaneous complete abortion within twenty-four hours.

A. The smear consists mainly of superficial squamous cells with a background of polymorphonuclear leucocytes and histiocytes. (×100.)

B. Despite the numerous polymorphonuclear leucocytes, the superficial squamous cells do not show evidence of inflammation, and their predominance in the smear is, therefore, a significant abnormality. (×200.)

Towards the end of the first trimester there may be an increase in the proportion of superficial cells in the smear, but even at that stage of pregnancy this type of smear with its overwhelming preponderance of superficial cells is not compatible with a normal pregnancy.

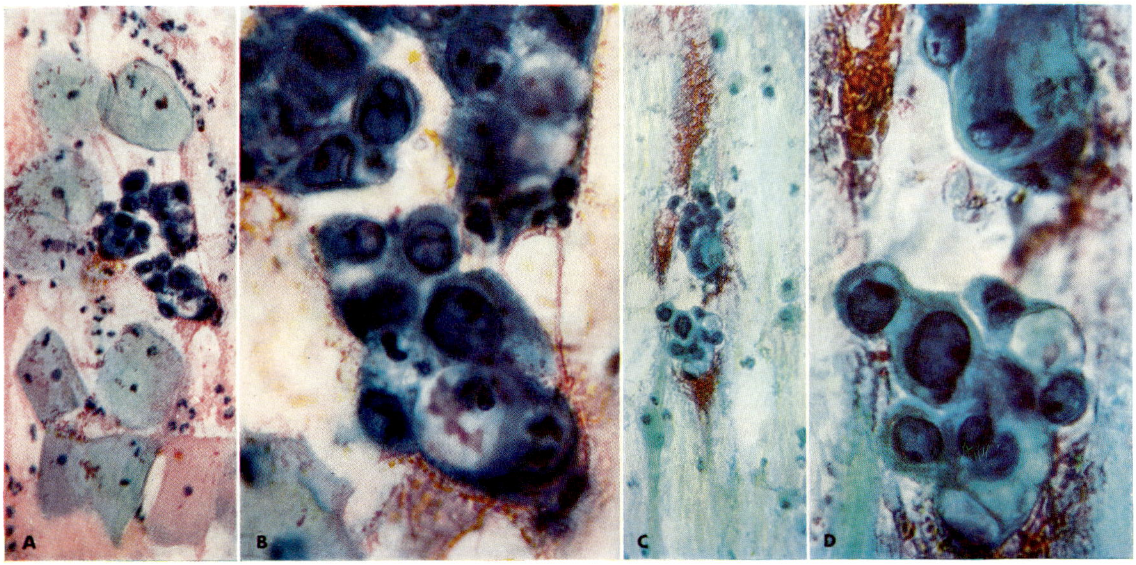

FIG. 74

A Post-abortion Vaginal Smear

This smear was taken from a patient who gave a history of an apparently complete, eleventh week, abortion in the previous twenty-four hours.

A. The squamous cells show a mixture of superficial and intermediate types; they lie singly and only the cell in the right upper quadrant shows minimal rolling of its cytoplasmic borders. A large group of cells with densely hyperchromatic nuclei is present in the centre of the field. Polymorphonuclear leucocytes are seen in the background. (×200.)

C. Further groups of cells with hyperchromatic nuclei are seen against a background of erythrocytes and histiocytes. (×200.)

B. and D. Both these high power fields reveal that the cell group consists of degenerate cells; infiltration by leucocytes is seen in field B. Degenerative changes make these cells unsuitable for definitive diagnosis; it seems probable that they are of decidual or trophoblastic origin and the hexagonal cellular outlines and adequate cytoplasm of the cells in the lower groups suggest a decidual origin. (×800.)

Abnormal Smears in Late Pregnancy

In view of the difficulty already discussed in determining clinically the actual maturity of a pregnancy and in relating maturity to the onset of labour, there is not such a clearly defined normal pattern to provide a basis for comparison in late pregnancy as there is in early pregnancy. In late pregnancy, too, great care must be taken that inflammatory changes seen in the smears, with the appearance of parabasal cells, are not credited with a hormonal significance. Occasionally, however, patterns are seen which differ strikingly from the normal pregnancy smear and these fall into two main types. One resembles the smear which may be seen in an abortion, with a striking predominance of superficial cells. A less common change is the disappearance of the normal clumps of folded intermediate cells and their replacement by single parabasal and rounded intermediate cells. In the absence of inflammation, an association has been described between this type of smear and the occurrence of placental insufficiency and post-maturity. Nevertheless, normal pregnancy patterns may persist after intra-uterine foetal death.

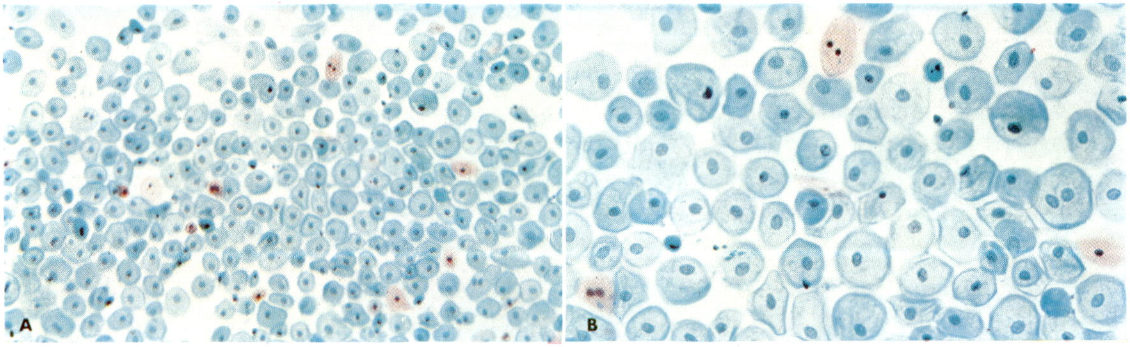

FIG. 75

An Atypical Vaginal Smear in Late Pregnancy

This vaginal smear was taken from a multiparous patient, with an apparently normal pregnancy, five days before her expected date of delivery. In fact she was delivered of a live female infant, weighing only five pounds two ounces, three days later.

A. The smear consists of discrete parabasal and intermediate squamous cells with rounded outlines. Note the absence of inflammatory changes. (×100.)

B. Higher magnification shows the pale-staining basophilic cytoplasm and thickened cellular borders of these cells. (×200.)

This type of smear has been described in association with post-maturity and placental insufficiency.

The Puerperium

After delivery marked changes occur in the smear, so that it consists largely of numerous erythrocytes, polymorphonuclear leucocytes and groups of decidual cells, which frequently show degenerative changes. The folded intermediate squamous cells rapidly decrease in number during the first five or six days and are replaced by rounded parabasal and intermediate cells with thickened cellular borders and glycogen-rich cytoplasm. Their nuclei are large, often eccentric, and frequently show degenerative changes. Sometimes there is eosinophilic staining of the cytoplasm of these cells.

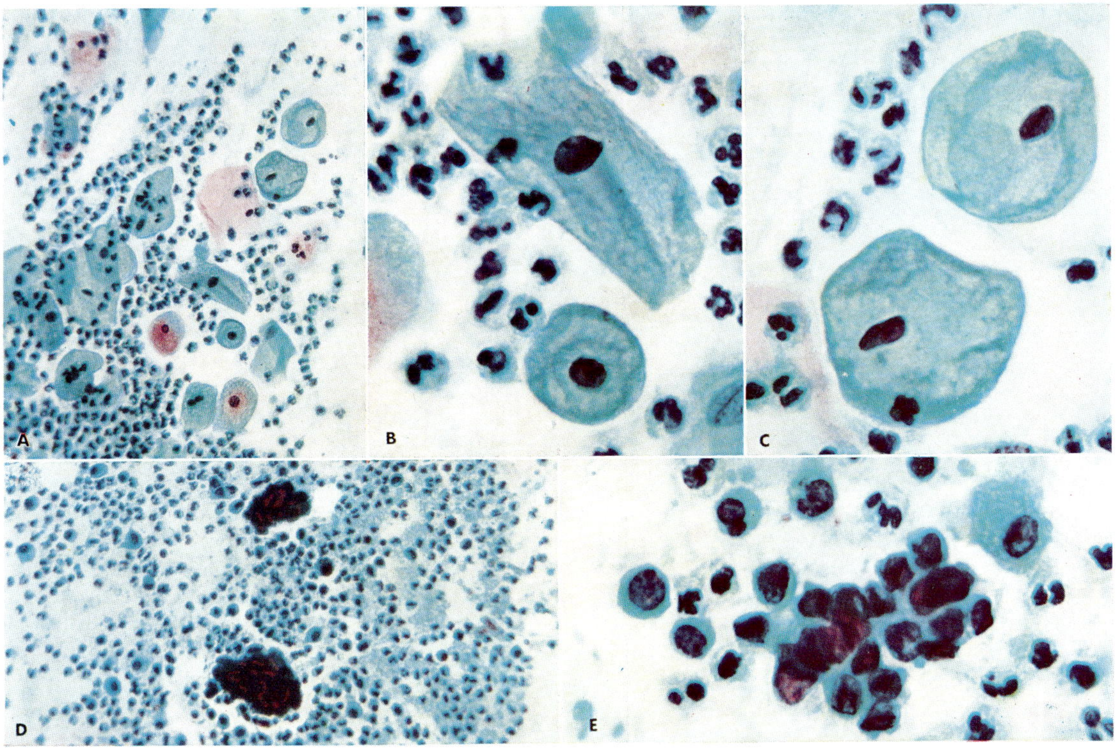

FIG. 76

The Vaginal Smear in the Early Puerperium

This smear was taken from a patient on the sixth day of the puerperium.

A. The smear shows numerous polymorphonuclear leucocytes and histiocytes. The squamous cells consist mainly of intermediate and parabasal types. A few intermediate cells still show folding of their cytoplasm but many have rounded outlines. (×200.)

B. This field shows an intermediate cell with folded cytoplasm lying above a parabasal cell which has a thickened cellular border and a large pyknotic nucleus. (×800.)

C. In this field two intermediate cells with rounded outlines, thickened cellular borders and yellowish-green glycogen-rich cytoplasm are shown. The general appearance of these cells is that of 'oyster' cells but their nuclei are no longer vesicular. (×800.)

D. Against a dense background of polymorphonuclear leucocytes and histiocytes two groups of cells and a few single cells with hyperchromatic nuclei are seen. (×200).

E. This group of cells and the adjacent single cells have degenerate pyknotic nuclei. However, the arrangement of the group and the hexagonal outlines of the single cells are compatible with a decidual origin. (×800.)

The Post-natal Smear

During the six weeks of the puerperium there is a gradual disappearance of non-epithelial and decidual cells from the smear. The changes in the epithelial cells vary according to whether the patient is lactating or not. In the lactating patient, at the end of the sixth week, the smear still consists mainly of intermediate and parabasal cells with rounded outlines, thickened cellular borders, glycogen-rich cytoplasm and eccentric nuclei. In the non-lactating patient the smears show a gradual increase in the numbers of superficial and intermediate cells so that by the end of the sixth week, the smear is similar to that usually seen in the proliferative phase of the menstrual cycle.

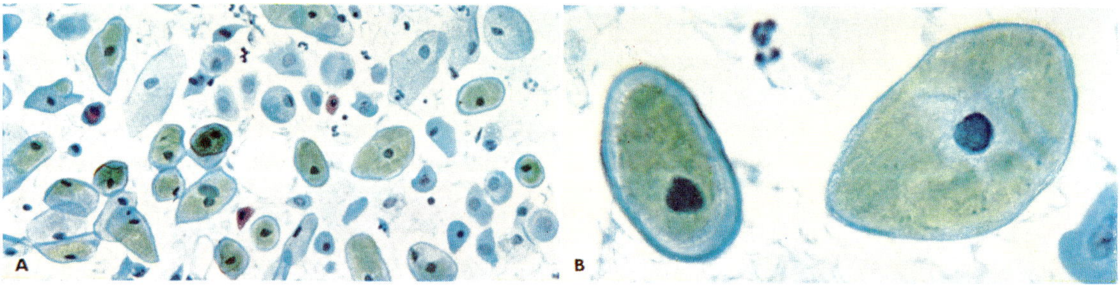

FIG. 77

The Vaginal Smear at the end of the Puerperium in a Lactating Patient

This smear was taken from a lactating patient, at a post-natal clinic, six weeks after delivery.

A. The smear consists of parabasal and intermediate cells, many of the intermediate cells have rounded outlines. Occasional parabasal cells exhibit eosinophilic staining of the cytoplasm. (×200.)

B. These smoothly rounded intermediate cells have thickened cellular borders. The glycogen within the cytoplasm gives a yellow tinge to the otherwise basophilic cytoplasm. (×800.)

SECTION II

RESPIRATORY TRACT

CHAPTER 6

THE NORMAL CELLULAR AND NON-CELLULAR PATTERNS

SPECIMENS of sputum and bronchial aspirates are usually collected either from symptomatic patients or elderly patients with a past history of respiratory disorder, so the 'normal' specimen shows considerable variations which often overlap pathological changes. The cellular content of specimens from the respiratory tract also varies according to the site from which the specimen has been collected.

THE UPPER RESPIRATORY TRACT

A specimen of saliva contains mainly superficial and intermediate squamous cells, similar to those seen in smears from the female genital tract, together with a few polymorphonuclear leucocytes. Occasional columnar cells may also be seen, but their presence is not sufficient evidence to indicate that the specimen contains sputum, as they may have originated in the nasal mucosa.

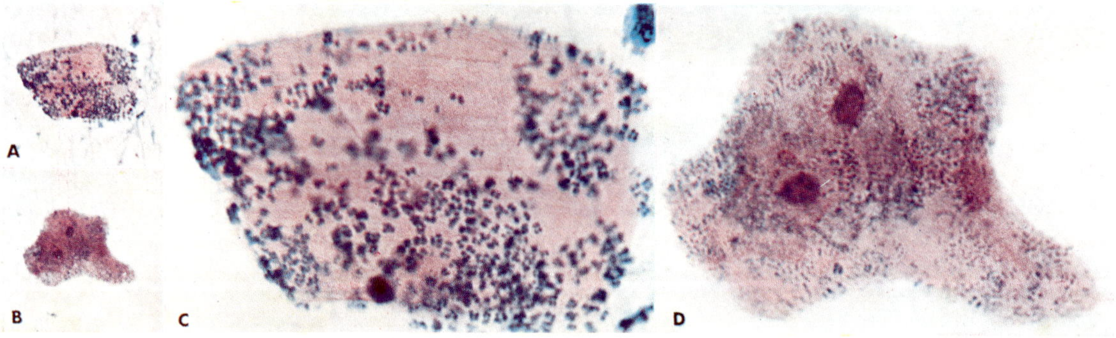

FIG. 78

Superficial and Intermediate Squamous Cells of Upper Respiratory Tract Origin in Saliva

A. and B. These squamous cells show a diffuse mottling of their cytoplasm. (×200.)

C. and D. On high-power examination, many coccoid micro-organisms are clearly seen on the cytoplasm of the cells. (×800.)

The presence, in a specimen of 'sputum', of significant numbers of squamous cells with this appearance suggests that the specimen consists mainly of saliva, containing cells largely of upper respiratory tract origin.

THE LOWER RESPIRATORY TRACT

Sputum, bronchial aspirates and bronchial washings are the specimens commonly collected from the lower respiratory tract. The cellular content of sputum consists of ciliated columnar and mucus-secreting columnar cells of lower respiratory tract origin, with histiocytes containing carbon particles, squamous cells of upper respiratory tract origin, polymorphonuclear leucocytes and relatively small numbers of lymphocytes. Bronchial aspirates and washings consist mainly of columnar cells of lower respiratory tract origin, histiocytes containing carbon particles, some polymorphonuclear leucocytes and occasional squamous cells.

COLUMNAR CELLS OF RESPIRATORY TRACT ORIGIN

These form two groups, ciliated and mucus-secreting columnar cells. The ciliated cells are of tall columnar type, with basal vesicular nuclei which show considerable variation in size with granular, but regular, chromatin patterns; nucleoli may be present but are not prominent. When the nucleus is large it distends the cell. The staining reaction of both cytoplasm and nucleus is basophilic. At the free edge of the cell eosinophilic staining cilia and a terminal plate may be seen; these two features are present only in benign columnar cells. Columnar cells are most often seen lying singly, but may be arranged in short parallel rows or in groups which on cross-section resemble a 'honeycomb'. Mucus-secreting columnar cells have basal vesicular nuclei and pale staining vacuolated cytoplasm. Mucus within the cytoplasm may distort the columnar shape of the cell and produce flattening of the superficial aspect of the nucleus. Columnar bronchial epithelial cells are usually readily recognisable, but may cause diagnostic difficulty if they occur in compact groups or when they are denuded of cytoplasm (see Fig. 89). The presence, in a group, of a cell with cilia or with a well-defined terminal plate is a useful indicator of the benign nature of that group of cells.

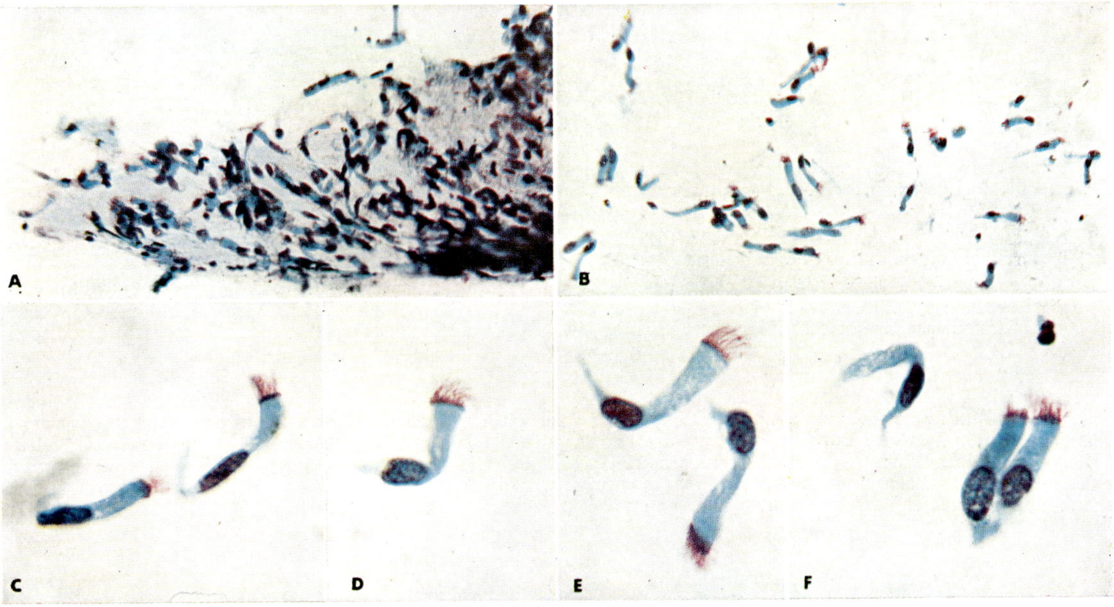

FIG. 79

Ciliated Columnar Epithelial Cells of Respiratory Tract Origin in a Bronchial Aspirate

A. This smear, prepared from a bronchial aspirate, is composed mainly of columnar cells with an occasional polymorphonuclear leucocyte. (×200.)

B. Even at this low-power magnification eosinophilic cilia can be seen on many of these elongated columnar cells. (×200.)

C. D. E. and F. In these cells cilia and terminal plates are clearly seen. The basal nuclei appear to distend the elongated cytoplasm in several of the cells. Note the granular chromatin pattern of the nuclei, which is especially well seen in the two cells to the right of field F. (×800.)

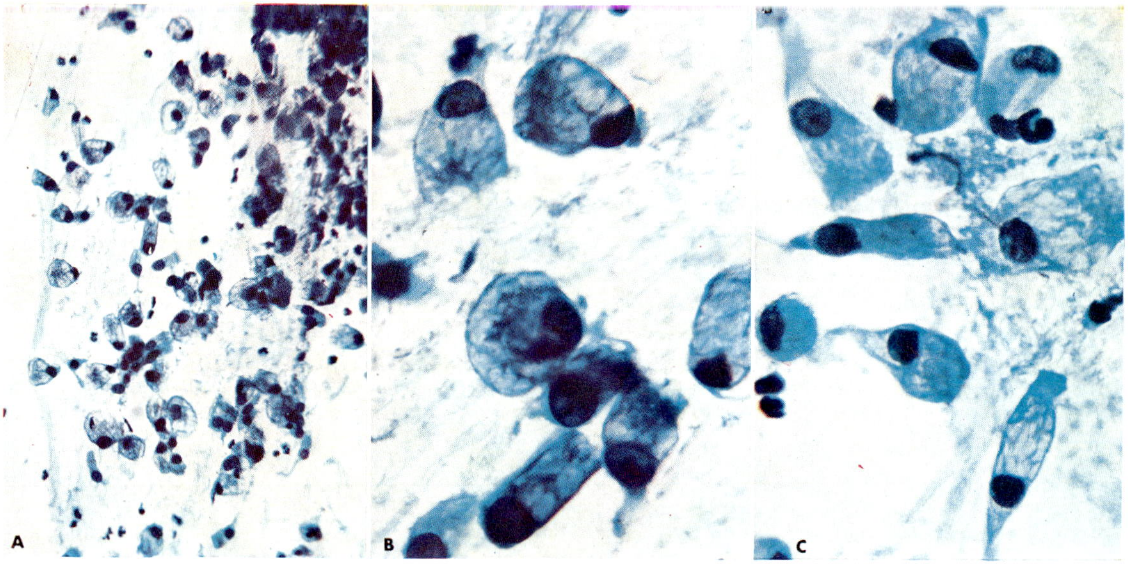

FIG. 80

Mucus-secreting Columnar Cells in Sputum

A. Initially the nuclei of these cells appear to be hyperchromatic and pleomorphic. Some cells exhibit an almost round cellular outline and contain eccentric nuclei. One columnar cell in the centre of the field, however, retains its ciliated border and confirms the bronchial epithelial origin of these cells. (×200.)

B. At higher magnification it is apparent that the rounded cellular outline of these columnar cells is due to distension of the cytoplasm by mucus which has also flattened the luminal aspect of the basal nuclei. The nuclear hyperchromasia is due to degeneration. (×800.)

C. These mucus-secreting, or 'goblet' cells, are typical in appearance and show less marked distortion of cellular and nuclear form by mucus. It is evident that these elongated cells, with almost basal nuclei and a fine proximal tail of cytoplasm, are of columnar type. Mucus, in varying amounts, is present in the cytoplasm of the cells. (×800.)

There may be considerable difficulty in distinguishing cells of this type from cells originating in an adenocarcinoma, especially when the bronchial epithelial cells are degenerate. Frequently the presence of a well-preserved ciliated columnar cell indicates the nature of the group. The nuclei of these benign cells, while uniform in size, may vary in shape, but this appearance is obviously due to distortion by mucus. Hyperchromatic nuclei indicate degeneration here. When the nuclear structure is still present it is finely stippled and does not show the abnormalities of chromatin pattern seen in adenocarcinoma. Nucleoli are not prominent in these bronchial epithelial cells, in contrast to the strikingly large nucleoli often seen in cells from an adenocarcinoma (cf. Figs. 104-107).

HISTIOCYTES

Histiocytes containing carbon particles are essential components of satisfactory specimens of sputum and of bronchial aspirates and washings. These round phagocytic cells have eccentric nuclei and a considerable amount of finely vacuolated cytoplasm, which contains ingested carbon particles. The nuclei are variable in shape, frequently round or oval, but typically reniform. Nuclear staining is basophilic with a coarse chromatin pattern and nucleoli may be seen. Multinucleation is often present and giant multinucleated forms occur. The cytoplasm is usually palely basophilic with a greyish-green or yellowish-green colour, but this colour may be obscured by the presence of black carbon particles.

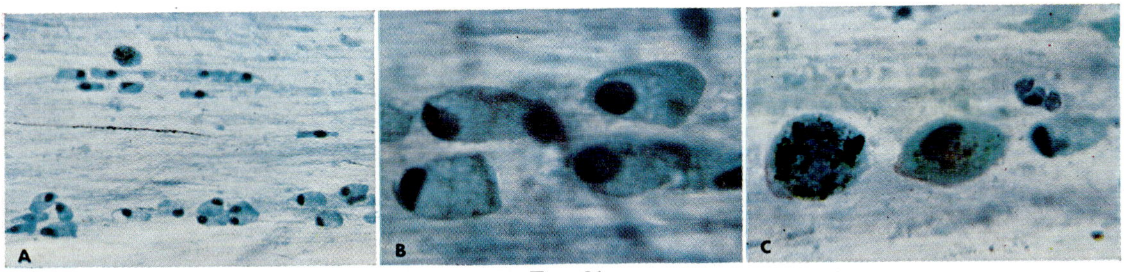

FIG. 81

Mucus-secreting Columnar Cells and Carbon-Laden Histiocytes in Sputum

A. Two elongated groups of columnar cells are seen against a background of streaks of basophilic mucus. A single histiocyte containing black granules within the cytoplasm lies above the upper group. The columnar cells in both groups are broad in comparison with the single non-mucus-containing columnar cell lying between them. (×200.)

B. The cytoplasm of these columnar cells is distended by mucus which has flattened the luminal aspect of the basal nuclei of the two cells to the left of the field. (× 800.)

C. Carbon particles within the cytoplasm partly obscure the nuclei of the histiocytes. (×800.)

The presence of columnar cells in a specimen is not sufficient evidence that it consists of sputum as these cells may originate in the upper respiratory tract from the nasal mucosa. It is essential to find histiocytes containing carbon particles in a smear before accepting it as a specimen of sputum. Further illustrations of histiocytes in sputum are shown in Fig. 102. Histiocytes stained by alternative techniques are shown in Figs. 82, 93, 95 and 162.

LEUCOCYTES

The predominant inflammatory cell in sputum and bronchial washings is the polymorphonuclear leucocyte. Occasionally bilobed nuclei in polymorphonuclear leucocytes suggest that eosinophils are present. To confirm this diagnosis specific staining of cytoplasmic granules by May-Grünwald-Giemsa technique or a similar Romanowsky stain is necessary as the Papanicolaou technique does not stain cytoplasmic granules. Lymphocytes are relatively uncommon and are identified by their densely basophilic crenated nuclei with a narrow pale basophilic rim of cytoplasm. When present in significant numbers, they should be carefully scrutinised to confirm their identity and to distinguish them from small undifferentiated malignant cells ('oat cells'). Monocytes and plasma cells are occasionally seen.

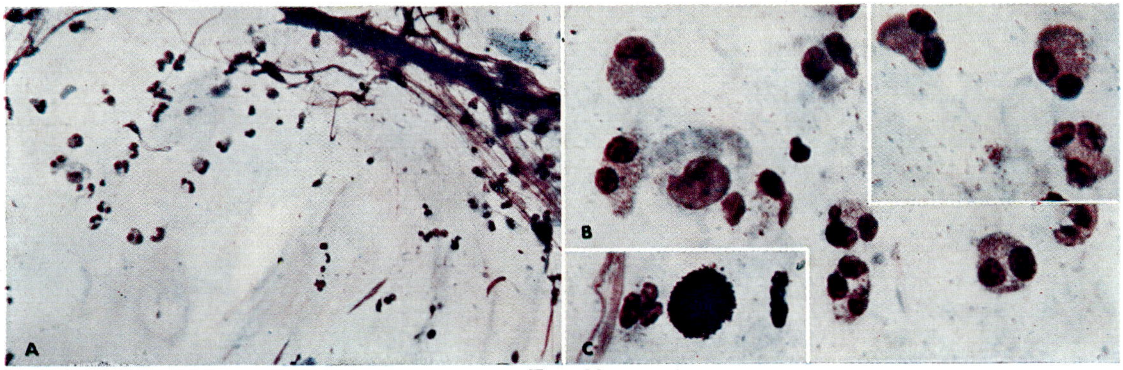

FIG. 82

Neutrophil, Eosinophil and Basophil Polymorphonuclear Leucocytes in Sputum

(May-Grunwald-Giemsa stain)

A. Many polymorphonuclear leucocytes are seen against a background of mucus. (×200.)

B. Three eosinophil polymorphonuclear leucocytes, with bilobed nuclei and eosinophilic cytoplasmic granules and four neutrophil polymorphonuclear leucocytes with trilobed nuclei are shown. Two further eosinophils and a neutrophil polymorphonuclear leucocyte are seen in the inset (top right). (×800.)

C. Dense basophilic cytoplasmic granules obscure the nucleus of the basophil polymorphonuclear leucocyte seen in the centre of the field. The nucleus of a polymorphonuclear leucocyte, probably of neutrophil type, is seen to the left of the basophil. (×800.)

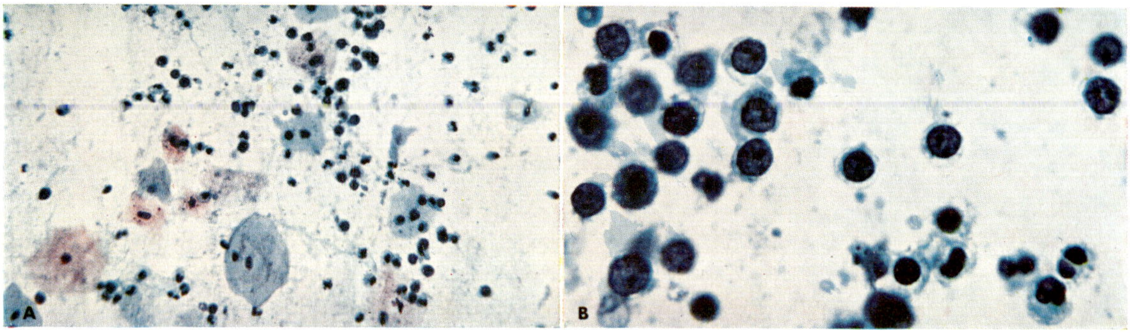

FIG. 83

Lymphocytes, Histiocytes and Polymorphonuclear Leucocytes in Sputum

A. This field shows benign squamous cells and a large number of small dark, round or oval cells. These are reminiscent of the appearance of malignant 'oat cells' in sputum, but even at this low-power magnification it can be seen that these dark cells are a mixture of several cell types, unlike the single cell type of an 'oat cell' group. The mixed inflammatory cells illustrated here have more diffuse outlines than the sharply defined punctate 'oat cells' seen in Fig. 102. (×200.)

B. At high-power magnification the mixture of inflammatory cells is now recognisable. Lymphocytes predominate, but a few small histiocytes and polymorphonuclear leucocytes are interspersed. The lymphocytes, with their dark nuclei and narrow rim of cytoplasm, most closely resemble 'oat cells', from which they can be differentiated by the absence of nuclear pleomorphism and chromatin clumping, and by their fairly uniformly rounded nuclei. (×800.)

NON-CELLULAR COMPONENTS

Mucus

In most specimens of sputum and bronchial washings mucus forms a blue background of uneven density (see Fig. 81). Greater diagnostic difficulty may be caused by inspissated mucus appearing as densely basophilic small round or oval masses. When superimposed on squamous cells these masses bear a superficial resemblance to the nuclei of malignant cells. Careful examina-ion, however, reveals the absence of a nuclear structure in these mucus plugs.

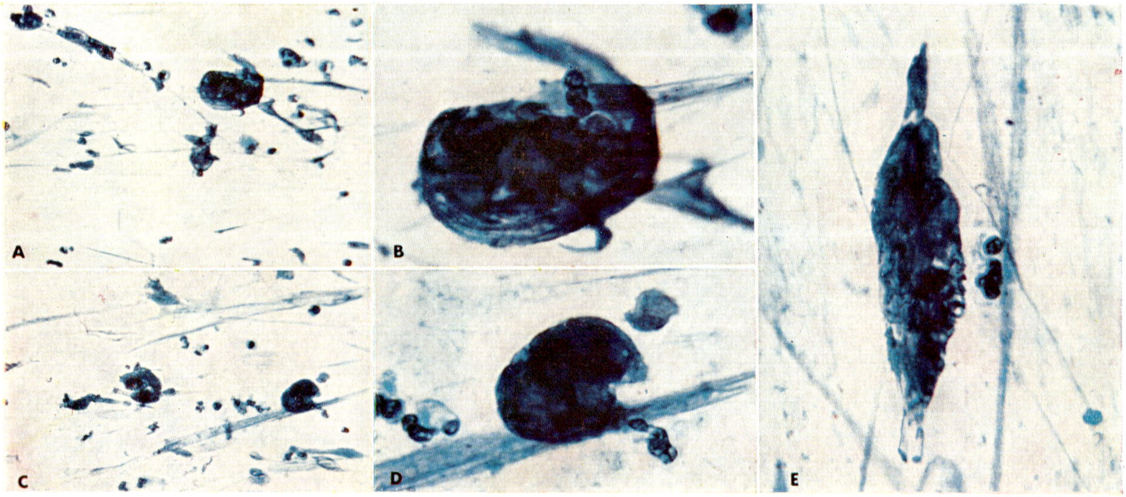

FIG. 84

Inspissated Mucus Mimicking the Appearance of Malignant Nuclei in Sputum

A. and C. On low-power examination, the densely hyperchromatic, basophilic, pleomorphic appearance of mucus plugs mimics that of undifferentiated malignant cells (cf. Fig. 103). (×200.)

B. D. and E. High-power examination, however, fails to reveal either nuclei or a cellular structure and confirms that these are mucus plugs. (×800.)

Micro-organisms

When a specimen consists only or mainly of saliva occasional squamous cells may be almost completely obscured by numerous bacteria. If this phenomenon is seen in many cells especial care should be taken to check that there is evidence that the specimen does contain sputum (see Fig. 78).

Bacteria may also cause diagnostic difficulty when closely clumped in the centre of a cell giving the appearance of an enlarged hyperchromatic nucleus. High-power examination, however, will identify the component micro-organisms.

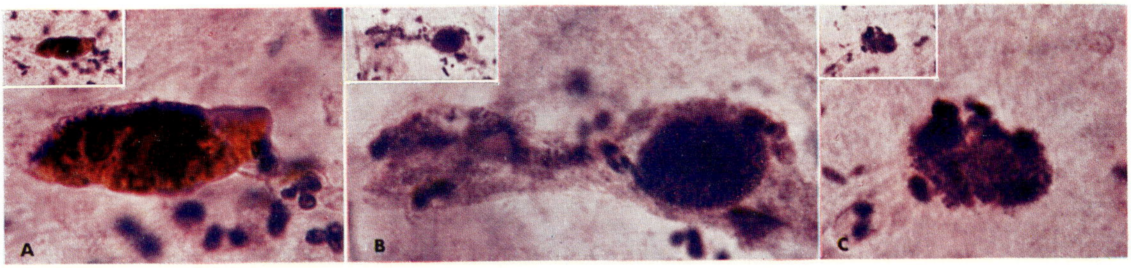

FIG. 85

Micro-organisms Simulating Malignant Nuclei in Sputum

A. The brightly eosinophilic opaque cytoplasm and apparently irregular densely hyperchromatic 'nucleus' of this cell, as shown on low-power examination (inset), suggest a differentiated malignant cell of squamous type. High-power examination, however, shows that this appearance is due to the presence of large numbers of micro-organisms overlying a squamous cell background. (Inset ×200) (×800.)

B. and C. The two low-power insets similarly suggest a poorly differentiated malignant cell in B, and an undifferentiated malignant cell in C. On high-power examination, in each case, it is apparent that the 'nuclei' are composed of densely packed micro-organisms. In these two examples epithelial cellular material is not present. (Insets ×200.) (×800.)

It is clear that it is of fundamental importance to examine critically any suspicious area of a smear under high-power magnification in order to establish that the abnormal appearance is of cellular origin and that a malignant 'nucleus' shows a recognisable nuclear structure.

Trichomonas buccalis is seen in sputum and saliva. The appearance of the protozoon is similar to that of *Trichomonas vaginalis* in smears from the female genital tract (q.v.).

The standard cytological techniques are not suitable for the recognition of the alcohol and acid-fast bacilli of tuberculosis.

Fungi are frequently seen in specimens consisting of saliva and also in sputum which is not freshly prepared, although even in fresh specimens of sputum, fungi may occasionally be present. This is usually *Candida albicans* and the characteristic eosinophilic hyphae and conidia are readily recognised. *Candida albicans* is most often seen as a secondary invader in patients who have had a course of treatment with an antibiotic.

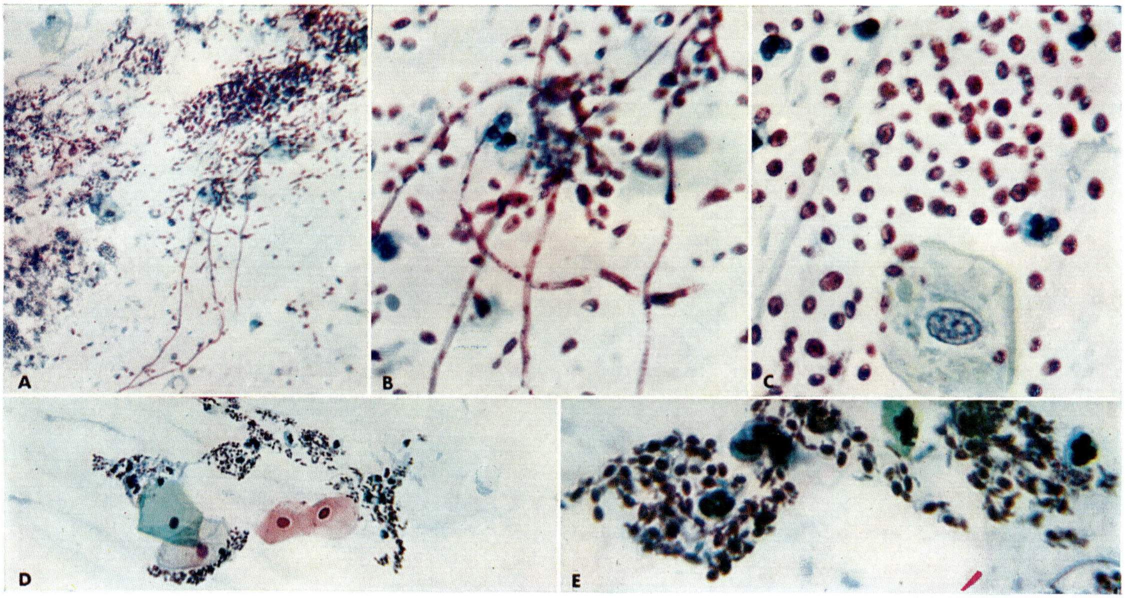

FIG. 86

Fungi in Sputum and in Saliva

A. This specimen consists of sputum, confirmed by the presence of histiocytes containing carbon particles as well as squamous cells and polymorphonuclear leucocytes. The fungus appears as long filaments of brightly eosinophilic material and small oval unicellular eosinophilic structures. (×200.)

B. and C. At high magnification septate hyphae and yeasts are seen in field B while many yeasts are present in field C. (×800.)

D. and E. These illustrations are from a specimen of saliva. The cellular content consists of squamous cells and polymorphonuclear leucocytes. While numerous yeasts appear as multiple eosinophilic ovoid unicellular structures, hyphae are not present. (× 200.) (× 800.)

When a specimen of sputum contains fungi, there are three probable explanations: (1) Saliva frequently shows the presence of fungi, but is readily recognised by the absence of histiocytes containing carbon particles. (2) A specimen of sputum, which has been delayed in reaching the laboratory, will often show fungi together with heavy bacterial contamination. (3) Fungi may be present as pulmonary pathogens. This last conclusion should only be reached after excluding the other two more common explanations, seeking to localise the fungus by identifying it in bronchial washings, confirming the diagnosis by culture, and examining radiologically for the appearances suggestive of a fungal infection.

Contaminants

The commonest contaminant in fresh sputum is food. The larger particles are usually readily recognisable as vegetable material by the cellulose membrane, but occasionally the single large dark nuclei may cause difficulty in diagnosis. Specimens contaminated by food should be discarded.

If sputum specimens contain a considerable proportion of saliva or if there is a significant delay in the preparation of smears fungi and bacteria may be present in large numbers as contaminants.

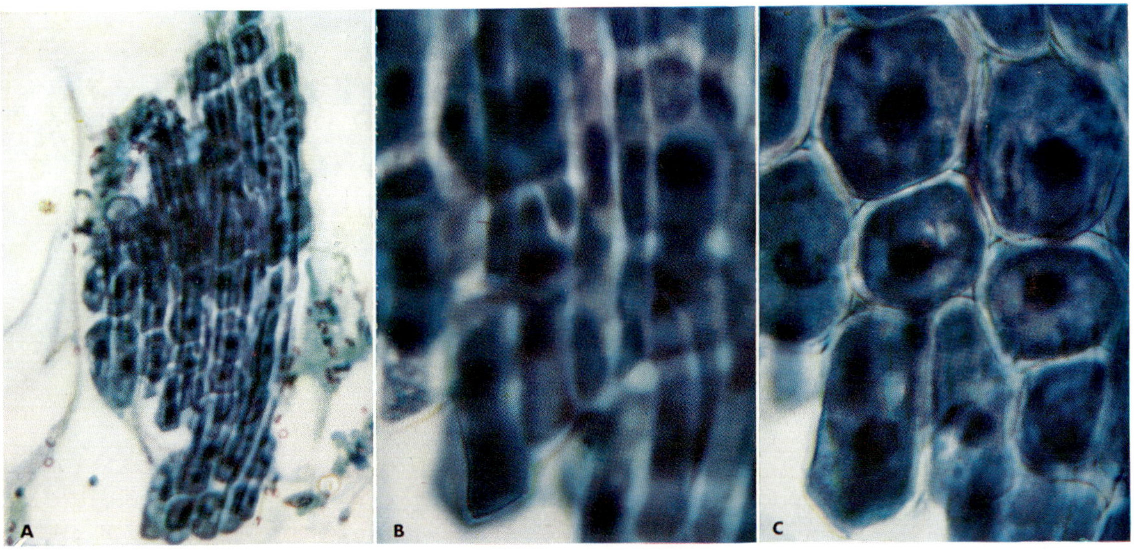

FIG. 87

Contamination of Sputum by Food Debris

A. This hyperchromatic linear group of cells, probably of vegetable origin, is seen against a background of squamous cells, erythrocytes and polymorphonuclear leucocytes. (×200.)

B. and C. Elongated fibres are apparent at this plane of focus (B), while at a deeper plane of focus on the same field (C) large hexagonal cells with central nuclei are seen. (×800.)

Initially food debris may require scrutiny because it appears as hyperchromatic material, but once it is recognised as vegetable matter it can be dismissed as a contaminant.

Contamination by food debris should be avoided by collecting *early morning specimens.* When food debris is present the specimen should be discarded, as single degenerate vegetable cells may appear to be densely hyperchromatic and require careful examination to avoid a false positive diagnosis of malignancy.

CHAPTER 7

NON-MALIGNANT DISORDERS

SINCE most patients subjected to cytological investigation of specimens from the respiratory tract are symptomatic or elderly, the changes associated with chronic infection or irritation are frequently apparent.

EPITHELIAL CELLS

Squamous cells show various changes in response to irritation or infection. Occasionally keratinised anucleated squames from the mouth may appear as a result of denture trauma and these cells are similar to the keratinised cells which are occasionally seen in cervical or vaginal smears. Their cytoplasm is eosinophilic or orangophilic but does not show the dense opacity and brilliant staining of the anucleated squames associated with squamous carcinoma. The shape of the benign anucleated squame is normal and contrasts with the bizarre forms seen in squamous carcinoma (cf. Figs. 98 and 111).

In response to inflammation 'deep' squamous cells may appear. They are small with relatively large nuclei, which may have heavy nuclear borders. Their uniformity indicates their benign character but occasionally degeneration of the nucleus may produce a suspicious appearance. These cells are nicknamed 'Pap' cells. It is said that Dr Papanicolaou noted similar cells in a specimen of his own sputum examined when he had an upper respiratory tract infection.

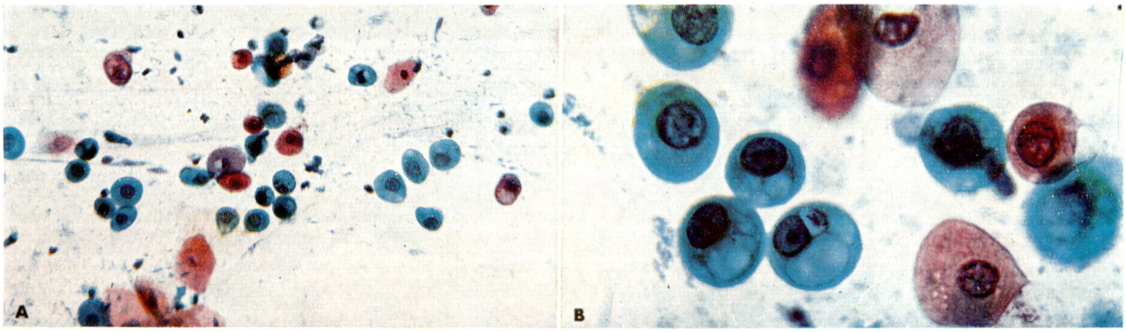

FIG. 88

Basal or 'Deep' Benign Squamous Cells in Sputum

A. Small rounded squamous cells with vesicular or oval nuclei are present. They are smaller than the superficial squamous cells seen in the lower left quadrant and their cytoplasm is more opaque. (×200.)

B. These basal squamous cells have rounded outlines. Only in the lowest cell on the right is slight 'peaking' of cytoplasm, suggestive of a cytoplasmic process, still seen. Several cells show degenerative changes with vacuolation of cytoplasm and nuclear pyknosis or karyolysis. Where the nuclear chromatin remains well preserved, in the cell in the upper left corner, it is finely stippled. The nuclei shown here are less uniform than the vesicular nuclei seen under low power, but the variation is due to degeneration, with karyorrhexis in the lower cell on the right and nuclear vacuolation in the highest cell on the right. The cytoplasm is eosinophilic in a number of the degenerate cells, but does not show the dense, brightly eosinophilic quality of the differentiated malignant cells of squamous type seen in Fig. 99. (×800.)

Bronchial epithelial cells may undergo considerable alteration in response to inflammation. Loss of cilia may occur and the cytoplasm may disappear. The resultant denuded nuclei could be confused with malignant 'oat cells', but the finely granular uniform chromatin pattern indicates their benign nature. Intact columnar cells are often found near a group of denuded nuclei and are an aid to their identification. Multinucleation may be seen in bronchial epithelial cells in inflammation, but the nuclei of any one cell are all similar in size, shape and chromatin pattern. Occasionally the chromatin pattern may become more coarsely granular and nucleoli appear more prominent. In groups of benign cells, however, the uniformity of cells and their nuclei is retained.

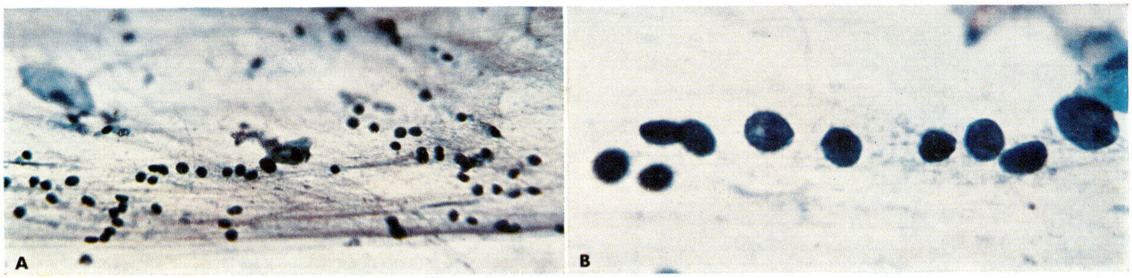

FIG. 89

The Differential Diagnosis of Denuded Nuclei of Benign Columnar Cells and Small Undifferentiated Malignant Cells in Sputum

A. Many small dark discrete oval nuclei are seen. They show some variation in size, but are uniform in shape. Below the squamous cell, on the left of the field, a cell has retained its cytoplasm and remains readily identifiable as a columnar cell. (×200.)

B. The nuclei vary in size but are uniformly oval in shape and, although they appear hyperchromatic, the chromatin is evenly distributed and clumping is not seen. (×800.)

These denuded nuclei are frequently seen, as columnar cells readily undergo cytolysis. In general, a few well-preserved columnar cells are present and identify the origin of the group. In their absence, however, a careful examination of the nuclei, which are uniform in shape and chromatin pattern, differentiates them from the small undifferentiated malignant cells ('oat cells') illustrated in Fig. 102.

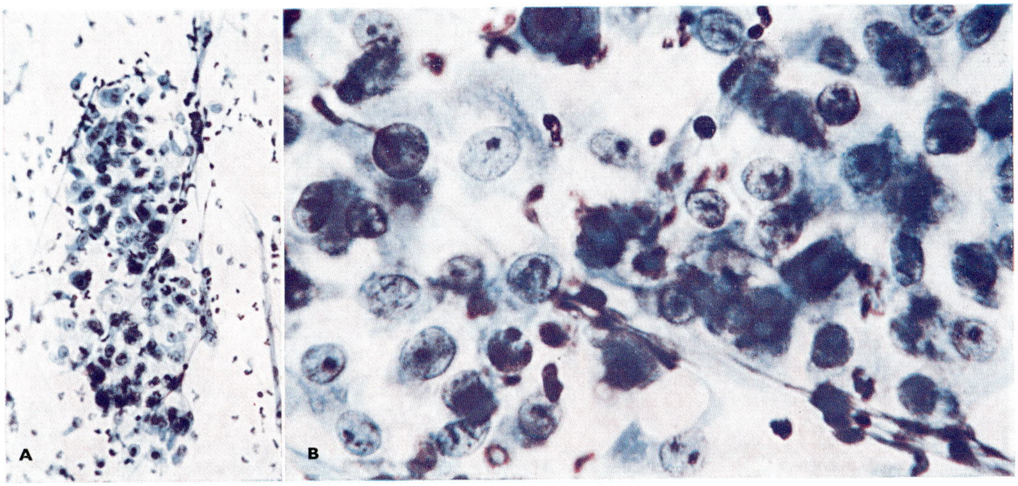

FIG. 90

Prominent Nucleoli in Bronchial Epithelial Cells from a Bronchial Wash

A. In the low-power view large nucleoli are a striking feature of this group of columnar bronchial epithelial cells. (× 200.)

B. At high-power magnification enlarged eosinophilic nucleoli are seen in most nuclei. This is an unusual feature in bronchial epithelial cells and led to careful scrutiny of the group. Although the nuclei vary in size they are uniformly oval in shape and they do not show any significant abnormalities of chromatin pattern. A few appear hyperchromatic, but their nuclei show degenerative changes. It was noted that bronchial epithelial cells at the edge of the group (beyond the high-powered field) showed well-preserved cilia. No other abnormality was seen in this specimen. (× 800.)

Bronchoscopy was performed on an elderly male patient because of a spontaneous pneumothorax. The bronchoscopy revealed only inflammatory changes. These smears were prepared from the bronchial washings. In view of these prominent nucleoli further specimens of sputum were examined and no significant abnormality was found. Subsequent clinical and radiological follow-up of this patient did not reveal any abnormality in the following two years.

Squamous metaplasia is frequently seen in specimens of sputum from patients with chronic respiratory infection. These sheets of metaplastic bronchial epithelial cells show good cohesion, in marked contrast with the tendency of malignant squamous cells to appear singly scattered throughout a specimen. Degenerative changes in metaplastic cells may give their nuclei a suspicious appearance, but careful examination permits recognition of pyknosis or karyorrhexis.

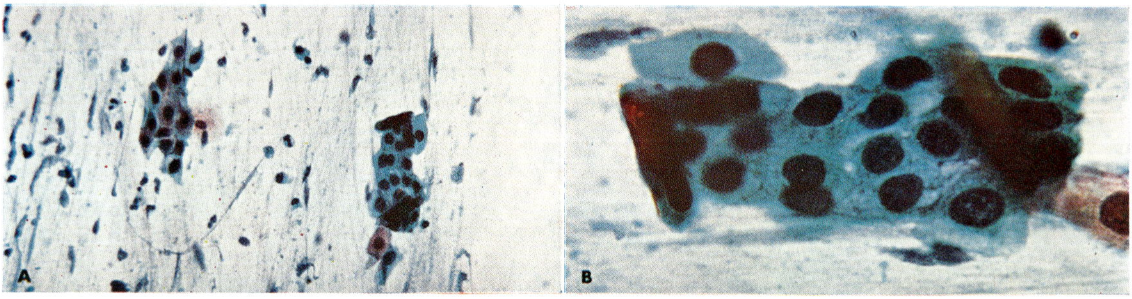

FIG. 91

Squamous Metaplasia in Sputum

A. Two groups of metaplastic cells are seen; both show good cellular cohesion and uniformity in cellular and nuclear size. (×200.)

B. The metaplastic cells seen in this compact group have oval nuclei surrounded by adequate cytoplasm. There is minimal variation in nuclear size, shape and staining reaction. (×800.)

Although groups of metaplastic cells in sputum require careful examination to exclude early malignant change, the type of cell group illustrated here, with good cellular cohesion and striking cellular and nuclear uniformity, is benign.

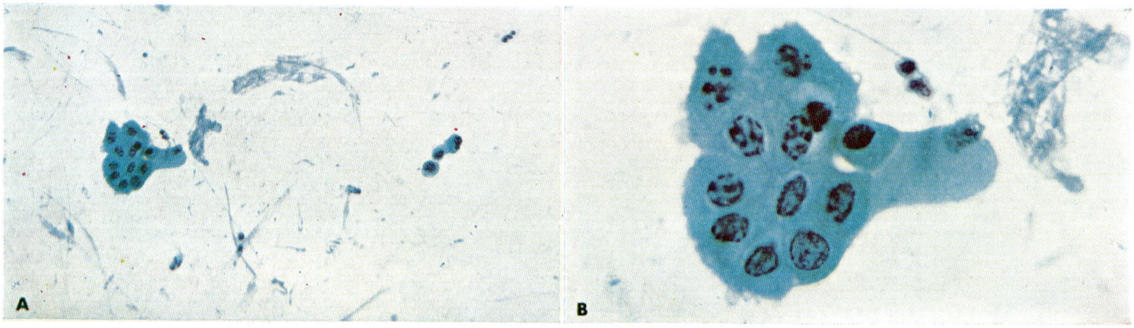

FIG. 92

Karyorrhexis in Metaplastic Cells in Sputum

A. The nuclei in the group of metaplastic cells on the left appear to be hyperchromatic and the chromatin seems to be irregularly arranged. The group, however, shows good cellular cohesion and the nuclei are fairly uniform in size. (×200.)

B. At high-power magnification it is evident that the apparent hyperchromasia and irregular chromatin patterns are due to degenerative changes. In the cell to the right of the group nuclear pyknosis is seen, while cells in the lower half show marginal condensation of chromatin, with breaks in the continuity of the nuclear border, and loss of the normal chromatin pattern, producing an 'empty' nucleus. The cells in the centre of the group show karyorrhexis, while those at the top also show karyorrhexis with a breakdown of the nuclear border, resulting in the appearance of coarse clumps of chromatin within the cytoplasm. All these cells are, therefore, benign metaplastic cells showing degenerative changes. (×800.)

NON-EPITHELIAL CELLS

In specimens of sputum and bronchial washings polymorphonuclear leucocytes are found in large numbers in the presence of infection. In allergic disorders, eosinophil polymorphonuclear leucocytes are often seen in significant numbers (see Fig. 82).

Lymphocytes and monocytes are rarely seen in large numbers. Their main significance lies in the possibility of confusion with 'oat cells' but the chronic inflammatory cells usually form groups of mixed cell types, while the tumour cell groups rarely show any admixture with non-neoplastic cells (Figs. 83 and 102).

Histiocytes, which are seen in all specimens of sputum, in bronchial aspirates and washings, show striking changes in response to irritation and inflammation; reactive and degenerative changes occur in their nuclei. Multinucleation is frequently seen and high multinucleation with giant-cell forms is not uncommon. Nuclei may appear 'active' with coarsely granular chromatin and prominent nucleoli. In well-preserved cells the characteristic nuclear shape and vacuolated cytoplasm, with frequent phagocytosis, usually makes identification easy. Difficulty may arise when degenerative changes in histiocytes have produced nuclear hyperchromasia, but the complete loss of nuclear structure and frequent evidence of phagocytosis distinguishes these degenerate cells from malignant cells.

Cytoplasmic changes occur as a result of phagocytosis. Black carbon particles are usually present within the cytoplasm (see Fig. 81) and, in inflammation, ingested polymorphonuclear leucocytes may be seen. Other ingested material found in histiocytes includes brown haemosiderin granules, in the so-called 'heart-failure' cells or in pulmonary haemosiderosis, and also lipoid material. Haemosiderin can be identified by using a staining method to demonstrate iron (e.g., the Prussian blue reaction) and lipoid material by fat stains (e.g., Sudan IV).

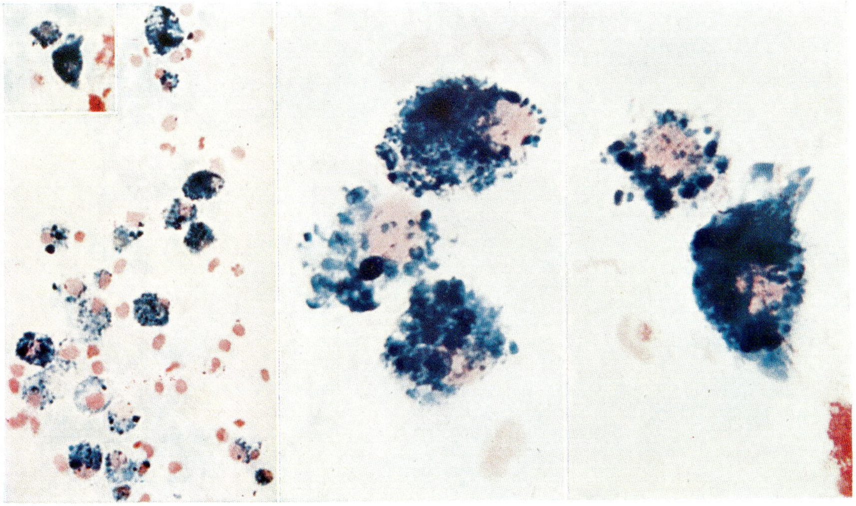

FIG. 93

Histiocytes in Sputum Stained by the Prussian Blue Reaction

This specimen of sputum from a case of pulmonary haemosiderosis shows, in the low-power fields on the left, a positive Prussian blue reaction in many histiocytes. The high-power fields (centre and right) show the granular cytoplasmic distribution of the haemosiderin. Its loosely bound ferric ion is responsible for the positive Prussian blue reaction. (×200.) (×800.)

NON-CELLULAR COMPONENTS

Mucus

Mucus may appear in a striking form as a Curschmann's spiral, which is a cast of inspissated mucus expelled from a small bronchus. The spiral has a dense coiled central core surrounded by thinner translucent material. Curschmann's spirals are frequently found in sputum from patients with asthma or with chronic respiratory tract infections.

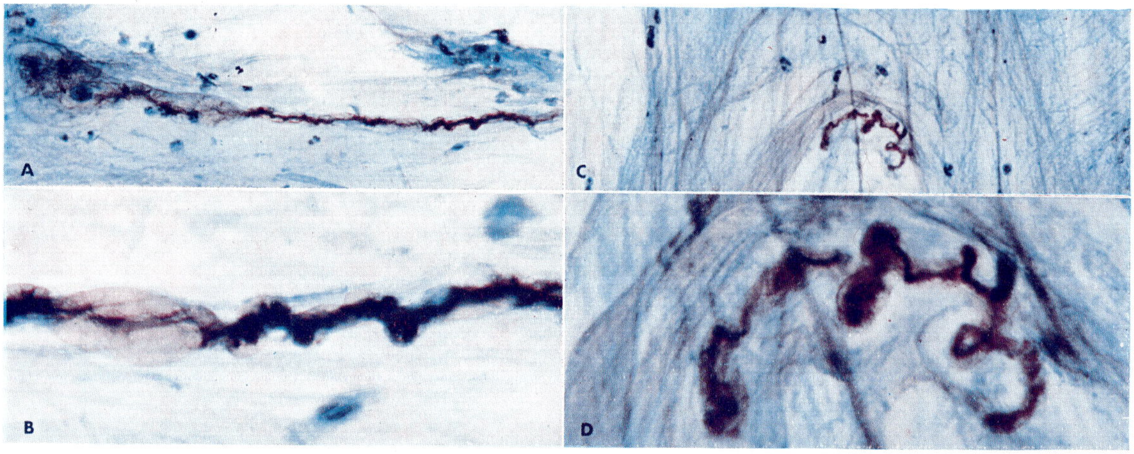

FIG. 94

Curschmann's Spirals in Sputum

A. and C. The characteristic hyperchromatic coiled appearance of two spirals is well demonstrated. (×200.)

B. and D. High-power examination shows the dense convoluted central core and the translucent appearance of the peripheral area of the spiral. The overall appearance of the spiral is that of a 'cast' of a small bronchus, which has been plugged by inspissated mucus; this now forms the dense central core. (×800.)

Asbestos Bodies

Asbestos bodies only occasionally appear in sputum in cases of asbestosis. Deposition of protein and iron salts around the silicate fibres produces yellowish-brown structures with a characteristic 'drumstick' appearance. The material deposited around the fibres gives a positive Prussian blue reaction. The presence of asbestos bodies in sputum specimens has considerable clinical significance because of the observed association between asbestosis and mesothelioma and bronchogenic carcinoma.

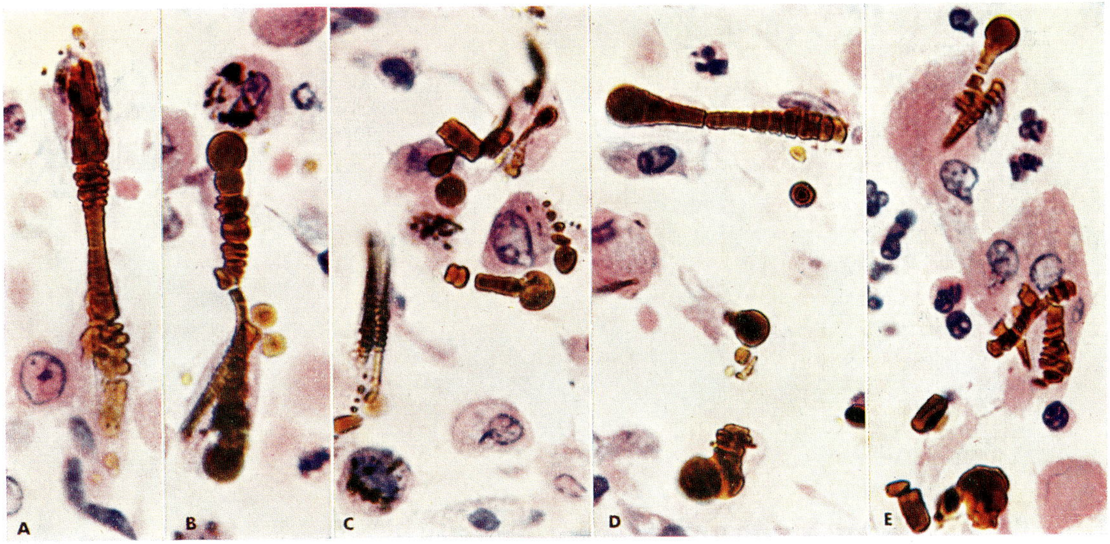

FIG. 95

Asbestos Bodies in Sputum Stained by Haemalum and Eosin

Segmented yellowish-brown asbestos bodies are present in each field of this composite illustration. The characteristic 'drumstick' appearance, produced by the deposition of protein and iron salts on the silicate fibres, is strikingly demonstrated in fields B and D, but smaller 'drumsticks' can be recognised in fields C and E. Polymorphonuclear leucocytes are obvious and carbon-laden histiocytes can be seen in fields B and C. Asbestos bodies may also be identified by their positive Prussian blue reaction. ($\times$800.)

CHAPTER 8

CANCER OF THE RESPIRATORY TRACT

THE most important application of cytologic diagnosis to the respiratory tract is its contribution to the diagnosis of cancer of the lower respiratory tract, but the method can also be usefully applied to the diagnosis of malignant lesions of the upper respiratory tract.

The general characteristics of malignant cells of respiratory tract origin are similar in many ways to those characteristics previously described with reference to cancer of the female genital tract. There are, however, certain modifications due to the relative inaccessibility of the lesions, the differing tumour types, a greater tendency to necrosis in the case of the squamous cell carcinomata in this area and to benign variations in the epithelial cells, which must all be reviewed in the differential diagnosis of malignant cells.

THE UPPER RESPIRATORY TRACT

The role of cytology in the diagnosis of malignant lesions of the upper respiratory tract is a comparatively small one, because most of the lesions are readily accessible for direct biopsy. However, cytologic diagnosis may be useful as a preliminary investigation, prior to biopsy of a suspicious lesion. In addition more frequent application of cytological methods in the diagnosis of lesions, which are not sufficiently clinically suspicious to warrant biopsy, may occasionally reveal an unexpected malignant lesion. Finally, it is important to remember that the upper respiratory tract may be the source of unexplained malignant cells in sputum and careful examination of the upper respiratory tract is, therefore, an essential step in the investigation of any patient with malignant cells in sputum and negative findings on radiological examination of the chest.

The Paranasal Sinuses

These may be investigated cytologically by the examination of sinus washings. The commonest tumour in this area is the squamous cell carcinoma; the features of the malignant cells are essentially similar to those already described (pages 37 to 41). In the presence of infection the specimen may consist mainly of polymorphonuclear leucocytes with relatively little epithelial cellular material and this may cause diagnostic difficulty.

The Larynx

Specimens suitable for examination for malignant cells from a squamous cell carcinoma of larynx may be obtained directly by scraping or swabbing the suspicious area or, indirectly, by the collection of sputum. The malignant cells of squamous type show the features previously described (see above).

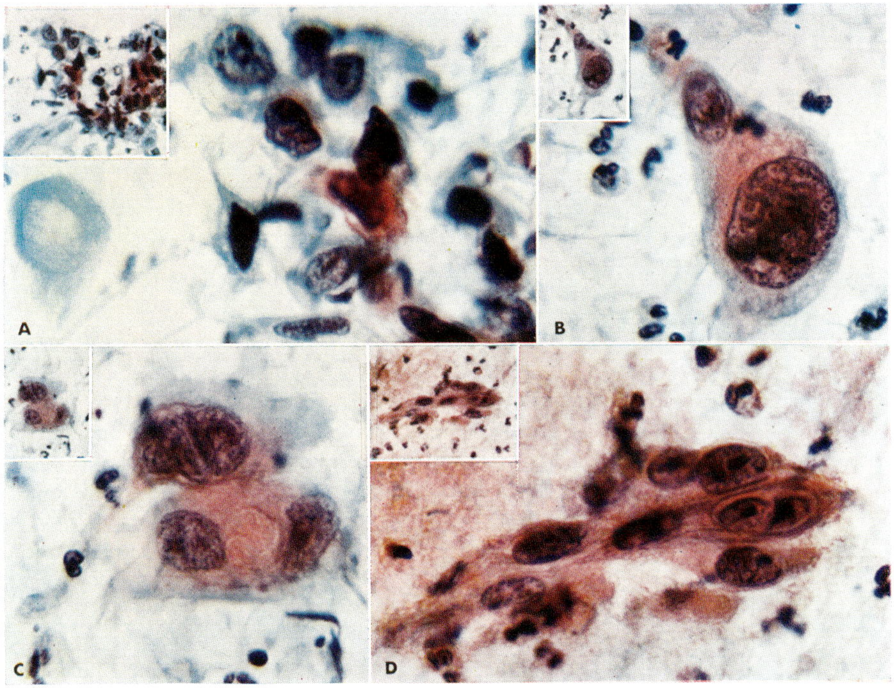

FIG. 96

Sputum Showing Undifferentiated Malignant Cells and Differentiated Malignant Cells of Squamous Type from a Squamous Cell Carcinoma of Larynx

A. This group of cells shows marked nuclear enlargement and hyperchromasia even in the low-power illustration (inset). High-power examination reveals that some of the hyperchromatic nuclei are degenerate, but others show the nuclear size variation, pleomorphism and abnormal chromatin patterns of undifferentiated malignant cells. (Inset ×200.) (×800.)

B. The larger cell displays the characteristic features of a differentiated malignant cell of squamous type, with a large central hyperchromatic nucleus showing an abnormal chromatin pattern. The smaller undifferentiated malignant cell has an irregular outline and an abnormal chromatin pattern. Note the 'dirty' background with several polymorphonuclear leucocytes. (Inset ×200.) (×800.)

C. The arrangement of this group of undifferentiated malignant cells with the nuclei at the periphery is suggestive of an origin from an adenocarcinoma, but on high-power examination it is apparent that these nuclei are not completely surrounded by cytoplasm. (Inset ×200.) (×800.)

D. In the low-power field (inset) this appears to be a group of 'fibre' cells, while high-power examination reveals that it is a mixture of differentiated malignant cells of squamous type ('fibre' cells) and undifferentiated malignant cells. Nucleoli are apparent in the elongated nuclei of these cells. (Inset ×200.) (×800.)

THE LOWER RESPIRATORY TRACT

The early diagnosis of carcinoma of the lower respiratory tract requires a combination of clinical alertness with investigation by radiological, cytological and bronchoscopic methods. The advantages of cytology as a method of investigation in this field are the minimal patient inconvenience in specimen collection, the opportunity to obtain samples which are fairly widely representative of the pathological changes occurring and the possibility of typing the tumour. The disadvantages are an inability to determine the site of a lesion and the fact that early peripheral single lesions frequently give negative results.

Two types of specimen can be examined, sputum and a bronchial aspirate or washing. Sputum is comparatively readily collected and several specimens may be examined, but occasional patients consistently fail to produce anything but saliva. A specimen of sputum represents material from a wider area than an aspirate or washing. A bronchial aspirate or washing from a suspected area causes more patient discomfort, but may yield a higher concentration of malignant cells and may be of help in localising a lesion.

The malignant cells from bronchial carcinomata can be divided into three groups: those arising from a squamous cell carcinoma, those from an adenocarcinoma and undifferentiated malignant cells, which may be either small or large. In practice, however, there may be a considerable degree of overlapping of cell types present, both in consecutive specimens from a patient and even in the same specimen; this reflects the variations in degree of differentiation which may be present in a tumour. Nevertheless the distribution pattern of malignant cells, together with their nuclear and cytoplasmic characteristics, will usually categorise a tumour with a considerable degree of accuracy and this may be useful to the clinician.

DIFFERENTIATED MALIGNANT CELLS OF SQUAMOUS TYPE

Squamous cell carcinoma of bronchus frequently undergoes necrosis, which produces a characteristic pattern in the smear. The malignant cells occur singly or in small groups of two or three against a background of polymorphonuclear leucocytes. Frequently they appear to be strung out along a strand of muco-pus. This contrasts sharply with the good cellular cohesion shown by groups of metaplastic cells (q.v.). The cytoplasm of the malignant squamous cells is often strikingly abnormal. The cells may be bizarre in shape with tadpole and fibre forms; amoeboid forms with the appearance of cytoplasmic pseudopodia are also seen. The consistency of the cytoplasm is also abnormal with an opaque quality and thick sharp cellular border, which gives the cell a curiously three-dimensional appearance. Cytoplasmic staining is usually brightly eosinophilic or orangophilic, often with a refractile quality. This refractile effect is enhanced if the substage condenser of the microscope is slightly lowered while examining the cell. The presence of bizarre cell shapes and abnormal cytoplasmic staining may be noted as suspicious evidence in cells in which the nuclei appear normal. Such cells should not be used as a basis for a definitive diagnosis, but their presence should be an indication for further search for malignant cells. The nuclei of differentiated malignant cells of squamous type may present considerable diagnostic difficulty because of degenerative changes. As a result many of the nuclei appear densely hyperchromatic with an irregular and often angular outline; frequently these nuclei are large and the cells show an abnormal nucleocytoplasmic ratio. However, careful examination of the specimen will usually reveal some cells in which the structure of the central malignant nucleus is more readily recognisable. The typical abnormalities of nuclear pleomorphism and chromatin patterns, with irregular strands or clumps of chromatin and uneven marginal condensation of chromatin are then seen. Occasionally prominent nucleoli are noted, but these are not such a striking feature as in the malignant cells of adenocarcinomatous type.

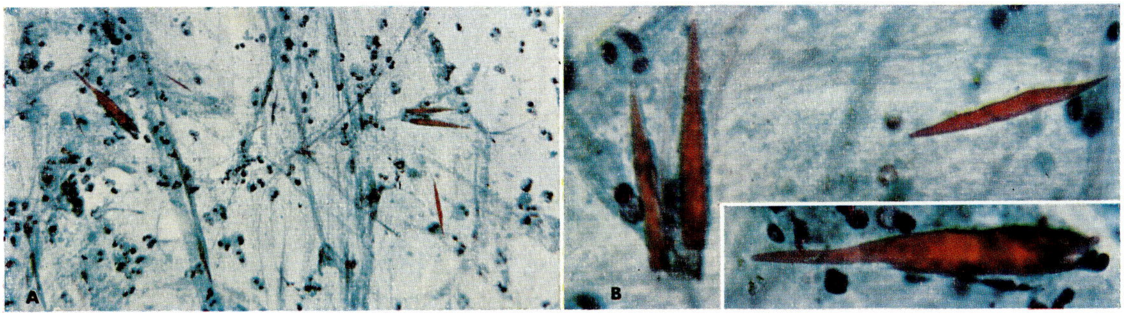

FIG. 97

Anucleated Keratinised Squamous Cells in Sputum

A. Several bright orangophilic elongated cells are seen against a background of mucus, polymorphonuclear leucocytes and histiocytes. (×200.)

B. The cytoplasm of these cells is opaque and shows bright orange staining. If, during microscopy, the substage condenser is lowered, these cells frequently appear refractile. It is not possible to recognise a nucleus in any of these cells. Note the tadpole form of the cell illustrated in the inset. (×800.)

In the absence of nuclei, these keratinised squamous cells are not diagnostic of squamous carcinoma, but when they are seen, a careful search is indicated for similar cells with nuclei showing the characteristic features of malignant change, and also for other less highly differentiated malignant cells of squamous type.

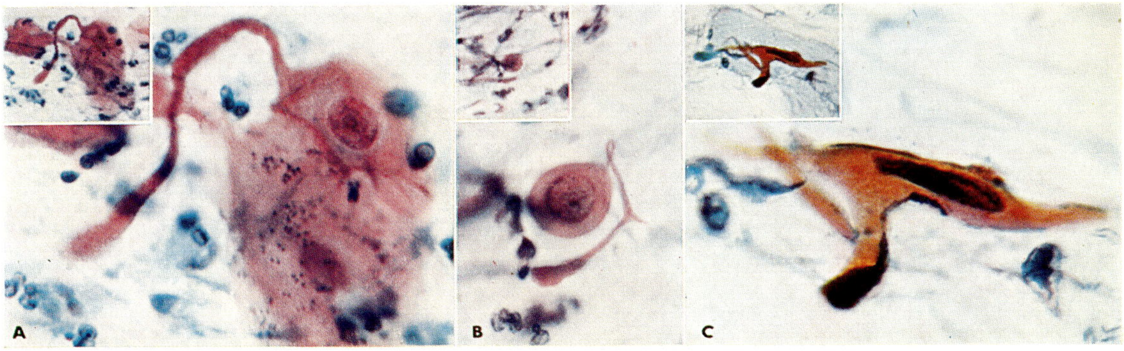

FIG. 98

Amoeboid Squamous Cells and a Differentiated Malignant Cell of Squamous Type in Sputum

A. The bizarre shape of this squamous cell (inset) leads to careful examination of the nucleus which shows only enlargement and is not sufficiently abnormal to justify a diagnosis of malignancy. Note especially the long blunt-ended cytoplasmic pseudopodium. (Inset ×200.) (×800.)

B. Again, an amoeboid outline with a long blunt-ended pseudopodium is seen in the cell illustrated in the inset. High-power examination, however, reveals only an enlarged nucleus and unusually opaque cytoplasm. (Inset ×200.) (×800.)

C. This opaque, brightly orangophilic squamous cell shows several amoeboid processes (inset). Higher magnification reveals two large hyperchromatic nuclei. The lower shows degenerative change, but the upper has an irregular outline, marginal condensation of chromatin and an abnormal chromatin pattern, thus identifying it as a differentiated malignant cell of squamous type. (Inset ×200.) (×800.)

The appearance of bizarre and especially of amoeboid cells should always lead to a careful search for differentiated malignant cells of squamous type.

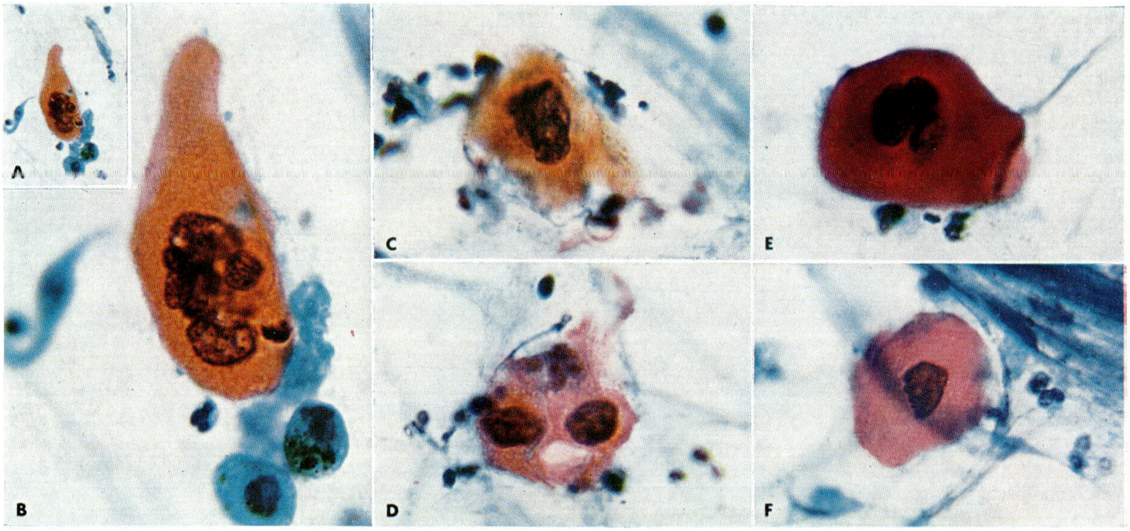

FIG. 99

Differentiated Malignant Cells of Squamous Type in Sputum Specimens

A. A giant multinucleated tadpole cell is seen. Its cytoplasm is brightly orangophilic and has a sharply defined border. The nuclei appear hyperchromatic. (×200.)

B. The bizarre tadpole shape, sharp cellular border and densely opaque, brilliantly orange stained cytoplasm, are characteristic features of a differentiated malignant cell of squamous type. The nuclei show variation in size and shape. Chromatin clumping is present and peripheral condensation of chromatin is seen in the lowest nucleus. (×800.)

C. This squamous cell shows degenerative changes with a loss of nuclear structure and is, therefore, unsuitable for final diagnosis. However, the appearance of this type of cell with orangophilic cytoplasm and a large bizarre irregular nucleus, should lead to a careful search for malignant cells of squamous type. (×800.)

D. Although degenerative changes are apparent in the cytoplasm of these cells, their nuclei, which are large, irregular in shape and show chromatin clumping, are recognisable as malignant nuclei. (×800.)

E. This multinucleated cell shows the striking and characteristic 'three-dimensional' cytoplasm of the differentiated malignant cell of squamous type. The cytoplasm has sharply defined edges, is densely opaque and brightly eosinophilic. The cytoplasm of such cells often appears to be refractile if the substage condenser is slightly lowered. The hyperchromatic nuclei vary in size and chromatin clumping is apparent in the nucleus on the right. (×800.)

F. This cell shows both the cytoplasmic and nuclear features characteristic of a differentiated malignant cell of squamous type. The cytoplasm is clearly defined, opaque and brightly eosinophilic. The nucleus is irregularly shaped, angular and hyperchromatic, with marginal condensation of chromatin and chromatin clumping. (× 800.)

Before differentiated malignant cells of squamous type are diagnosed it is essential to see both the nuclear and cytoplasmic features. However, the presence of cells with either the typical nuclear or cytoplasmic characteristics should arouse suspicion and lead to a prolonged search for cells suitable for definitive diagnosis.

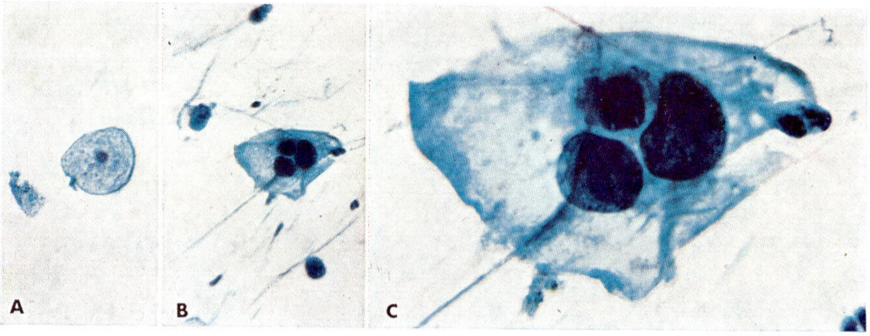

FIG. 100

A Benign Squamous Cell and a Multinucleated Differentiated Malignant Cell of Squamous Type in Sputum

A. This illustration shows a benign squamous cell with a large amount of translucent cytoplasm and a small central nucleus. (× 200.)

B. In this multinucleated malignant cell marked nuclear hyperchromasia and enlargement are obvious even on low-power examination. (× 200.)

C. High-power examination reveals a striking variation in nuclear size and shape. Marginal condensation of chromatin produces sharp irregular nuclear borders and chromatin clumping is also seen. Note the moulding of adjacent nuclei. (× 800.)

Three types of benign cell may cause difficulty in the differential diagnosis of differentiated malignant cells of squamous type. Degenerate deep squamous cells of upper respiratory tract origin may have opaque eosinophilic or orangophilic cytoplasm, but they are fairly uniform in size; their nuclei, although hyperchromatic, have smooth round or oval outlines and their nucleo-cytoplasmic ratios are normal (see Fig. 88). These cells usually predominate in those areas of a specimen which consist mainly of saliva. Histiocytes are usually readily recognisable, but may present a problem if their nuclei are pyknotic as a result of degeneration. Since malignant squamous cells may appear to show 'phagocytosis' with polymorphonuclear leucocytes or even another squamous cell within their cytoplasm, producing the typical 'bird's-eye' appearance, the phagocytic activity of histiocytes cannot be used as a diagnostic criterion. However, the finely vacuolated cytoplasm of histiocytes and the presence of reniform nuclei will usually identify them. Squamous metaplasia presents diagnostic difficulty only when the nuclei show nuclear atypia. The good aggregation of these cells in sheets or groups and the regularity of their arrangement help to distinguish them from malignant squamous cells, occurring singly or in completely irregularly arranged groups. Although the fact that malignant squamous cells occur singly is often a striking feature, the diagnosis of malignancy should not be based on such a single cell; most problems in diagnosis can be resolved only by an evaluation of all the abnormal cells present in a specimen.

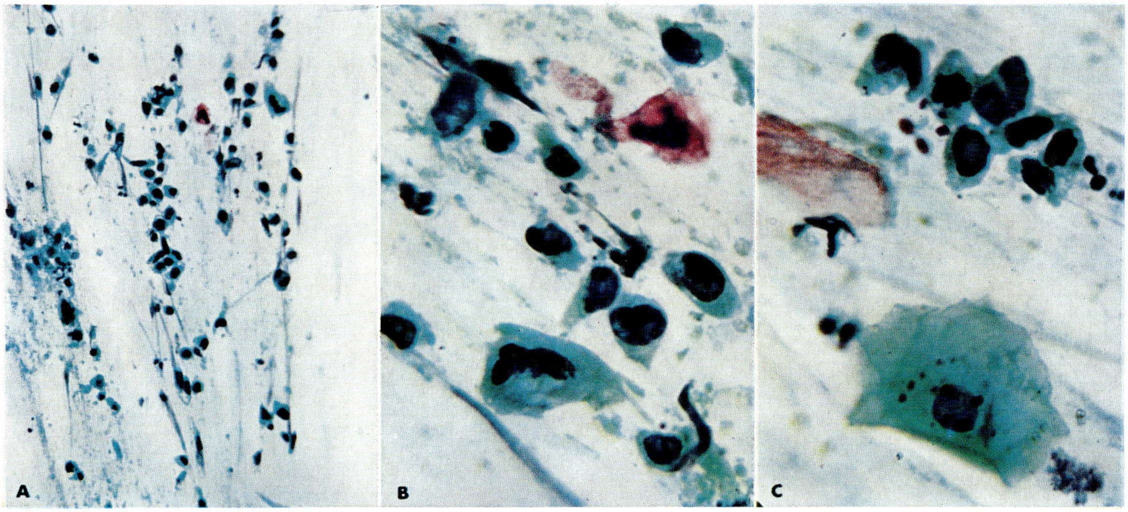

FIG. 101

Differentiated Malignant Cells of Squamous Type in Sputum

A. These differentiated malignant cells of squamous type show strikingly abnormal cellular and nuclear features. Many bizarre cells are present including a number of 'tadpole' forms. The cytoplasm is opaque with sharply defined margins and a single brightly eosinophilic cell is seen. The nuclei are densely hyperchromatic (the 'ink-dot' appearance), and are irregular in shape. (×200.)

B. The opaque, dense cytoplasm, which in one cell shows bright eosinophilia, forms sharply defined cells with irregular outlines, and 'tails' of cytoplasm are seen on many cells. The nuclei are intensely hyperchromatic and their irregular outlines are frequently angular or notched. The cells in the group are loosely arranged in contrast to the close cohesion of the benign metaplastic cells seen in Figs. 91 and 92. (×800.)

C. The nuclei in the group of malignant squamous cells are again densely hyperchromatic and irregular in size and shape. Although this is a more compact group, there is still separation of individual cells. Note the characteristic 'dirty' background in this field and in field B. (×800.)

These groups of differentiated malignant cells of squamous type, which are loosely arranged and show gross nuclear abnormalities contrast sharply with the metaplastic cells, showing good group cohesion and comparative nuclear uniformity, illustrated in Fig. 91.

UNDIFFERENTIATED MALIGNANT CELLS

Undifferentiated malignant cells may be the only type of malignant cell identified in a specimen, but they are also seen in specimens which include differentiated malignant cells. Two types of undifferentiated malignant cell, distinguished by size, are found; these are small undifferentiated malignant cells (oat cells) and large undifferentiated malignant cells.

Small Undifferentiated Malignant Cells (Oat Cells)

These small undifferentiated malignant cells occur in elongated groups or streaks, usually on a clear background and without any admixture of inflammatory cells. Although the individual cells are mainly discrete, cells in close contact show moulding of adjacent nuclei. The cells are strikingly small, similar in size to lymphocytes. Their cytoplasm is absent or appears as a narrow, often incomplete, basophilic rim surrounding nuclei which show slight variations in size and shape with a predominance of round or oval forms. The chromatin patterns show considerable diversity; some nuclei appear densely hyperchromatic, while in others abnormal chromatin clumping is recognisable and, occasionally, pale staining nuclei are seen. Condensation of chromatin at the nuclear border occurs and gives the nuclei a punctate appearance on low-power examination.

Those benign cells which may be confused with oat cells are columnar bronchial epithelial cells and inflammatory cells. Columnar bronchial epithelial cells cause difficulty only when their nuclei are denuded of cytoplasm. These benign nuclei are usually degenerate and although they may appear hyperchromatic the nuclear structure is lost and the whole nucleus is opaque; their oval or round forms are retained and they do not show moulding of adjacent nuclei (see Fig. 89). Elongated groups of inflammatory cells may appear suspiciously similar to groups of oat cells on low-power examination but high-power examination reveals that the groups consist of a mixture of cells, usually mainly lymphocytes and small histiocytes (see Fig. 83). The group does not show the complete preponderance of a single-cell type which is usually seen in groups of oat cells. The main diagnostic hazard with oat cells is the risk of overlooking them entirely, because of their small size or alternatively accepting them as groups of lymphocytes. The observer should carefully examine all groups of small hyperchromatic cells.

Although undifferentiated malignant cells often occur alone in a specimen, it is not uncommon to find that, in a group, an occasional cell has a margin of opaque cytoplasm and is in fact a miniature differentiated malignant cell of squamous type. Occasionally a specimen which contains oat cells will also include a few larger differentiated malignant cells of squamous type or the two types of cell may be found in separate specimens in a series from the same patient.

Large Undifferentiated Malignant Cells

Large undifferentiated malignant cells may be found without other malignant cells or seen in a specimen which also contains differentiated malignant cells of adenocarcinomatous or squamous type. Large undifferentiated malignant cells appear in groups, which are usually more compact than groups of oat cells. Individual cells, however, are easily recognisable. Their cytoplasm has either disappeared completely or is present only over a part of the nuclear surface. The nuclei show variations in size and their shapes are irregular with angular outlines; adjacent nuclei in a group frequently show moulding. Their chromatin patterns are strikingly abnormal with irregular condensation of chromatin both at the nuclear borders and in strands and angular clumps within the nuclei; enlarged nucleoli are noted in some of these nuclei. Although most nuclei are hyperchromatic occasional pale staining nuclei with abnormal chromatin patterns are seen.

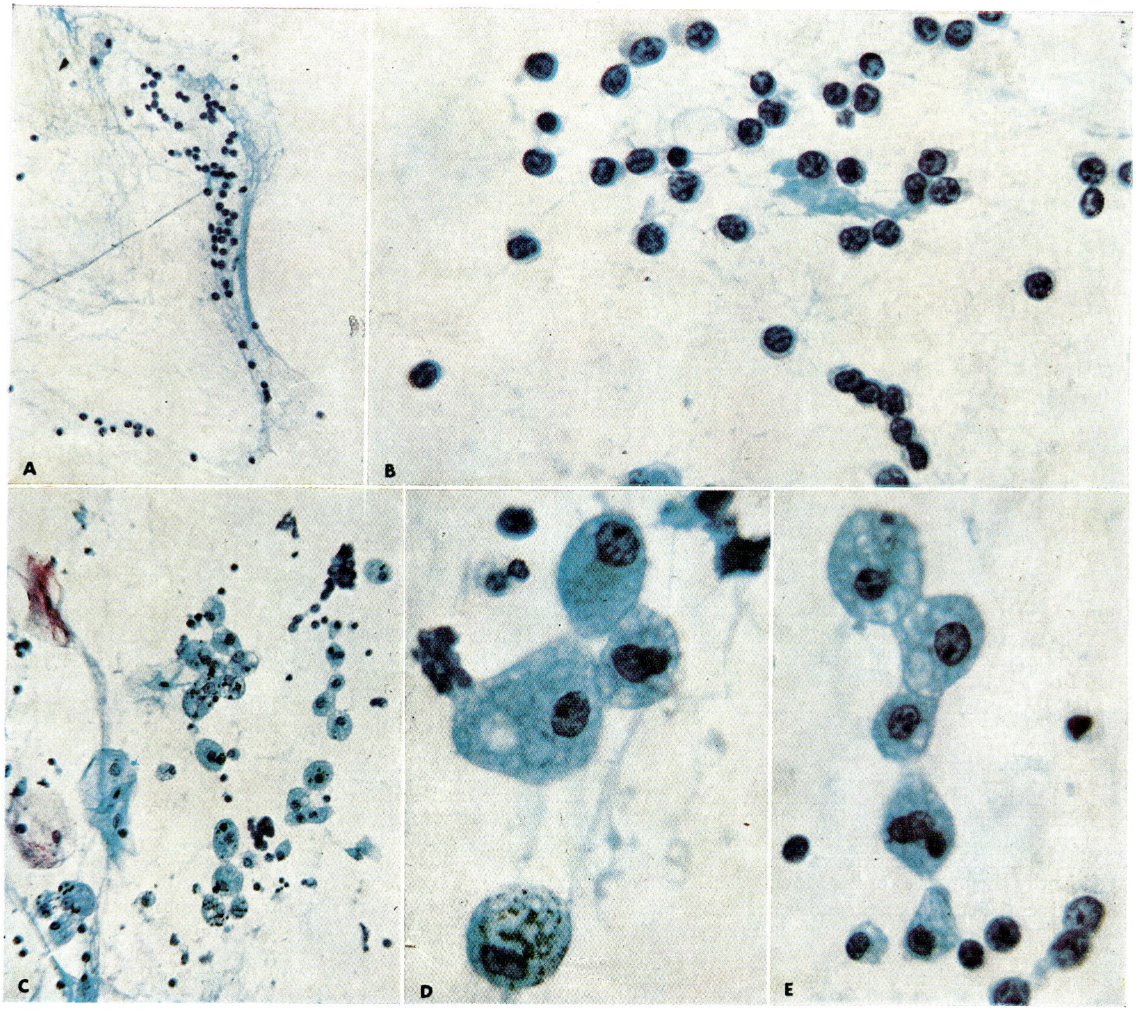

FIG. 102

Small Undifferentiated Malignant Cells, Oat Cells, in Sputum

A. This elongated group of loosely arranged discrete cells, surrounded by strands of mucus, consists of small dark cells closely resembling lymphocytes. These cells are all of the same type and there is no admixture of inflammatory cells. (×200.)

B. At high-power magnification a similar group is seen to consist entirely of small undifferentiated malignant cells with round or oval nuclei which show only slight variation in size. They show coarse clumping of chromatin with clear spaces separating the irregularly arranged and shaped chromatin wedges. Many of these nuclei are seen to have an incomplete narrow rim of cytoplasm, while the single cell in the centre of the lower half of the field shows a nucleus wholly surrounded by cytoplasm. With this central hyperchromatic nucleus its appearance is that of a small malignant cell of squamous type. The linear group of cells in the lower right quadrant shows moulding of nuclei by their neighbours, and moulding is also seen in the group immediately above. (×800.)

C. Two groups of tiny undifferentiated malignant cells are seen in the right half of the field. Compare the size of the individual cells spraying out from the upper group with that of the adjacent carbon-laden histiocytes. (×200.)

D. The total area of the group of undifferentiated malignant cells in the upper left quadrant of this field is much less than that of the single adjacent histiocyte. Note the moulding of the nuclei of this compact group of oat cells which abuts on thc largc vacuolated histiocyte. (×800.)

E. The tiny nuclei in the lower right corner show characteristic chromatin clumping and have a narrow incomplete margin of cytoplasm. Notice again their small size in comparison with the histiocytes. (×800.)

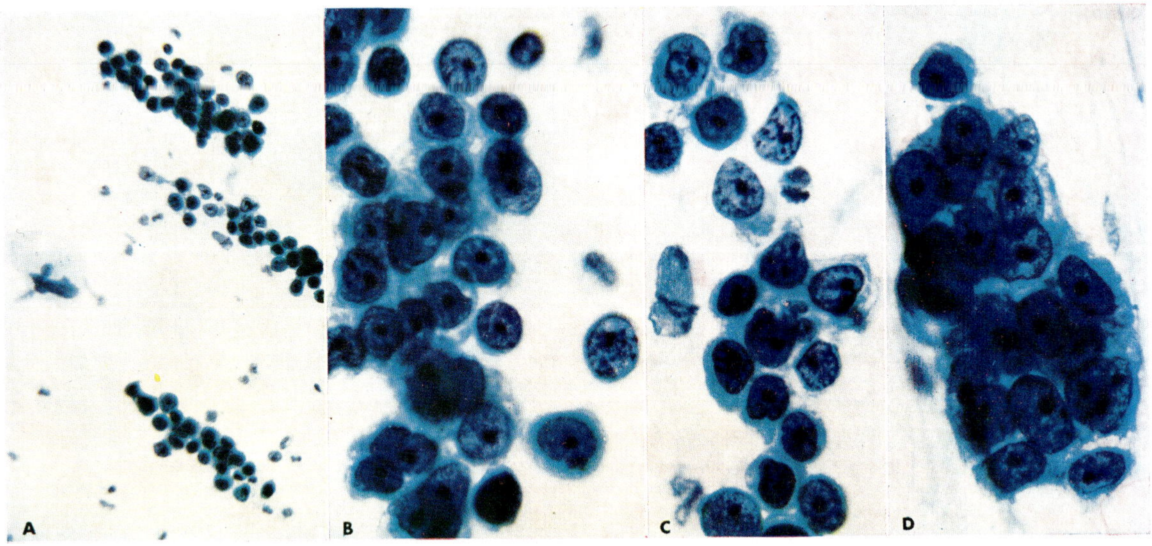

FIG. 103

Large Undifferentiated Malignant Cells in Sputum

A. Three large groups of undifferentiated malignant cells are present. Although single cells are easily discernible they are more compactly grouped than the minute cells of oat cell type seen in Fig. 102. Occasional overlapping of cells occurs. (×200.)

B. The cytoplasm is seen only as an ill-defined background to these cells. The nuclei vary in size and show some irregularity of shape; their chromatin is coarsely clumped and prominent angular eosinophilic nucleoli are seen. (×800.)

C. The nuclei of these cells show striking variation in shape, with irregular, angular and notched nuclear outlines and their chromatin patterns show great diversity. The large nuclei in the upper half of the group are comparatively pale but still show clumping and peripheral condensation of chromatin. The smaller nuclei in the lower half of the field are densely and unevenly hyperchromatic. Most of the nuclei in the upper half of the field are devoid of cytoplasm while those in the lower half are surrounded by cytoplasm, although it is difficult to identify the outlines of individual cells. (× 800.)

D. This crowded overlapping group of cells shows a poorly defined cytoplasmic background, with nuclei which vary in size, in shape and in chromatin pattern. Peripheral condensation of chromatin and prominent nucleoli are striking features. (×800.)

This type of malignant cell group in sputum must be distinguished from a group of histiocytes, which also shows variation in nuclear shape and may have prominent nucleoli. Histiocytes usually retain their characteristic foamy cytoplasm, and careful examination will frequently show ingested carbon particles. Some histiocyte nuclei will show the typical 'kidney-bean' shape and they do not have the marked abnormalities of chromatin pattern seen in the malignant cells.

Difficulty in diagnosis is unlikely to occur when cells showing the characteristic chromatin abnormalities are seen, but certain appearances in the smear may initially arouse a false suspicion that they are large undifferentiated malignant cells. Mucus may simulate irregular hyperchromatic 'cells' on low-power examination, but careful scrutiny reveals the absence of a cellular or nuclear structure (see Fig. 84). A group of histiocyte nuclei, denuded of cytoplasm, may show considerable nuclear pleomorphism (see Fig. 129). However, some nuclei in the group usually display the typical reniform shape peculiar to the histiocyte and, although the nuclei may have granular chromatin patterns with prominent nucleoli, the abnormalities of chromatin pattern seen in malignant nuclei are absent. Errors can be minimised by basing the diagnosis on groups of cells rather than on single cells and by seeking the characteristic nuclear pleomorphism and chromatin abnormalities of malignant cells.

DIFFERENTIATED MALIGNANT CELLS FROM AN ADENOCARCINOMA

These cells usually appear in compact, often elongated groups in which they are crowded together. Often these groups appear in clear areas and on low-power examination have a raspberry-like appearance. The scalloped edge of the group is often due to the fact that it is bordered by nuclei rather than by cytoplasm. In the individual cells the nuclei are eccentric in position, with an abnormal nucleocytoplasmic ratio due to nuclear enlargement. The amount of cytoplasm present varies with the degree of differentiation of the cells. Nuclei in a group vary in size and in shape, with occasional angular irregular outlines. The chromatin patterns are strikingly abnormal with uneven condensation of chromatin in clumps and strands. Nucleoli are enlarged, prominent, angular and often multiple.

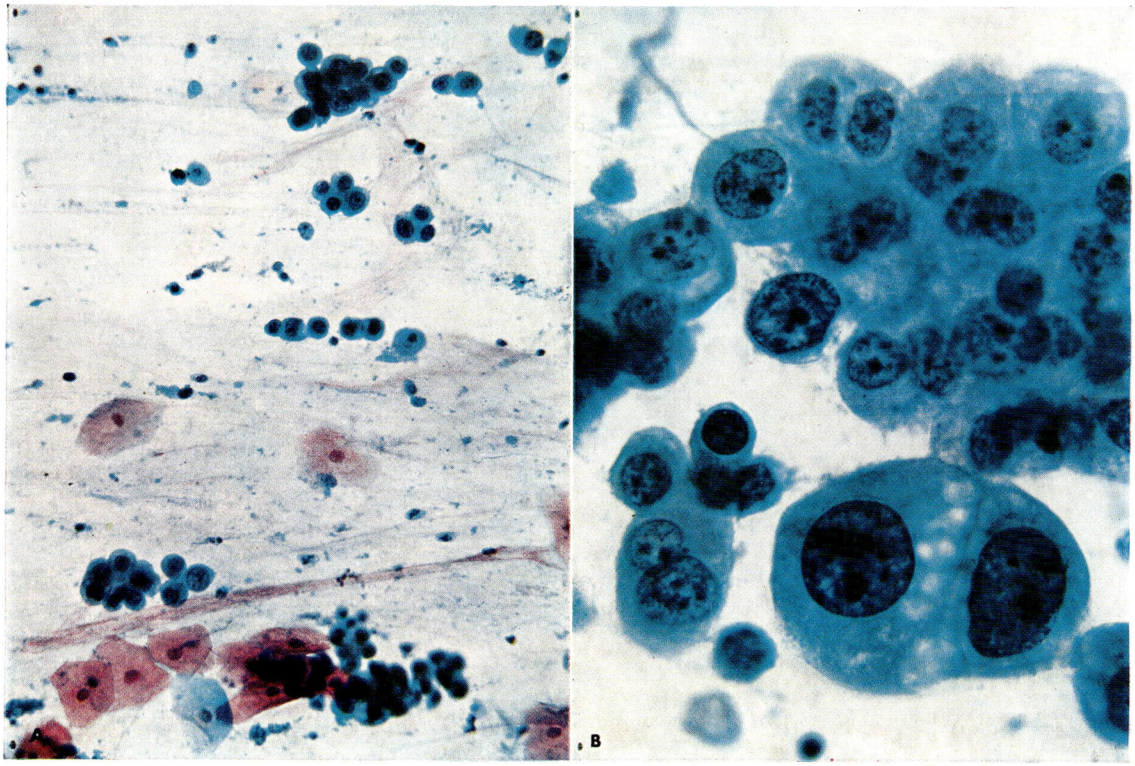

FIG. 104

Sputum Showing Differentiated and Undifferentiated Malignant Cells from an Adenocarcinoma

A. Several raspberry-like groups of malignant cells are seen. It is characteristic for cells originating in an adenocarcinoma to appear in clumps. Note the irregular borders of the groups; a linear edge is not seen. In the lowest group of cells, the nuclei are close to the border, but in the better differentiated cells of the upper groups this feature is absent. Striking variation in nuclear size is apparent and a coarse chromatin pattern is recognisable in the larger nuclei. (×200.)

B. Most of these cells show the features of differentiated malignant cells arising from an adenocarcinoma, although one undifferentiated malignant cell is present, just above the centre of the field. The differentiated malignant cells have foamy vacuolated cytoplasm with eccentric nuclei, which show marked variation in size, with some irregularity of shape. Their chromatin patterns are grossly abnormal with marked clumping and marginal condensation of chromatin. Prominent nucleoli are present in many of the nuclei. (×800.)

These malignant cells must be differentiated from histiocytes, which also have foamy cytoplasm and eccentric nuclei. Histiocytes, however, show neither the striking variation in nuclear size nor the bizarre chromatin patterns of these malignant cells.

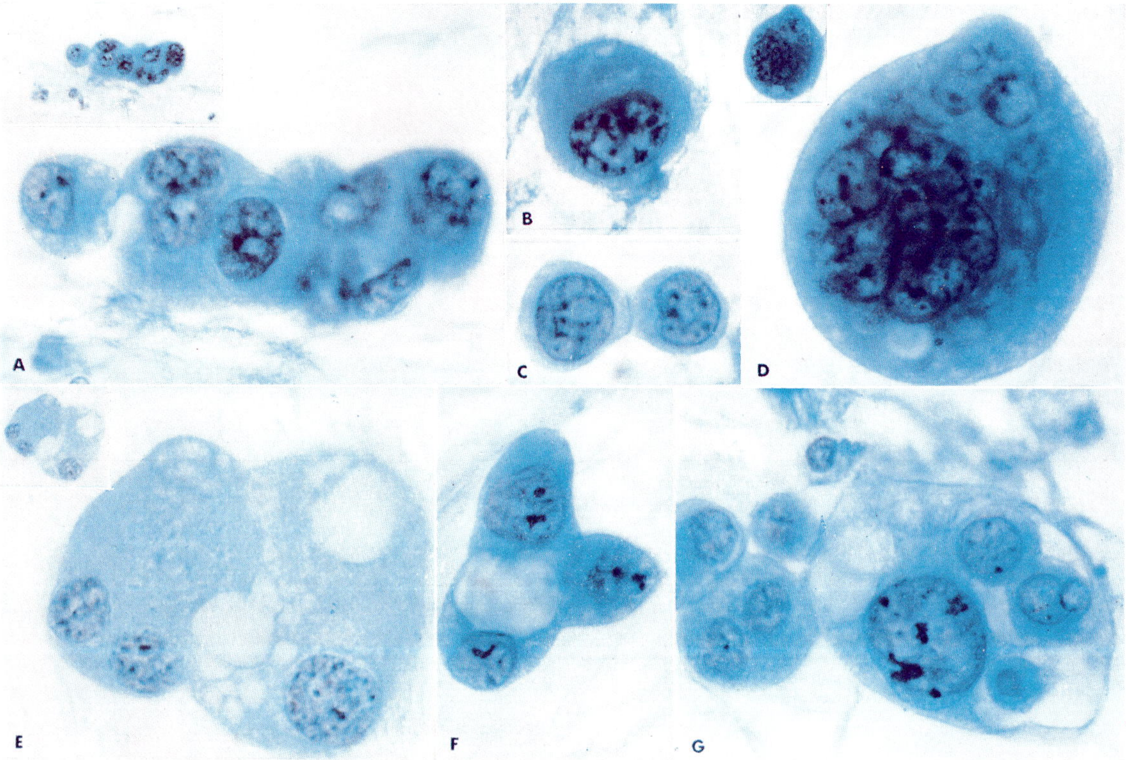

FIG. 105

Sputum Showing Differentiated Malignant Cells from an Adenocarcinoma

A. This group of malignant cells, shown under low power (×200) in the inset, shows the characteristic eccentric nuclei arranged mainly around the edges of the group. The nuclei vary in size and shape and show clumping of chromatin. (×800.)

B. Chromatin clumping is a striking feature of this differentiated malignant cell. The eccentric nuclear position is compatible with an origin in an adenocarcinoma. (×800.)

C. This pair of differentiated cells shows an abnormal nucleocytoplasmic ratio, variation in nuclear size and shape, marginal condensation of chromatin and chromatin clumping, but there is insufficient evidence in these cells to indicate the type of tumour from which they originate. (×800.)

D. This giant multinucleated malignant cell, shown at low-power (×200) in the inset, shows marked nuclear abnormalities, with variation in size and shape, uneven marginal condensation and irregular clumping of chromatin. (×800.)

E. F. and G. The malignant cells shown in these fields have essentially similar nuclei to those described above. In these cells, however, in addition to the eccentric position of the nuclei, there is also coarse cytoplasmic vacuolation; both features are compatible with an origin from an adenocarcinoma. Note that in these groups of malignant cells the nuclei form part of the border of the group. (Inset E.× 200.) (× 800.)

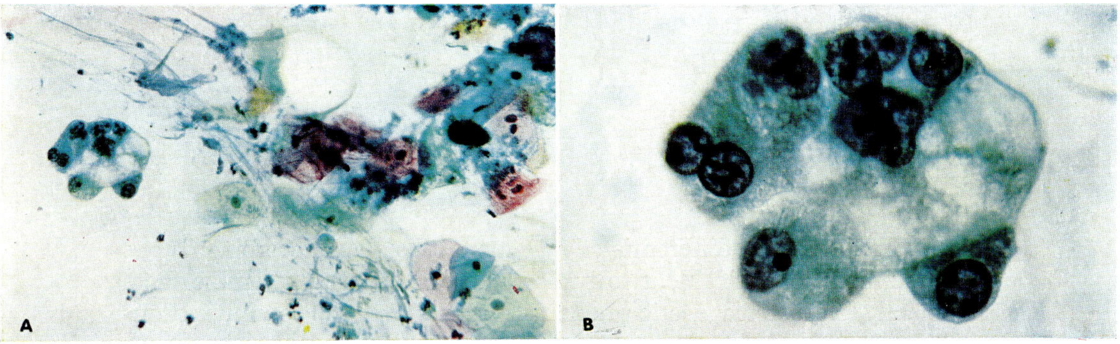

FIG. 106

Sputum Showing Differentiated Malignant Cells from an Adenocarcinoma

A. The group of cells on the left of the field shows striking vacuolation of cytoplasm and the hyperchromatic nuclei are peripherally arranged. The overall appearance suggests a pseudo-acinar pattern. (×200.)

B. Cytoplasmic vacuolation in the cells on the right of the group is so marked that it has distorted and blurred the individual cellular outlines. The nuclei lie close to the outer margin of the group and protrude irregularly on its surface. There is only slight variation in nuclear size and shape, but all the nuclei show grossly abnormal chromatin patterns with marginal condensation and coarse irregular chromatin clumping. Prominent nucleoli are seen in several of the nuclei. (×800.)

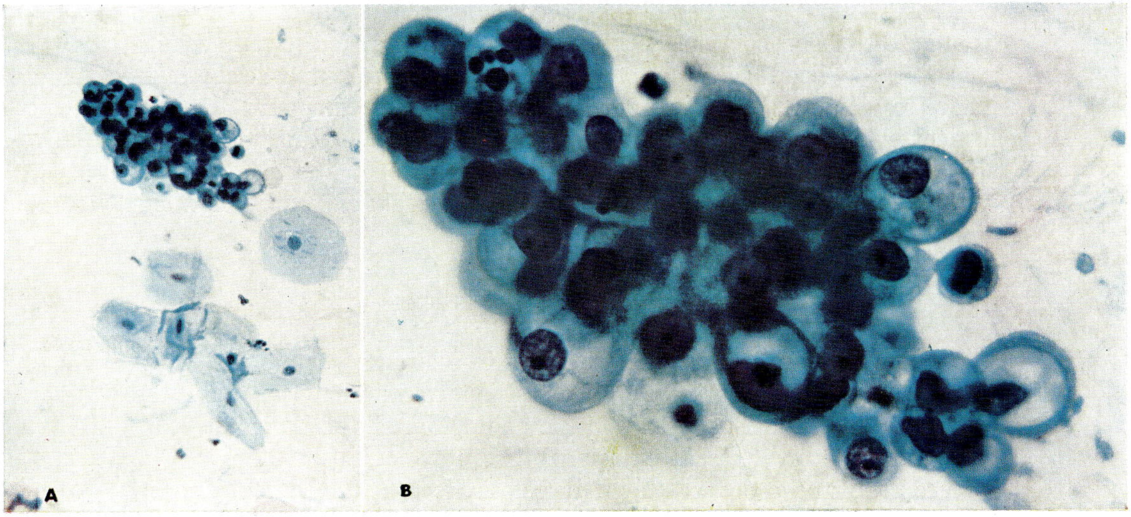

FIG. 107

Sputum Showing Differentiated Malignant Cells from an Adenocarcinoma

A. A compact group of overlapping cells is seen. The outline of the group is irregular. The nuclei are hyperchromatic and cytoplasmic vacuolation is present. (×200.)

B. The cells show many of the characteristic features of well-differentiated malignant cells from an adenocarcinoma. In this group, however, the nuclei do not form the border of the group. The cells vary in size and are irregularly arranged; their cytoplasm shows coarse vacuolation. The eccentric nuclei vary in size and shape and some have sharply angular outlines. Hyperchromasia is seen in all the nuclei and those on the periphery of the group show clumping and marginal condensation of chromatin. Prominent nucleoli can be identified in several nuclei. (×800.)

Groups of benign columnar bronchial epithelial cells may closely resemble these highly differentiated malignant cells. The variation in nuclear size, shape and chromatin pattern, and the prominent nucleoli of these malignant cells, however, confirm their identity.

Although typical differentiated malignant cells of adenocarcinomatous type are readily identifiable, two types of benign cell may resemble them sufficiently to cause diagnostic problems. Individual single columnar bronchial epithelial cells are easily recognised, but when they appear in groups they can cause diagnostic difficulty. In a group of bronchial epithelial cells the presence of cilia and/or well-defined terminal plates confirms the benign nature of that particular group. The border of a group of benign cells usually appears smoother than the outline of a group of malignant cells; in the benign group there is often a margin of cytoplasm uniformly separating the nuclei from the edge of the group. Nuclei of benign columnar cells vary in size, but show little variation in shape unless distorted by mucus secretion (see Fig. 80). In benign nuclei the chromatin may appear more coarsely granular in the presence of inflammation, but the abnormal chromatin patterns of malignant change are absent. Occasionally nucleoli may be enlarged in benign columnar cells, but this is less marked than in malignant cells (see Fig. 90). Groups of histiocytes may show a superficial resemblance to differentiated malignant cells from an adenocarcinoma, but the histiocyte group is more loosely arranged than the group of malignant cells and careful examination reveals the typical reniform nuclei and finely vacuolated cytoplasm of the histiocytes, which will frequently show ingested carbon particles (see Fig. 81).

METASTATIC CARCINOMA

There is often considerable difficulty in securing a cytological diagnosis in a case of carcinoma metastasising to the lung, because the metastasis does not usually broach the epithelium of the bronchus at an early stage. Nevertheless, in occasional cases, cytological methods can be used to confirm the presence of a metastasis, to type the tumour and, in some circumstances, be of value in determining whether a particular lesion in lung is likely to be a primary carcinoma or a metastatic deposit.

SECTION III

GASTRO-INTESTINAL TRACT

CHAPTER 9

THE NORMAL CELLULAR CONTENT

THE cellular content varies according to the site from which the specimen is collected.

THE ORAL CAVITY

The predominant cells are superficial or intermediate squamous cells and polymorphonuclear leucocytes. Occasionally, tall ciliated columnar cells from the nasal mucosa may be seen.

OESOPHAGEAL WASHINGS

Superficial squamous cells predominate in such specimens, with occasional polymorphonuclear leucocytes. If the specimen is contaminated by swallowed sputum ciliated columnar cells and carbon-containing histiocytes of respiratory tract origin may also be seen.

GASTRIC WASHINGS

In a normal specimen cells from other areas may be more striking than cells from the gastric mucosa. Squamous cells of oral and oesophageal origin, ciliated columnar cells, mucus-producing goblet cells and carbon-containing histiocytes of respiratory tract origin, together with small numbers of polymorphonuclear leucocytes and lymphocytes, may predominate. Often these cells show early degenerative changes, with pale staining cytoplasm and poor definition of cellular borders and nuclear structure; this may be most marked in normal specimens when there is high acidity. The columnar cells of gastric mucosal origin may appear as small groups or sheets of cells, or as single cells. The groups of cells are evenly spaced in a regular pattern, characteristically producing a 'honeycomb' appearance. When surface cells are still present in a group they appear as a regularly arranged 'picket-fence' or 'palisade'. Single columnar cells have a vesicular nucleus with a granular chromatin pattern; an eosinophilic nucleolus may be identified, but will be of very small size. These columnar cells appear shorter and broader than those of respiratory tract origin. The nucleus, although eccentric in position, may be less obviously basal than in the respiratory tract columnar cell and this contributes to their somewhat cuboidal appearance. Degenerative changes with complete loss of cytoplasm are frequently seen in gastric mucosal cells. Although these denuded nuclei may initially appear suspicious, their uniformity of size, shape and chromatin pattern indicate their benign character. Occasionally degenerative change in the nucleus produces a hyperchromatic appearance, but further examination reveals the loss of the normal nuclear structure. Contamination of the specimen by food particles may be seen if the patient has not been suitably prepared for investigation and this can cause diagnostic difficulties.

COLONIC AND RECTAL WASHINGS

These specimens consist mainly of cellular debris, polymorphonuclear leucocytes and histiocytes. Only a few epithelial cells are seen; these are squamous cells of anal origin and occasional columnar cells of colonic or rectal origin. Frequently, the epithelial cells are obscured by debris and inflammatory cells.

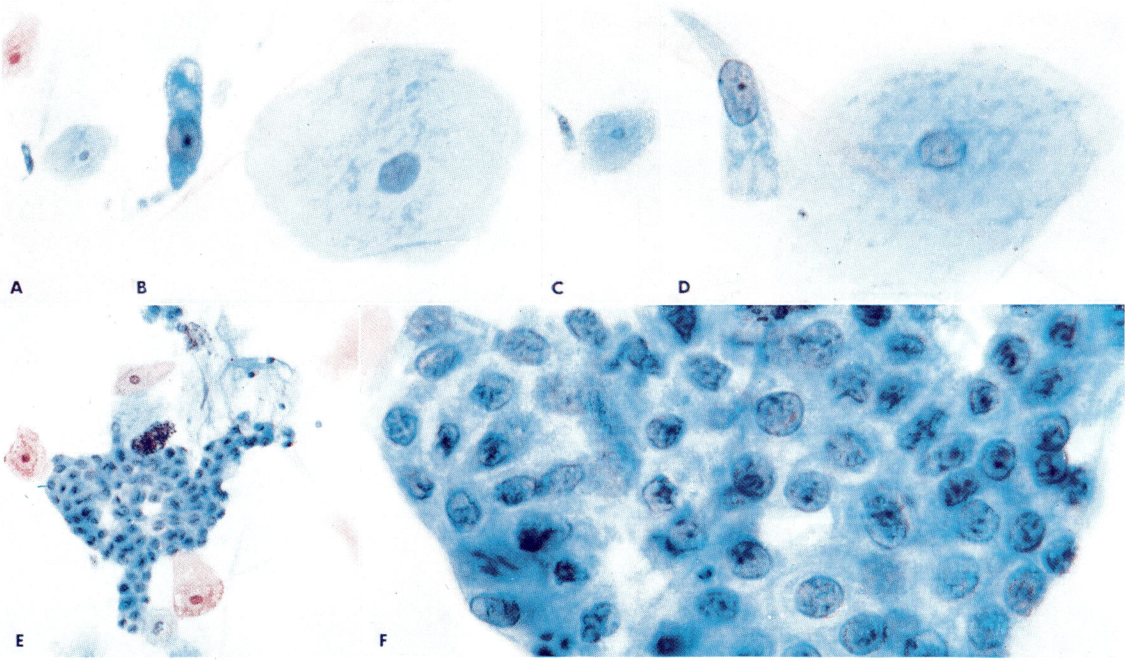

FIG. 108

Benign Columnar and Squamous Cells in Gastric Washings

A. B. C. and D. Columnar and squamous cells are shown at low-power (A and C) and high-power (B and D) magnifications. The overall pale staining of the squamous cells illustrated here is frequently seen in gastric washings. The gastric columnar cells illustrated are well preserved and readily recognisable, but when compared with columnar cells of lower respiratory tract origin they appear shorter and broader; their nuclei, although eccentric, are situated nearer to the cell centre. Although the nuclei of the two columnar cells show some size variation, they are uniformly oval in shape. Note their finely granular chromatin pattern and the small round single eosinophilic nucleoli. (A and C ×200.) (B and D × 800.)

E. A large sheet of benign columnar cells is seen against a background of pale staining squamous cells and mucus. These columnar cells are regularly arranged with evenly spaced nuclei well separated by cytoplasm. Two clumps of polychromatic staining micro-organisms are also present in the field. (×200.)

F. Most of the nuclei are evenly spaced without crowding or overlapping, and where an individual cell border is identifiable an adequate margin of cytoplasm is seen around the nucleus. Note the finely granular chromatin pattern of the nuclei; where nucleoli are present they are small and occur singly. (×800.)

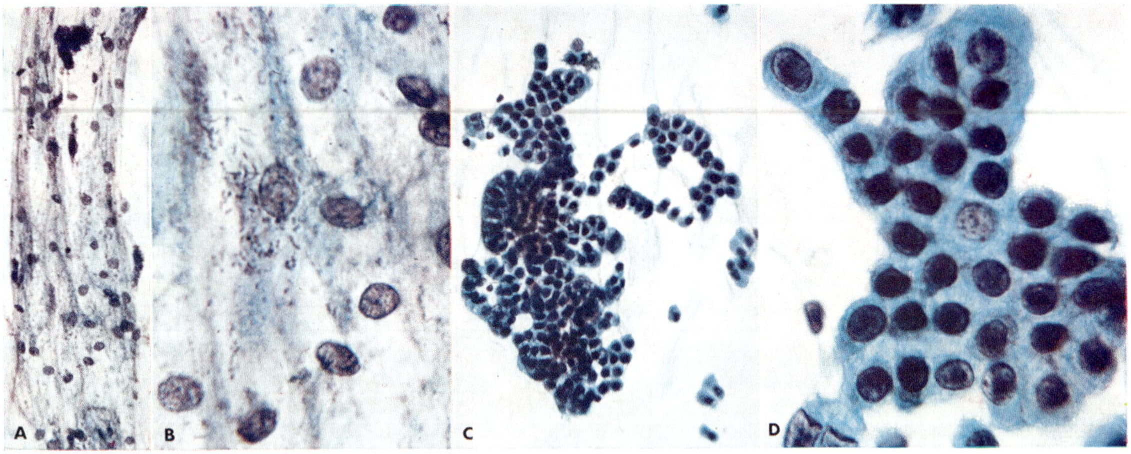

FIG. 109

Benign Denuded Nuclei of Cytolysed Columnar Cells and Groups of Intact Benign Gastric Columnar Cells in Gastric Washings

A. Cytolysis frequently produces the appearance seen in this illustration in which only the denuded nuclei of columnar cells are identifiable lying enmeshed in a background of mucus and bacteria. (×200.)

B. In spite of the loss of cytoplasm, the benign character of these nuclei is apparent. They show slight variation in size, but are uniformly oval in shape, and their chromatin pattern is finely granular. When a nucleolus is observed it is single and small. It is not possible to determine the origin of these nuclei with certainty as columnar cells of nasal and bronchial origin may be present in a gastric wash, but these denuded nuclei are so numerous that a gastric origin seems probable. (×800.)

C. A large sheet and several smaller groups of columnar cells are present. In the middle sector of the left border of the large group, a smooth surface is noted where the superficial cells show a 'palisade' arrangement. The rest of the large sheet of cells shows a 'honeycomb' pattern with regularly arranged and spaced nuclei. Several of the smaller groups also show 'palisade' arrangements with parallel rows of columnar cells. (×200.)

D. This group shows regularly arranged cells with evenly spaced nuclei separated by adequate margins of cytoplasm, which also separate the nuclei from the free border of the sheet at its upper edge. The nuclei are fairly uniform in size and shape. Many are hyperchromatic but abnormalities of chromatin pattern are absent. (× 800.)

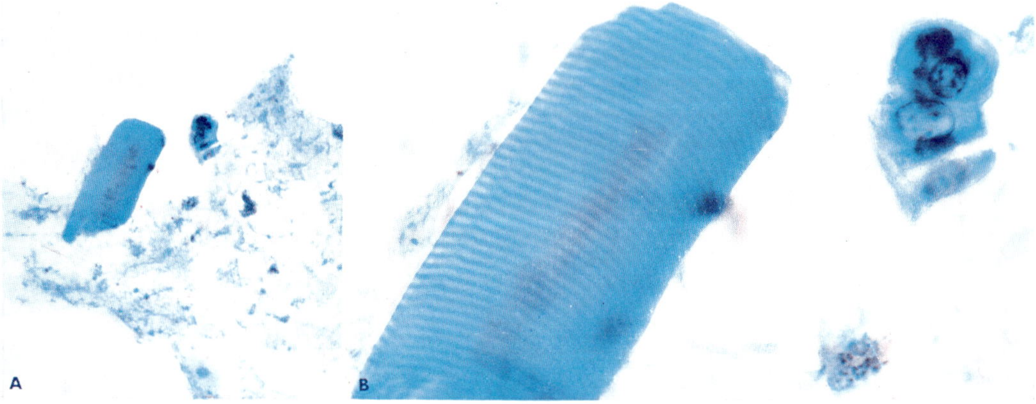

FIG. 110

Food Contamination of a Gastric Aspirate

A. and B. In the low-power illustration a single group of columnar cells (upper centre) is seen lying against a background of debris. The large body to the left of the cells is a fragment of tissue, which at high-power magnification (B) is identified, by its cross striations, as muscle tissue of animal origin. The columnar cells show degenerative changes with a loss of nuclear detail. (×200.) (×800.)

Food contamination of gastric aspirates and washings can be minimised by a period of starvation for twelve hours prior to the collection of the specimen but may still occur in patients who have some degree of obstruction. When it does occur to any significant degree the specimen should be discarded.

CHAPTER 10

NON-MALIGNANT DISORDERS

THE ORAL CAVITY

SIMPLE local lesions can produce considerable changes in the cytological pattern. The pressure of ill-fitting dentures may cause the appearance of orangophilic keratinised anucleated squames. These retain normal shapes and cytoplasmic translucency and can be readily distinguished from the bizarre opaque anucleated squames seen in association with squamous carcinoma.

INFLAMMATION

In benign non-specific ulcerative conditions parabasal squamous cells appear and these may show nuclear enlargement or, occasionally, degenerative changes in the nucleus. The enlarged nuclei retain a normal chromatin pattern. There is an increase in the number of polymorphonuclear leucocytes.

LEUKOPLAKIA

After adequate biopsy has been performed initially, cytology may be of value in following this lesion, as numerous repeat biopsies of persistent lesions may be impracticable. The cells from an area of leukoplakia are superficial squamous cells and keratinised anucleated squames, and usually these cells are remarkably uniform in appearance. If intermediate or parabasal squamous cells are present in significant numbers they should be carefully scrutinised. If nuclear pleomorphism or hyperchromasia is seen, further investigation by biopsy is indicated as the thick keratinised epithelium may prevent the exfoliation of fully diagnostic malignant cells.

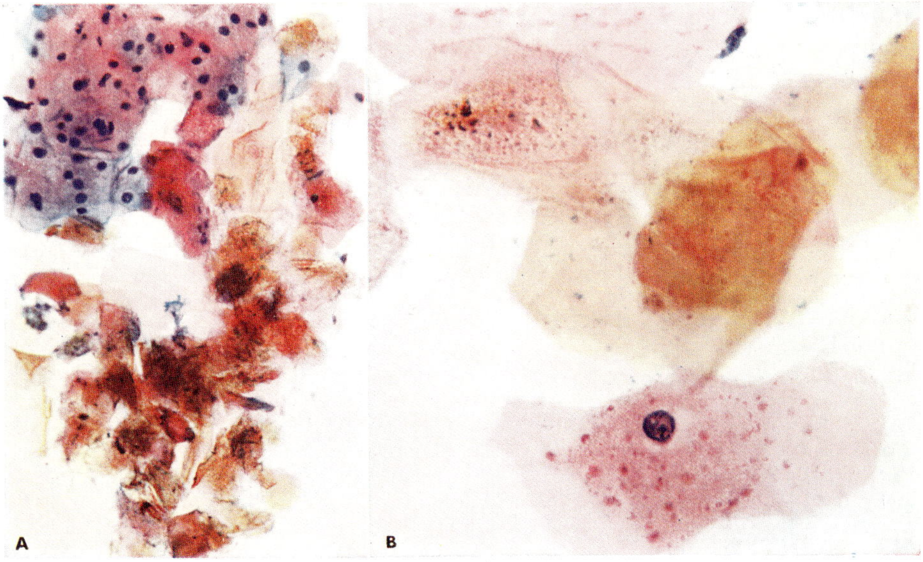

FIG. 111

Benign Squamous Cells and Anucleated Squames from an Area of Lingual Leukoplakia

A. This low-power field consists mainly of orangophilic anucleated squames accompanied by a sheet of benign superficial and intermediate squamous cells (upper left quadrant). A 'ghost' nucleus is apparent in one squame lying to the right margin of the group of squamous cells. (×200.)

B. At high-power magnification the wrinkled translucent orange-brown cytoplasm of the anucleated squames is well demonstrated. An intermediate squamous cell is seen at the lower margin. Keratohyalin granules are present in the cytoplasm. (× 800.)

THE OESOPHAGUS

INFLAMMATION

Parabasal squamous cells appear as a result of the loss of the more superficial layers of epithelium in oesophagitis. These cells have relatively large nuclei with granular chromatin, but occasional rounded clumps of chromatin are seen. There is good cellular cohesion in these atypical benign cells. Polymorphonuclear leucocytes are present in increased numbers.

THE STOMACH

ACUTE GASTRITIS

The high acidity of the specimen in acute gastritis causes rapid cytolysis of the cells. Although smears show an increase in the number of columnar cells, these usually appear as partly or wholly denuded nuclei. The benign character of these denuded nuclei is recognised by their uniformity of size and shape (see Fig. 109). Accompanying squamous cells and polymorphonuclear leucocytes, which are numerous, also show degenerative changes, with pale staining cytoplasm and swollen nuclei with poorly defined chromatin patterns.

CHRONIC ATROPHIC GASTRITIS

In chronic atrophic gastritis, gastric washings are often strikingly cellular with numerous groups of columnar cells and many inflammatory cells. Micro-organisms are usually present in considerable numbers and cocci frequently predominate. The columnar cells may show nuclear enlargement and prominent, occasionally multiple, nucleoli but they retain a normal chromatin pattern. The surface mucosal cells are often short and broad, with a distinctly cuboidal appearance. Frequently cells from areas of intestinal metaplasia (q.v.) are present. The inflammatory cells are mainly polymorphonuclear leucocytes, but there may also be a considerable number of plasma cells, which may cause diagnostic difficulty if their characteristic eccentric nuclei, with characteristic 'cart-wheel' chromatin patterns, are not recognised (see Fig. 14). Erythrocytes are present in small numbers.

INTESTINAL METAPLASIA

If areas of intestinal metaplasia are present, typical signet ring cells are seen. These are large rounded cells in which the nucleus is situated at the periphery; the cytoplasm is vacuolated and may contain one, or more, polymorphonuclear leucocytes. The cytoplasm has an affinity for the mucicarmine stain (Schade 1960). There may be considerable difficulty in distinguishing these cells from those of an adenocarcinoma. However, although the nuclei of intestinal metaplastic cells are densely stained, they are usually fairly uniform in size and do not show prominent nucleoli. If they occur in groups cellular overlapping is not marked and the boundaries of the individual cells are apparent. Nevertheless great care should be exercised in the diagnosis of gastric washings from cases of chronic atrophic gastritis showing areas of intestinal metaplasia.

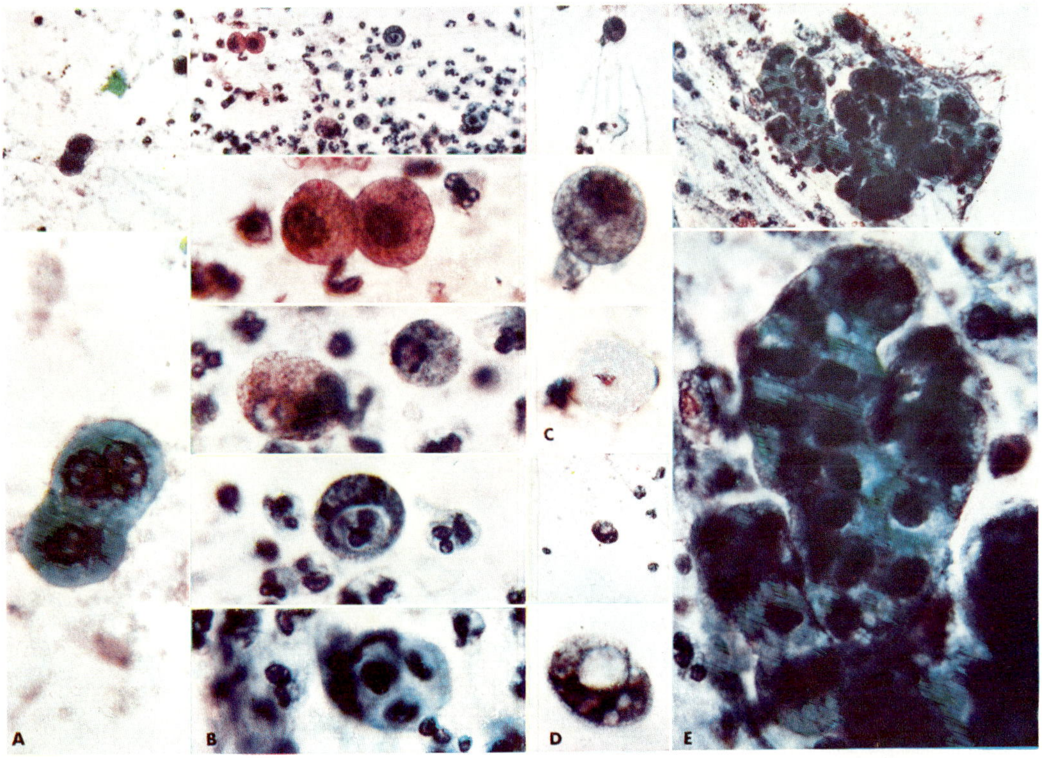

FIG. 112

Gastric Washings in Chronic Atrophic Gastritis with Intestinal Metaplasia

A. Two well preserved surface mucosal cells are shown at low magnification in the upper field and at high magnification in the lower field. Note the large overall size of these cells and of their nuclei; these cells nevertheless have normal nucleocytoplasmic ratios. Although there is slight variation in nuclear size the two nuclei of the upper cell and the single nucleus of the lower cell are uniform in shape and chromatin pattern. (×200.) (×800.)

B. The top field is a low-power view of the typical picture of gastric washings from a case of chronic atrophic gastritis with intestinal metaplasia. Many polymorphonuclear leucocytes and a few plasma cells are present in the background. The larger epithelial cells arise from the surface of the gastric mucosa and from areas of intestinal metaplasia.
The four fields below show selected cells at high-power magnification. The upper field shows two cuboidal surface mucosal cells, while the three lower fields all show cells with signet ring forms from areas of intestinal metaplasia. These cells have eccentric nuclei and granular vacuolated cytoplasm, in which polymorphonuclear leucocytes are apparent in the two lower fields. Note that, in addition to the many polymorphonuclear leucocytes shown in these high-power fields, a plasma cell is present to the right of the upper of the three signet ring cells. (× 200.) (× 800.)

C. The top field shows, at low magnification, a surface mucosal cell and a starch granule. Below this at high-power magnification, the large size and cuboidal shape of the surface mucosal cell are apparent, while the characteristic appearance of a starch granule is illustrated in the lowest field. (×200.) (×800.)

D. Another signet ring cell from an area of intestinal metaplasia is shown at low magnification (upper field) and high magnification (lower field). Note the striking cytoplasmic vacuolation. (×200.) (×800.)

E. The upper field, illustrated at low-power magnification, shows a group of mucosal cells from a case of chronic atrophic gastritis. The large size of the cells and their dark nuclei are readily apparent. In the left half of the group both the regular arrangement of the cells and their normal nucleocytoplasmic ratios are easily recognised. This area is shown below at high-power magnification, when the regular arrangement of the evenly spaced cells, their uniform oval nuclei and the adequate amount of cytoplasm surrounding individual nuclei help to differentiate these benign cells from the malignant cells arising from an adenocarcinoma. (See Figs. 118 and 119.) (× 200.) (× 800.)

GASTRIC POLYPI

Unless a fragment of a polyp has become detached it is not possible to make a cytological diagnosis but when such a fragment is seen the uniformity of the columnar cells and their regular arrangement indicate their benign nature.

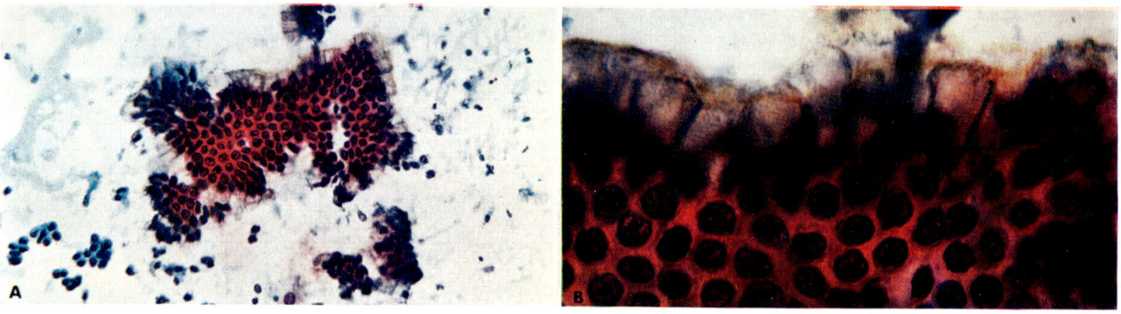

FIG. 113

A Polypoid Group of Benign Gastric Columnar Cells in a Gastric Wash

A. This polypoid group of benign gastric columnar cells shows a 'honeycomb' arrangement in the centre with the typical 'palisade' or 'picket-fence' arrangement on the surface. The eosinophilic staining of the cytoplasm is unusual and is probably due to the thickness of the cellular material. The smaller groups to the left of the field show the usual basophilic staining reaction. (×200.)

B. At high-power magnification the deep cells have uniform nuclei which are regularly arranged and separated by clear cytoplasm. The surface cells show the typical columnar appearance. (×800.)

THE COLON AND RECTUM

INFLAMMATORY CHANGES

Inflammatory conditions of the colon or rectum cause an increase in the number of inflammatory cells, especially polymorphonuclear leucocytes, and in the amount of cellular debris present. These features, combined with the small amount of epithelial cellular material present, make the specimens unsuitable for diagnosis.

POLYPI

It is not possible to diagnose the presence of polypi on cytological evidence alone. When the presence of a polyp has been demonstrated by other methods cytological evidence is not an adequate or safe substitute for investigation by biopsy.

CHAPTER 11

MEGALOBLASTIC ANAEMIA DUE TO A DEFICIENCY OF VITAMIN B12 AND/OR FOLIC ACID

CHANGES in the erythrocyte and myeloid series in megaloblastic anaemia due to a deficiency of vitamin B12 and/or folic acid have been recognised for a long time. Recently, however, it has become apparent that changes can also be identified in epithelial cells. Moreover, certain of these changes can cause errors in the diagnosis of malignancy.

In megaloblastic anaemia straightforward cellular and nuclear enlargement, with the maintenance of a normal nucleocytoplasmic ratio, are seen both in squamous and columnar cells. Alternatively the enlarged cells may show multinucleation; this is seen especially in squamous cells. When these changes occur alone, without abnormalities of either nuclear chromatin pattern or nucleoli, they present little diagnostic difficulty. This type of cellular and nuclear alteration is seen in megaloblastic anaemia due to a deficiency of vitamin B12 and/or folic acid. This is not an uncommon finding in cervical smears taken from pregnant patients, many of whom have megaloblastic anaemia due to a deficiency of folic acid.

In severe megaloblastic anaemia due to a deficiency of vitamin B12, epithelial cells may display other abnormal features, which can give rise to considerable diagnostic difficulty. These changes are especially significant when they occur in the columnar cells of the gastric mucosa, because the cellular and nuclear features may closely mimic the appearance of the malignant cells from a carcinoma of stomach. In this site, too, there are major practical problems involved in confirming or refuting the presence of an early gastric carcinoma.

There has been considerable discussion on the specificity of these changes in gastric mucosal cells; opinion is divided between those who relate the changes directly to the deficiency of vitamin B12 and those who consider the cellular changes are satisfactorily explained by the chronic atrophic gastritis, which is present in this condition. In practice problems arise from the possible relationship between gastric carcinoma and chronic atrophic gastritis and/or megaloblastic anaemia due to vitamin B12 deficiency. As a result, especial care is required in the interpretation of cells in gastric washings from patients with untreated vitamin B12 deficiency anaemia, to minimise the risk of a 'false-positive' diagnosis of malignancy or the even graver risk of failing to recognise cells arising from a gastric carcinoma, simply because their abnormal appearance is attributed to the fact that the patient is known to suffer from this type of anaemia. The latter hazard can be decreased by prolonged and detailed follow-up, after treatment, of patients with this type of anaemia, particularly the group of patients in whom gastric epithelial abnormalities have been observed. It should be realised, however, that some abnormal features may persist in epithelial cells, especially columnar cells, for a considerable period after successful treatment of the anaemia.

Gastric washings from cases of megaloblastic anaemia due to vitamin B12 deficiency show good cell preservation due to the associated achlorhydria. The columnar cells show nuclear enlargement with considerable variation in nuclear size and shape, while enlarged nucleoli may be strikingly prominent. Occasionally nuclei, which show marked variation in chromatin pattern, may be seen; it is cellular material of this type which presents diagnostic problems. In addition squamous cells of respiratory tract or oesophageal origin, which are present in the smear, may show cellular and nuclear enlargement or multinucleation.

Although abnormal features associated with severe vitamin B12 deficiency megaloblastic anaemia present the greatest practical difficulty in gastric washings, similar appearances, with cellular and nuclear enlargement, associated with multinucleation and abnormalities of nuclear chromatin pattern and nucleoli, have occasionally been observed in squamous cells in sputum and in cervical smears. However, the possibility that the presence of a megaloblastic anaemia, due to a deficiency of vitamin B12 and/or folic acid, is the explanation of abnormal cells and nuclei, especially in cervical and vaginal smears, in which they are only rarely attributable to this cause, should not be accepted until every effort has been made to exclude a malignant lesion.

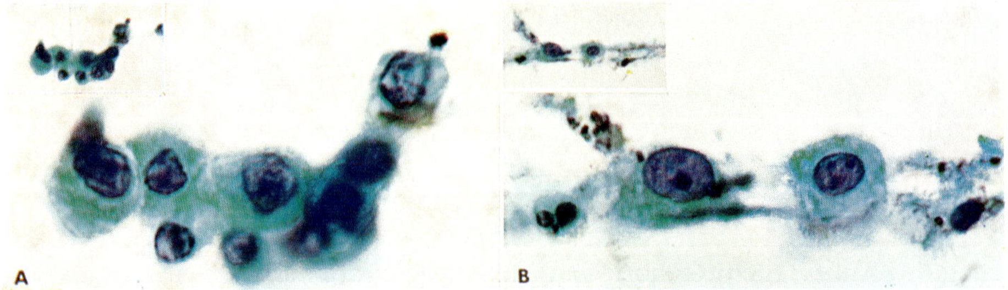

FIG. 114

Abnormal Columnar Cells in Gastric Washings from a Patient with an Untreated Vitamin B12 Deficiency Megaloblastic Anaemia

A. This group of cells shows considerable variation in nuclear size and shape with abnormal clumping and marginal condensation of chromatin. (Inset ×200.) (×800.)

B. In these cells the most striking feature is the prominence of the nucleoli, especially the large eosinophilic nucleolus seen in the cell on the left and the multiple nucleoli in the cell on the right. There is also considerable enlargement of the nucleus in the cell on the left. (Inset ×200.) (×800.)

At the time at which this specimen was reported the diagnosis of a vitamin B12 deficiency anaemia had not been made in this patient, who complained of nausea, vomiting and weight loss. The appearance of the cells illustrated above and other similar groups was therefore thought to be due to malignant change. Gastroscopy showed only a chronic atrophic gastritis and on further investigation the diagnosis of vitamin B12 deficiency megaloblastic anaemia was made. In view of this finding and negative radiological and gastroscopy investigation it was decided to treat the anaemia and repeat the gastric washings in six weeks time. After six weeks, when the anaemia had responded to treatment, gastric washing was again performed; atypical cell groups obtained at that time are shown in Fig. 115.

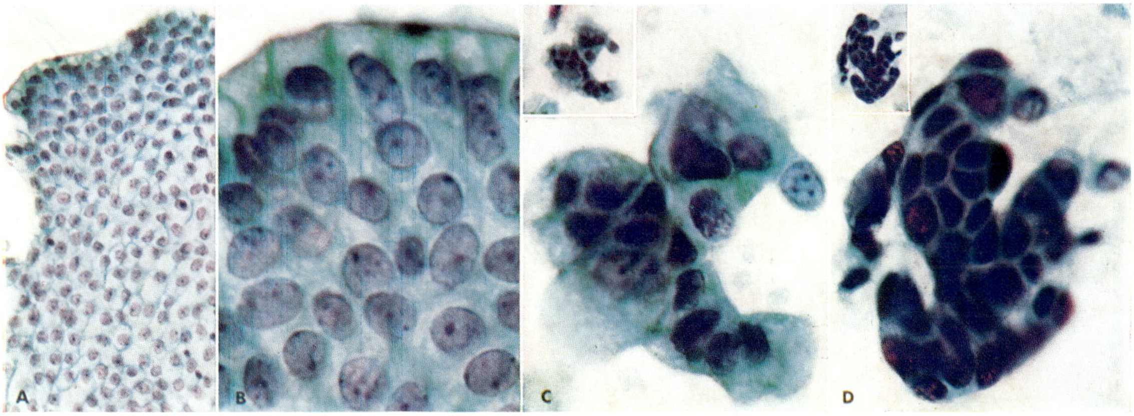

FIG. 115

Atypical and Persistently Abnormal Columnar Cells in Gastric Washings after Treatment of a Case of Vitamin B12 Deficiency Megaloblastic Anaemia

A. Most of the groups of columnar cells now show this appearance. The individual cells and nuclei, although enlarged, are regularly arranged. (×200.)

B. High-power examination shows that the large nuclei are now fairly uniform in size, shape and chromatin pattern, but eosinophilic nucleoli are still a prominent feature. (×800.)

C. and D. Occasional groups of cells, however, still show strikingly abnormal features. In both these groups the nuclei are crowded together and show moulding of adjacent nuclei. In plate C the nuclei show marked variation in size, shape and chromatin pattern; these features are also seen in plate D although the chromatin pattern is not so readily apparent. (Insets ×200.) (×800.)

The changes illustrated here are frequently seen in vitamin B12 deficiency anaemia. They should not present diagnostic difficulty in view of the uniformity with which the changes occur. The abnormal features seen in Fig. 114, however, may cause considerable diagnostic difficulty from their close similarity to malignant cells. A further problem arises when these abnormal cells persist, as in this case, after the patient has responded to treatment. Repeat gastroscopy again revealed chronic atrophic gastritis. Six months later gastric washings showed only uniformly enlarged columnar cells, similar to those shown in A and B above. The patient has now been carefully watched for two years and remains well and free from gastric symptoms.

CHAPTER 12

CANCER OF THE GASTRO-INTESTINAL TRACT

In most sites in the gastro-intestinal tract, technical difficulties in the collection of suitable diagnostic specimens impose limitations on the widespread application of cytological methods to the early diagnosis of cancer. Nevertheless, good results have been reported from careful techniques, especially in the diagnosis of cancer of the stomach.

THE ORAL CAVITY

Most lesions of the mouth are so readily accessible for direct biopsy that it might appear that cytological investigation would not make any significant contribution, whereas in fact, cytology can be usefully employed in two different ways. First, direct cytological smears can form part of the follow-up of patients with previously treated malignant lesions and of patients with leukoplakia. Second, good results have been reported from surveys involving the examination of large series of specimens obtained by mouth-washing techniques from patients with a wide range of oral lesions. By this method a significant number of unsuspected early malignant lesions have been diagnosed.

The commonest type of malignant tumour in this area is squamous cell carcinoma. The characteristic features of malignant cells from a squamous cell carcinoma are essentially similar to those previously described (pp. 34-43).

THE OESOPHAGUS

The role of cytology in the diagnosis of malignant lesions of the oesophagus is a relatively small one, as direct biopsy of the suspicious area is the procedure of choice. Cytology may prove useful, however, in the investigation of symptomatic patients in whom no localising lesion is seen.

Specimens for the diagnosis of carcinoma are collected by oesophageal washings. It should also be remembered that many of the cells in gastric washings originate in the oesophagus, so the oesophagus should not be overlooked as the possible source of malignant cells in a gastric wash. The commonest type of oesophageal carcinoma is the squamous cell carcinoma. The malignant cells from this tumour show the features previously described (pp. 34-43). In oesophageal washings malignant cells are only present in small numbers; this makes careful processing of specimens essential to achieve satisfactory results. Adenocarcinoma occurring mainly at the lower end of the oesophagus is most likely to be diagnosed by the malignant cells found in gastric washings. The general features of these cells are similar to those of cells from an adenocarcinoma of stomach (q.v.).

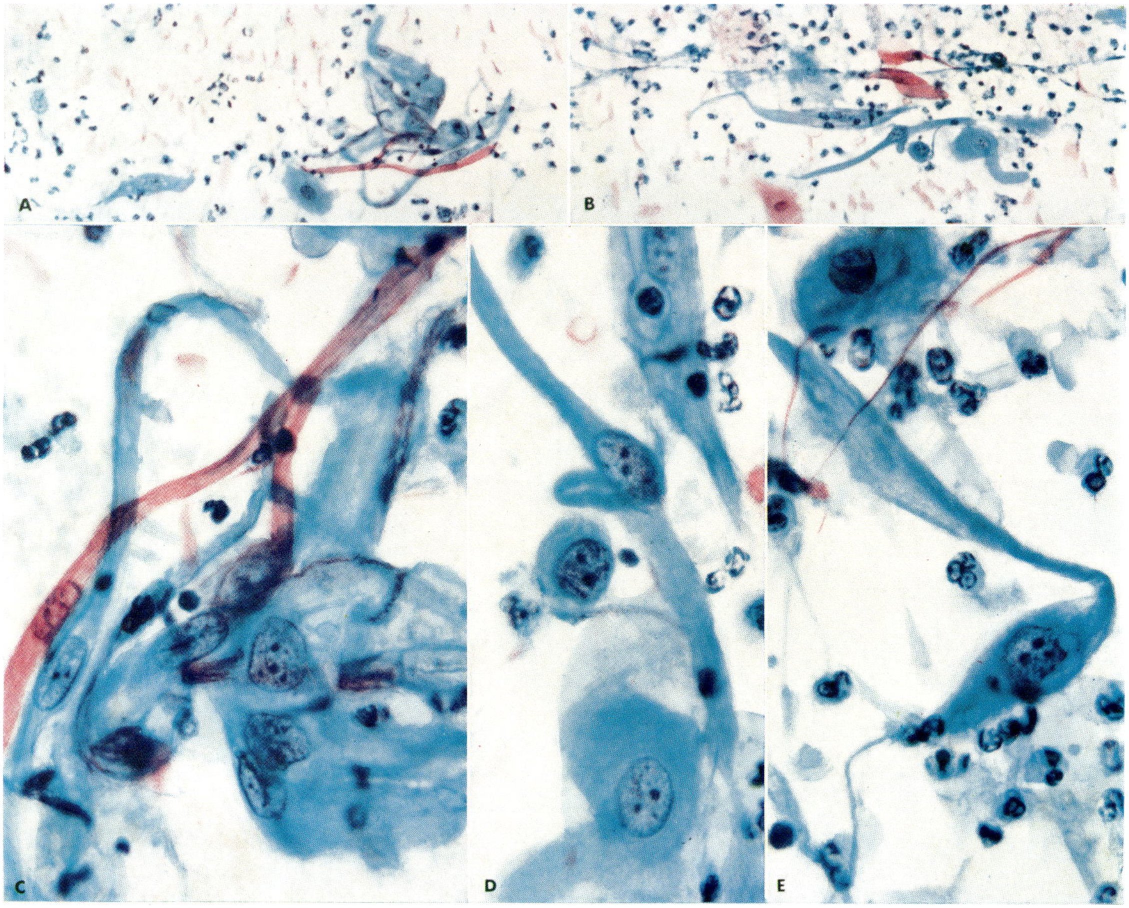

FIG. 116

Differentiated Malignant Cells of Squamous Type from a Squamous Cell Carcinoma of Tongue

A. and B. These low-power fields show typical differentiated malignant cells of squamous type, with characteristic fibre and tadpole shapes, together with an occasional round third-type malignant cell. (×200.)

C. At high-power magnification it is apparent that differentiated maiignant cells of fibre type predominate, but a large multinucleated malignant cell is also seen. Their nuclei show only slight variation in size, but those seen in the giant cell are irregular in shape. The chromatin patterns are abnormal with marginal condensation and uneven clumping of chromatin. (×800.)

D. A fibre cell and two third-type malignant cells are seen. In all the cells the nuclei are large and show abnormal chromatin patterns, with irregular clumps varying in size and shape. (×800.)

E. This malignant squamous cell shows the characteristic shape of a tadpole cell. The irregular nucleus has a strikingly abnormal chromatin pattern, in which large clumps of chromatin are easily seen against an almost clear background. (×800.)

These smears were obtained by direct scraping of an ulcer of the tongue. Subsequent biopsy and histological examination confirmed the presence of a squamous cell carcinoma.

THE STOMACH

The value of cytology in the diagnosis of early carcinoma of the stomach is restricted by technical problems. First, some degree of patient discomfort is unavoidable during the gastric intubation necessary for the collection of the specimens. Second, there may be only minimal exfoliation of epithelial cells, whether benign or malignant; this can lead to negative reports even in the presence of a carcinoma. However, this particular difficulty can be partly surmounted either by using an abrasive collection technique or, more simply and satisfactorily, by the use of a large volume gastric lavage method. Third, the cellular material present in a specimen readily undergoes cytolysis if there is any delay in processing the specimen; these cytolysed cells may be unsuitable for diagnosis. This hazard can be minimised by rapid specimen processing.

For these reasons it is not at present practicable to consider the routine use of gastric cytology as a screening procedure in well patients. However, the application of the method to symptomatic or high risk groups of patients is another matter. As an early diagnosis of gastric carcinoma provides the best opportunity for cure, it is essential to employ every diagnostic method available, especially as different methods show considerable variation in the types and sites of tumour which they can most successfully demonstrate. When cytological investigation is combined with other methods it is valuable in two different ways. Often it may indicate or confirm the malignant nature of a lesion, which has been localised by radiotherapy or visualised directly by gastroscopy or indirectly by the camera. Less commonly it may demonstrate the presence of a malignant lesion, which has not been identified at all by other methods, and thus lead to further attempts to find the lesion. It is, of course, in this second group that cytology makes its most important contribution, as this category includes very early malignant lesions, as well as those established malignant lesions, which have not been recognised by other methods because of their particular site in the stomach.

Gastric lavage specimens from cases of adenocarcinoma and undifferentiated carcinoma of stomach or distal portion of oesophagus show considerable variation both in the overall cellular pattern of the smears and in the features of the malignant cells. The background of the smears in established lesions consists mainly of erythrocytes and polymorphonuclear leucocytes with a small number of lymphocytes and plasma cells. As a result of the achlorhydria, often found in association with gastric carcinoma, the polymorphonuclear leucocytes are usually well preserved. In early lesions this background of erythrocytes and inflammatory cells may not be seen. The malignant cells show considerable diversity, from case to case, both in degree of differentiation and in cell size. In any individual case, however, the features of the malignant cells are fairly consistent.

Malignant cells appear in groups and singly, often among inflammatory cells. Frequently the malignant cells are large and form readily recognisable irregularly arranged groups, but single small undifferentiated or differentiated malignant cells occur and may only be identified after careful scrutiny. Differentiated malignant cells, from an adenocracinoma, with well preserved cytoplasm and eccentric nuclei, occasionally contain mucus. More often only the eccentric position of the nucleus in the differentiated malignant cell or the pseudo-acinar arrangement of a cell group suggest an origin from an adenocracinoma. Frequently groups of cells are seen in which individual cell borders are not identifiable. Undifferentiated malignant cells, in which the nucleus is incompletely surrounded by cytoplasm, occur both in association with differentiated malignant cells and also as the only type of malignant cell present in a specimen. The nuclei of differentiated and undifferentiated malignant cells show the characteristic features: variation in size, with a preponderance of nuclear enlargement producing abnormal nucleocytoplasmic ratios, variation in shape, although this is less striking than the size variation, abnormal chromatin patterns with irregular angular clumping and marginal condensation of chromatin, and enlarged, prominent, irregular and often multiple nucleoli.

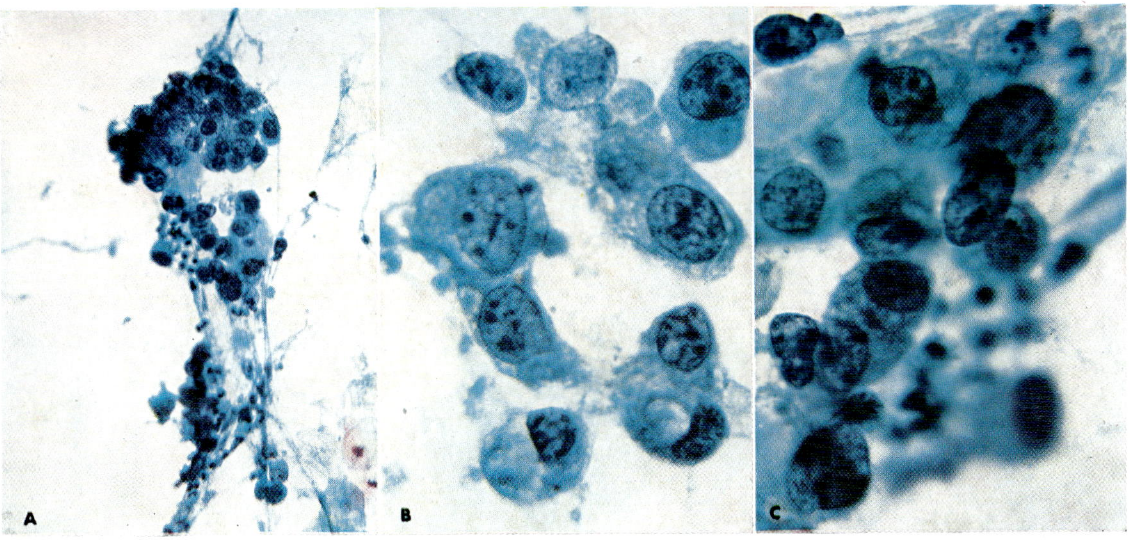

FIG. 117

Gastric Washings Showing Differentiated and Undifferentiated Malignant Cells from an Adenocarcinoma of Stomach

A. At the top of this low-power field a sheet of large malignant cells is shown; their hyperchromatic nuclei are seen against a poorly defined cytoplasmic background in which individual cell borders are not recognisable. The smaller groups seen lower in the field show large malignant cells with a background of inflammatory cells. In the small group of cells in the right lower quadrant individual malignant cells of the type arising from an adenocarcinoma display the characteristic eccentric nuclei, with clumping and marginal condensation of chromatin. (×200.)

B. At higher magnification the characteristic features of differentiated malignant cells arising from an adenocarcinoma are identifiable in most of these cells. Their large eccentric nuclei display marked variation in size with some irregularity of shape. All the nuclei show abnormal chromatin patterns with clumping and uneven marginal condensation of chromatin; in some nuclei large nucleoli are recognisable. Note the variation in differentiation in this field; two undifferentiated malignant cells are present at the upper border of the group of differentiated malignant cells, while the differentiated cell at the right lower border shows a cytoplasmic vacuole, suggestive of the presence of mucus. (× 800.)

C. The cytoplasm of these cells appears as an indefinite background and individual cell borders are only occasionally identifiable. The nuclei, however, display the variation in size and shape and the chromatin abnormalities characteristic of malignant cells. (×800.)

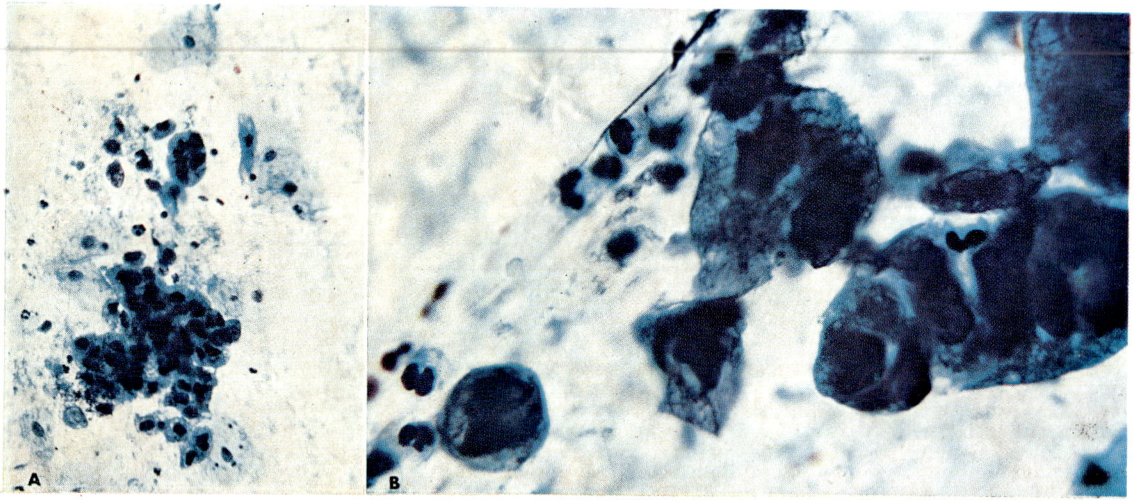

FIG. 118

Gastric Washings Showing Differentiated Malignant Cells from an Adenocarcinoma of Stomach

A. In this low-power field groups of large malignant cells with densely hyperchromatic nuclei, showing marked variation in size and shape, are seen. A few single undifferentiated malignant cells are also present scattered in the background of the cell groups. (×200.)

B. In this high-power field it is apparent that individual cell borders are not identifiable in the groups of cells on the right. Their finely vacuolated cytoplasm forms a diffuse background for the large nuclei which show striking irregularity of shape. Note also the arrangement of the nuclei, which are crowded together and show moulding of adjacent nuclear borders. The nuclei of the two single differentiated malignant cells, seen in the left lower quadrant of the field, are eccentric in position; this is compatible with an origin from an adenocarcinoma. (×800.)

The differential diagnosis of these malignant cells involves distinguishing them from benign gastric columnar cells and from inflammatory cells, especially histiocytes of respiratory tract origin and plasma cells. Benign gastric columnar cells in groups are characterised by the regularity of the group pattern, which differs markedly from the irregular, often overlapping or moulded arrangement of the malignant cell group. Single benign columnar cells may show slight variation in nuclear size, but are fairly uniform in shape; their nuclei display normal chromatin patterns, and nucleoli, if seen, occur singly and are small and round. As a result of cytolysis denuded benign nuclei may appear, but these again retain normal chromatin patterns. Occasionally histiocytes of respiratory tract origin may cause difficulty because of their eccentric nuclei. Usually, however, they can be readily identified by their finely vacuolated cytoplasm, reniform nuclei and absence of chromatin pattern abnormalities. Often the presence of carbon particles in their cytoplasm confirms their identity. Plasma cells, which may be found in significant numbers in cases of chronic atrophic gastritis, are only likely to present a problem if they are unusually large or if their presence has not been considered in the differential diagnosis.

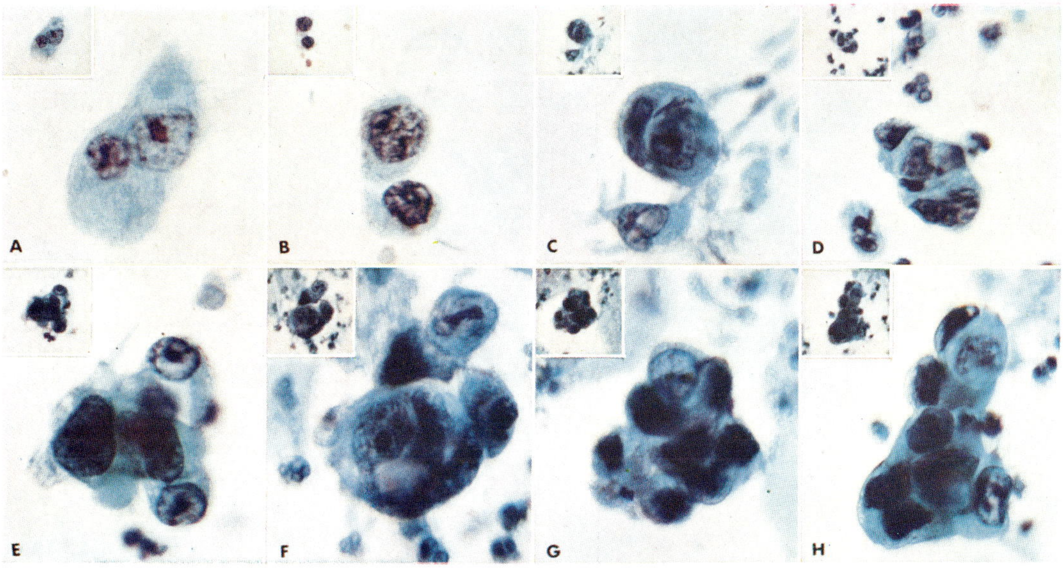

FIG. 119

Differentiated and Undifferentiated Malignant Cells seen Singly and in Groups in Gastric Washings from Four Cases of Adenocarcinoma of Stomach

A. This large multinucleated malignant cell shows a disparity in the size, shape and chromatin patterns of the two nuclei. In the larger nucleus an irregular eosinophilic nucleolus is a striking feature. (Inset ×200.) (×800.)

B. Although these two undifferentiated malignant cells show almost complete loss of cytoplasm, their nuclear pleomorphism and abnormally clumped chromatin indicate their malignant character. Note the resemblance to the nuclei of the differentiated malignant cell from the same patient shown in field A. (Inset ×200.) (×800.)

C. In this small group of malignant cells there is marked nuclear pleomorphism and hyperchromasia. The upper pair of cells shows the appearance of phagocytosis of one malignant cell by another. (Inset ×200.) (×800.)

D. Despite their small size the pleomorphism and abnormal chromatin patterns of these almost denuded nuclei are characteristic of undifferentiated malignant cells. Note also the well preserved polymorphonuclear leucocytes in the background. (Inset ×200.) (×800.)

E. F. G. and H. These larger clumps show the typical grouping of malignant cells. In each case the cells are irregularly arranged with overlapping of adjacent cells and moulding of adjacent nuclei. The edges of the groups are irregular and in many cases the actual border of the group is formed by nuclei. The individual nuclei show the size variation, pleomorphism, abnormal chromatin clumping, marginal chromatin condensation and prominent nucleoli characteristic of malignant cells. (Inset ×200.) (×800.)

These small groups and single malignant cells of gastric origin should be compared with the groups and single benign cells of gastric origin shown in Figs. 108 and 109, with the histiocytes of respiratory tract origin shown in Figs. 81 and 102 and with the plasma cells shown in Fig. 14.

THE COLON AND RECTUM

Technical difficulty in obtaining suitable specimens for diagnosis from these areas is considerable, as contamination by cellular debris often persists in spite of adequate patient preparation, while there may also be difficulty in obtaining exfoliated cells from a lesion. For these reasons biopsy is preferable as a method of investigation of lesions which are suspicious on clinical or radiological evidence. However, when malignant cells are present in colonic washings their appearance is essentially similar to the appearance of malignant cells from an adenocarcinoma of stomach.

SECTION IV

URINARY TRACT

CHAPTER 13

NORMAL CELLULAR PATTERNS

THE cellular content of a specimen of urine varies according to the site from which it is collected and the method of collection. In a voided specimen of urine only a few cells are present. In the male these are mainly transitional epithelial cells from the bladder and proximal urethra with an occasional squamous cell of distal urethral origin. In the female, squamous cells of vaginal origin may outnumber the transitional cells. When a specimen is collected by catheterisation transitional cells are more numerous and may appear in groups, probably detached from the bladder by the catheter tip. Ureteric catheter specimens contain multinucleated giant cells from the renal pelvis in addition to transitional cells.

TRANSITIONAL EPITHELIAL CELLS

Transitional epithelial cells show considerable variation in size, in shape and in nuclear characteristics. They are often similar in size and shape to parabasal squamous cells, but can be cuboidal or columnar. A basal tail of cytoplasm is occasionally seen on these columnar cells. Variation in the size of the cells and their nuclei is often present, but normal nucleocytoplasmic ratios are retained. The appearance of the nuclei shows greater diversity than is seen in parabasal cells. The vesicular nucleus is usually oval or round, but may be irregular in shape with a sharply defined nuclear border. The chromatin pattern is more coarsely granular than that of a squamous cell nucleus and clumping of chromatin may be seen; the chromatin masses, however, are of small size and rounded shape. Multinucleation is often seen and, when it occurs, the nuclei in a cell may vary in size. Unless the specimen is fresh, degenerative changes are seen in which the cytoplasm is frequently lost and the nuclei may show karyorrhexis or karyolysis. Degenerate nuclei, denuded of cytoplasm, may require careful scrutiny for identification, in order to distinguish them from undifferentiated malignant cells.

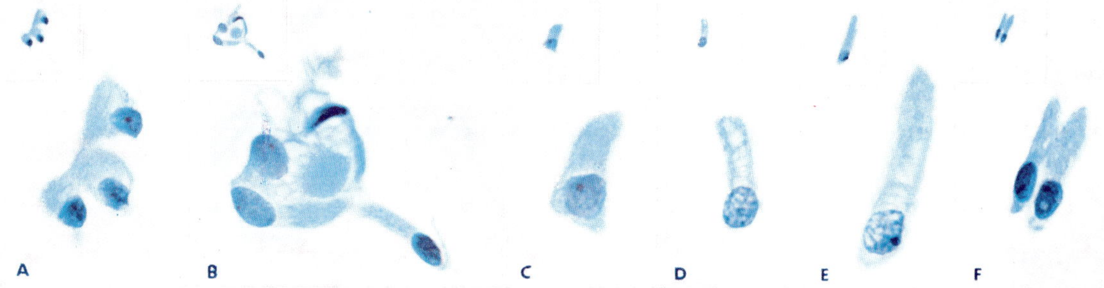

FIG. 120

Benign Transitional Cells of Cuboidal and Columnar Type from a Voided Specimen of Urine

Transitional cells are usually seen as small flat rounded or quadrangular cells, similar in shape to parabasal cells (see Fig. 122). Frequently, however, cuboidal or columnar forms appear. Cuboidal shapes are seen in A, B and C, and columnar forms in B, D, E and F. Note that the columnar form on the right in B shows a cytoplasmic 'tail' process, suggesting a point of attachment to the basement membrane. The nuclei of these benign transitional cells show some variation in size, shape and chromatin pattern. Quite coarsely granular chromatin patterns are seen in the cells illustrated in D and E. Single small eosinophilic nucleoli are identifiable in several of the cells in A, B, C, E and F. (Insets ×200.) (×800.)

MULTINUCLEATED GIANT CELLS

Multinucleated giant cells from the renal pelvis and ureter are most commonly seen in ureteric catheter specimens, but may also appear in other specimens of urine. These cells vary considerably in size, but may be strikingly large and irregular in shape; they often display an amoeboid outline. The cytoplasm shows both coarse and fine vacuolation. The number of nuclei ranges from two to thirty or forty; in a single cell there may be striking differences in the size of these nuclei. The chromatin pattern is mainly finely granular, but rounded clumps of chromatin are seen.

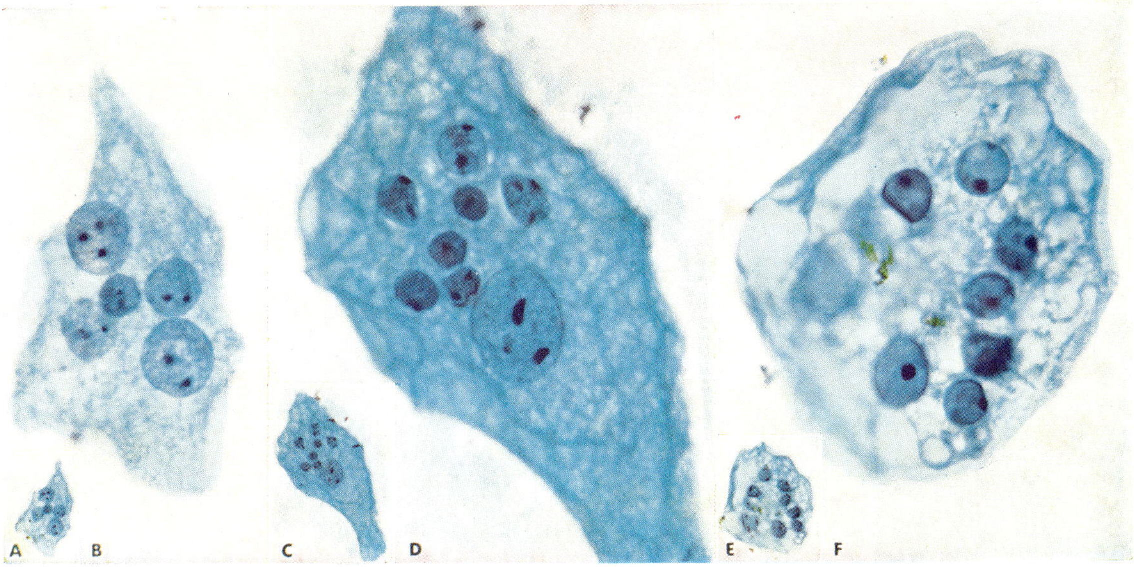

FIG. 121

Benign Multinucleated Giant Cells from the Renal Pelvis or Ureter in a Specimen of Urine Collected by Ureteric Catheterisation

These giant cells show features frequently seen in specimens of urine collected by ureteric catheterisation. The cells have a large amount of foamy, coarsely and finely vacuolated cytoplasm and multiple nuclei. There is a striking degree of variation in nuclear size, especially marked in the cell illustrated in D. However, the nuclei are fairly uniformly round or oval in shape. Rounded clumps of chromatin are seen, but the background chromatin pattern is evenly granular. Some of the chromatin clumps may be associated with and partly obscure nucleoli, as eosinophilic nucleoli cannot be identified. (A. C. E. ×200.) (B. D. F. ×800.)

SQUAMOUS EPITHELIAL CELLS

The squamous cells of distal urethral origin show no significant difference from the benign squamous cells of genital tract origin (q.v.). In a female patient it is not possible to distinguish squamous epithelial cells of urethral origin from those of genital tract origin.

CHAPTER 14

NON-MALIGNANT DISORDERS

As a result of inflammation changes occur in the transitional cells, and non-epithelial cells also appear in the urine in significant numbers. The exfoliation of transitional cells increases and these cells are present in larger numbers in the urine; sheets or groups of these cells may be seen even in the voided specimen. Both single and groups of transitional cells show degenerative changes, with loss of cytoplasm or cytoplasmic vacuolation and nuclear pyknosis or karyorrhexis. Occasionally nuclei become enlarged or appear hyperchromatic, but they retain a normal chromatin pattern.

In acute inflammation the non-epithelial cells are predominantly erythrocytes and polymorphonuclear leucocytes. These may be present in such large numbers that they cause diagnostic difficulty by obscuring the transitional cells. In chronic inflammation small histiocytes may be seen, with occasional plasma cells, as well as polymorphonuclear leucocytes and some erythrocytes.

'BENIGN' TUMOURS

The papilloma of the bladder presents considerable diagnostic difficulty. It is rarely possible to make this diagnosis on cytological evidence, but it should be considered if a voided specimen of urine contains erythrocytes and groups or sheets of transitional cells. Occasionally a papillary arrangement may be seen in a group of these cells. Although the individual cells appear to be completely benign the patient should be fully investigated, including cystoscopy and biopsy. Even when the histological diagnosis appears to confirm the benign nature of the lesion, careful follow-up is advisable as many papillomas recur. Cytology may play a useful role in the prolonged follow-up which is essential for these patients.

It must, however, be remembered that transitional cell groups may appear for other reasons. When considerable numbers of transitional cells are present in a specimen, for example in the presence of inflammation, cell grouping may be produced, as an artefact, by centrifugation. If a specimen is collected by catheterisation small groups of transitional cells may be detached by the tip of the catheter and subsequently appear in the specimen of urine.

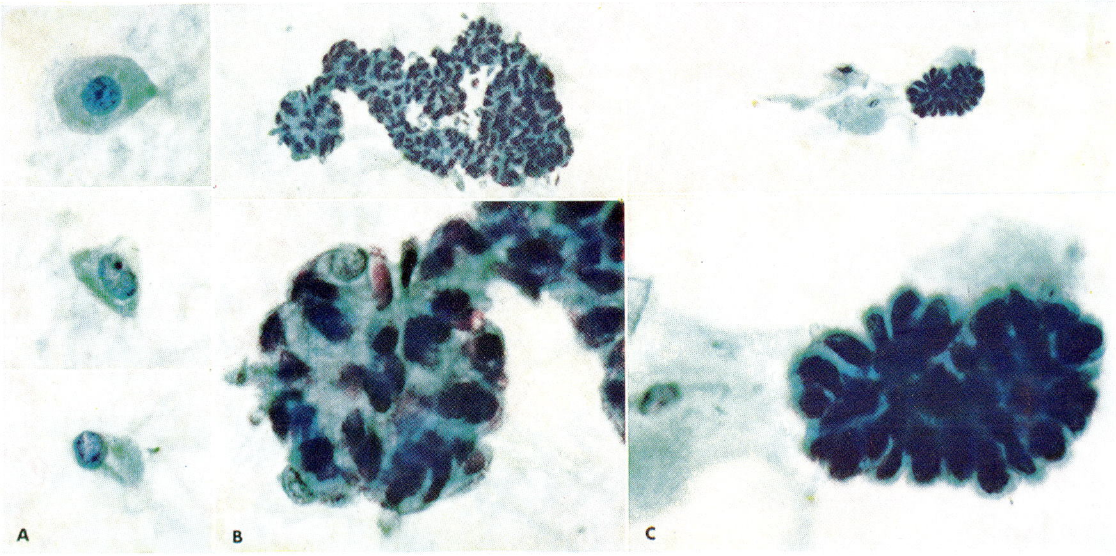

FIG. 122

Benign Transitional Epithelial Cells in a Specimen of Urine Collected by Catheterisation

A. In normal urine the transitional cells, which frequently resemble parabasal cells, usually appear singly. A typical example of such a cell is illustrated in the upper field. These benign single cells may have a cuboidal form, as seen in the centre field, or even a columnar form, as shown in the lower field. These cells show slight variation in nuclear size, while retaining their normal nucleocytoplasmic ratios. Note the coarsely granular nuclear chromatin pattern of the uppermost cell and the single eosinophilic nucleolus present in each of the three cells. (×800.)

B. This large group of irregularly arranged benign transitional cells may have been produced as an artefact by centrifugation or detached by the catheter tip. If large groups of cells are seen in a voided specimen further investigation is advisable, even though the individual cells appear to be within normal limits. (Inset ×200.) (×800.)

C. The arrangement of this large group of transitional cells suggests a papillary structure. The individual cells show no significant abnormality and they are regularly arranged in a radial pattern. This appearance is compatible with the tip of a detached papillary lesion and would require further investigation. (Inset ×200.) (×800.)

HORMONAL ASSESSMENT

The effect of oestrogens, both in female patients and in male patients receiving oestrogen therapy for carcinoma of the prostate, can be assessed by the examination of the epithelial cells in urine, since the epithelium overlying the trigone of the bladder appears to respond to oestrogens. The main advantages claimed for this method of assessment of oestrogen activity in female patients, compared with vaginal specimens, are the absence of contamination by cells of cervical origin and the avoidance of the inflammatory changes which often confuse the pattern in vaginal smears.

CHAPTER 15

CANCER OF THE URINARY TRACT

The most important application of cytological methods to the urinary tract is in the diagnosis of cancer of the bladder, but cytology can also be of value in the diagnosis of malignant lesions of the ureter and renal pelvis.

THE BLADDER

Cytological methods of diagnosis of carcinoma of the bladder can be applied in the investigation of patients in three categories: the scrutiny of asymptomatic patients in high-risk groups, the investigation of symptomatic patients and the follow-up of patients with treated bladder carcinomata. In all these patients the simplicity of the investigation and the complete absence of patient discomfort make repeated checks possible; this is especially valuable in the high-risk and follow-up groups of patients, who must be investigated at regular intervals. Cytological methods may reveal the presence of malignant cells, not only before symptoms occur, but also prior to recognisable changes in the bladder on cystoscopy. In fact the localisation of the site of origin of the malignant cells can pose a considerable practical problem. In this context it should not be forgotten that the lesion may lie in a bladder diverticulum, or alternatively, that the primary lesion may lie outside the bladder, in cervix or body of uterus or in the prostate. In the case of extra-vesical lesions malignant cells may be found in the urine either because the lesion has spread to involve the bladder or, in female patients, because of contamination of a voided specimen.

Cytological examination of urine specimens has long been employed as a method of screening patients with a history of industrial exposure to 'carcinogens' in the 'aniline' dye industry. Recent awareness of a similar hazard to some workers in the rubber industry has extended the high-risk group of workers, who should be investigated in this way at six-monthly intervals. In these patients cytology should be directed to the early recognition of any abnormality of transitional cells and the selection of those patients, who should undergo full investigation in a urological department. It is, therefore, justifiable in this group to advise further investigation of patients when cellular abnormalities, which are not in themselves fully diagnostic of malignancy, are identified in the urine. Moreover negative results on cytological examination of the urine should not be accepted as excluding significant disease in patients in this group, who complain of relevant symptoms, such as haematuria. For these symptomatic patients cytological examination of the urine should only form part of a full urological investigation.

In known cases of carcinoma of the bladder cytology can be used to study the response of the tumour to radiotherapy or cytotoxic drugs, but its main importance is in the follow up of treated patients, whether they have been treated by radiotherapy or surgery. Cytological examination of the urine at each return visit is advisable in view of the risk of recurrence of even the most 'benign' bladder tumours.

For the diagnosis of early carcinoma of the bladder it is essential that the specimen for investigation should consist of the whole volume of an early morning specimen of urine, as this provides a maximum amount of exfoliated cells. In established carcinoma, however, where cytological evidence is merely confirming the diagnosis, the smaller volume mid-stream specimen of urine will usually contain sufficient cellular material for diagnosis. A catheter specimen of urine is not required for cytological investigation and should be avoided, especially as many of these patients require repeated investigations.

The malignant cells arising from bladder carcinoma show great variation in their degree of differentiation. Thus, at one extreme, the highly differentiated cells from some bladder tumours may show little cellular evidence of malignancy although the tumour ultimately behaves as a malignant neoplasm. For this reason a proportion of 'false' negative results seems unavoidable. This risk can only be diminished by full investigation of patients, in whom even apparently 'normal' transitional cells are found in large numbers or in groups in voided specimens of urine. More often, however, the malignant cells lie at the other extreme and are undifferentiated cells; their nuclei are not surrounded by cytoplasm, which is frequently seen only as a poorly defined background. The malignant nuclei may appear in groups, but are often seen singly; they vary in size and shape, but all display striking abnormalities of chromatin pattern with clumping and often marginal condensation of chromatin. These cells present little diagnostic difficulty, but there may be appreciable problems in the recognition of differentiated malignant cells, which have the characteristic malignant nuclei surrounded by cytoplasm. These differentiated malignant cells should only be diagnosed where the nuclei display the typical features, especially the abnormal chromatin patterns, of malignant cells, as benign transitional epithelial cells show considerable variation in nuclear size and cellular shape, although they do not have abnormal chromatin patterns.

The cells which must be considered in the differential diagnosis of malignant cells are benign transitional epithelial cells, especially those showing degenerative changes, and the large multinucleated cells arising from the renal pelvis. Well-preserved benign transitional cells are only likely to be confused with differentiated malignant cells if insufficient attention is paid to the nuclear abnormalities of the malignant cells and the normal chromatin patterns of the benign cells. Degenerate transitional epithelial cells may lose their cytoplasm and their nuclei may appear hyperchromatic, thus producing a superficial resemblance to undifferentiated malignant cells. Again, however, the nucleus of the degenerate cell, which may be enlarged but usually retains a normal shape and has an opaque structureless appearance, distinguishes this cell from the undifferentiated malignant cell with its characteristic abnormal nuclear chromatin pattern. Degenerative changes in malignant cells may occasionally cause a loss of nuclear detail and lead to a 'false' negative diagnosis. This is especially liable to occur with advanced malignant lesions, as a result of surface necrosis and infection. The large multinucleated cells, which arise from the renal pelvis, are occasionally seen in voided specimens of urine. Initially their appearance may be considered suspicious, as their nuclei show considerable variation in size and rounded clumps of chromatin are recognised. However, their vacuolated cytoplasm, multinucleation and absence of significant nuclear chromatin abnormalities serve to identify them.

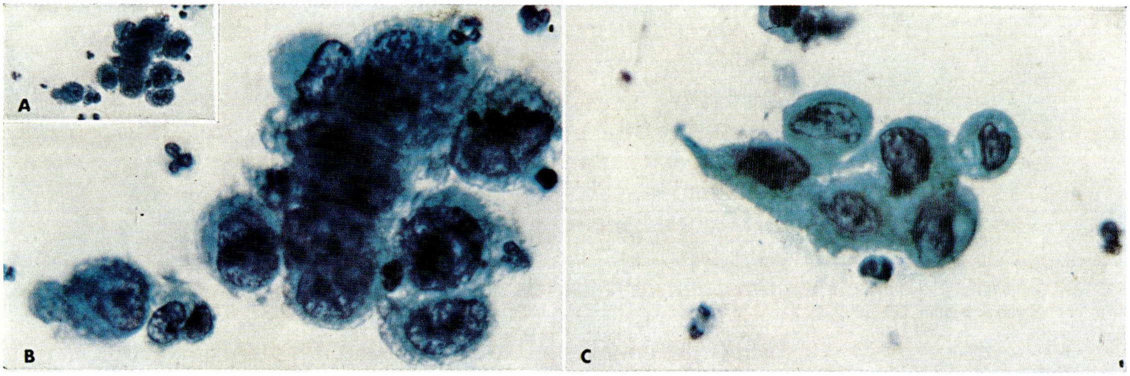

FIG. 123

Undifferentiated and Differentiated Malignant Cells in Urine from a Case of Transitional Cell Carcinoma of Bladder

A. This group consists mainly of undifferentiated malignant cells; their cytoplasm is present only as an ill-defined background for their hyperchromatic nuclei. (×200.)

B. At high-power magnification it is apparent that the group consists mainly of undifferentiated malignant cells, but single differentiated malignant cells are seen at the edges of the group. These malignant cells show the characteristic nuclear abnormalities; the nuclei vary considerably in size and are irregular in shape. The chromatin is coarsely clumped and, in some nuclei, it is irregularly concentrated at the nuclear margin. Cytoplasm is present only as a diffuse background to the nuclei in the centre, but surrounds the single nuclei on the periphery of the group. (×800.)

C. This group of differentiated malignant cells shows the typical irregularity of nuclear size, shape and chromatin pattern. However, although the nuclei are enlarged and the nucleocytoplasmic ratios are abnormal, the individual nuclei are surrounded by cytoplasm. Differentiated malignant cells of this type are less often seen than undifferentiated malignant cells; they should only be used as a basis for diagnosis if the typical nuclear features of malignant cells are present, as benign transitional cells often show considerable variation in nuclear pattern. (×800.)

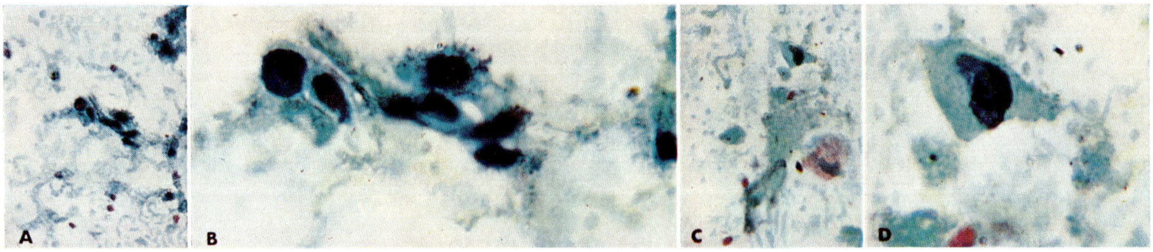

FIG. 124

Differentiated Malignant Cells in Urine from a Case of Transitional Cell Carcinoma of Bladder Showing 'Squamoid' Changes

A. A group of elongated cells with hyperchromatic nuclei is shown at low-power magnification. Note the erythrocytes in the background. (×200.)

B. At higher magnification this group of spindle-shaped malignant cells shows marked nuclear hyperchromasia with considerable variation in nuclear size and shape and also abnormal nucleocytoplasmic ratios. (×800.)

C. In the upper half of this low-power field a single differentiated malignant cell with a large hyperchromatic central nucleus is seen. (×200.)

D. At higher magnification it is apparent that this differentiated malignant cell has a large bizarre hyperchromatic nucleus surrounded by cytoplasm. Its overall appearance is 'squamoid', but it does not show keratinisation of cytoplasm. It resembles the 'third-type' differentiated malignant cell seen in squamous cell carcinoma of the cervix. (×800.)

THE KIDNEY AND URETER

Cytological investigation is far more valuable in the diagnosis of malignant tumours of the ureter and renal pelvis than in the diagnosis of malignant tumours of the renal parenchyma. Malignant cells exfoliated from tumours of the pelvis or ureter can be identified in voided urine and also in urine collected by ureteric catheterisation. In contrast, tumours arising from the renal parenchyma rarely exfoliate malignant cells into the urine until a late stage of the disease. Although urine collected by ureteric catheterisation may yield a greater concentration of malignant cells, which warrants the use of this technique, the findings on examination of ureteric catheterisation specimens should not be used alone to locate the site of a carcinoma. Occasionally the method can be of value in identifying the right or left kidney or ureter as the source of malignant cells. However, it is possible for cellular material, including malignant cells, of vesical origin to be carried into the ureter by the catheter tip and in this way an early bladder carcinoma may be misdiagnosed as a carcinoma of ureter or renal pelvis.

The features of the malignant cells arising from a transitional cell carcinoma of ureter or renal pelvis are essentially similar to those of malignant cells from a transitional cell carcinoma of bladder (q.v.). In ureteric catheterisation specimens especial care is necessary to distinguish the giant multinucleated cells, of renal pelvis origin, from malignant cells. However, their greatest resemblance is to malignant cells from a renal carcinoma of clear-cell type (see Fig. 142) rather than to the malignant cells from a transitional cell carcinoma. As the clear-cell carcinoma of renal parenchymal origin rarely exfoliates malignant cells in the urine until a late stage in the disease, when haematuria is also likely to be present, it is highly probable that multinucleated giant cells in asymptomatic patients and patients free from haematuria will prove to be benign. It is extremely inadvisable to make a diagnosis of malignancy based on cells of this type unless the nuclei show the characteristic chromatin abnormalities of malignant cells.

THE PROSTATE

The routine use of cytological methods for the detection of carcinoma of the prostate in asymptomatic patients has proved disappointing, although excellent results have been reported in studies of symptomatic patients. The difficulties encountered fall into two main groups. The first difficulty arises in specimen collection; urine does not normally contain a significant number of cells of prostatic origin. It is, therefore, necessary to collect a specimen after prostatic massage; this raises the problem of a possible risk of dissemination of the disease. The second problem is a pathological one; frequently the pathologist bases the diagnosis of carcinoma in a prostate biopsy on histological rather than cytological evidence. In this type of case the individual cells may show few characteristic features to provide a basis for a cytological diagnosis of malignancy from exfoliated cells. The cytologist, therefore, frequently has to combine the appearance of the individual cells and the arrangement of the cell groups to achieve a diagnosis. Often there is not sufficient cellular material present in the specimens to enable the cytologist to make a diagnosis in this way. However, when otherwise unexplained malignant cells are encountered in the urine of a male patient the prostate should be considered as their possible source.

METASTATIC CARCINOMA

Malignant cells may be found in the urine as a result of direct spread from malignant tumours of adjacent organs, such as the cervix and body of uterus, and also from secondary deposits from distant tumours.

SECTION V
THE BREAST

CHAPTER 16

NON-MALIGNANT DISORDERS

CYTOLOGICAL methods can be employed in the investigation of benign breast lesions in two different ways; the examination of discharge from the nipple or of fluid aspirated from cystic swellings.

DISCHARGE FROM THE NIPPLE

In the healthy breast, discharge from the nipple does not normally occur except during late pregnancy, lactation and the post-lactation period, although a small amount of secretion can occasionally be expressed from the nipple for a prolonged period once lactation has occurred. In the presence of infection, duct dilatation or duct papilloma, discharge from the nipple may be the presenting symptom; occasionally this discharge is blood-stained.

Smears prepared from discharge from the nipple contain two main cell types; these are foamy macrophages and ductal lining cells. The foamy macrophages, which usually predominate, are large cells with eccentric nuclei and finely vacuolated cytoplasm. Their nuclei are usually round or oval, although occasionally a reniform nucleus is seen. The chromatin pattern of the nuclei is evenly granular and single nucleoli can often be identified. The nuclei are similar to those of the ductal lining cells, from which these foamy macrophages originate. The ductal lining cells, in the absence of a papilloma, are usually present in small numbers; they appear as round or oval cells with opaque cytoplasm. Although their nuclei resemble those of the foamy macrophages they often appear larger because of their relatively scanty cytoplasm. Ductal cells may appear singly, but if they are present in appreciable numbers the cell groups show good cohesion. Occasional ductal cells from areas of apocrine metaplasia appear; these are larger cells with more abundant cytoplasm which often shows eosinophilic staining. Their nuclei frequently appear hyperchromatic, but this only represents a degenerative change. The striking uniformity of these cells, however, precludes a diagnosis of malignancy.

In the presence of a duct papilloma there is an increase in the number of ductal lining cells, which often appear in groups. These groups tend to retain their good cellular cohesion. Although there may be some variation in cell size within these groups, abnormalities of nuclear structure are absent.

ASPIRATES FROM BENIGN BREAST CYSTS

The cellular content of aspirates from benign breast cysts is essentially similar to that of nipple secretion and consists of foamy macrophages and ductal lining cells, occasionally including cells from areas of apocrine metaplasia. The cells, however, are often better preserved than those in nipple secretion which often show some degenerative changes. As a result the details of nuclear chromatin patterns and nucleoli may initially appear abnormally prominent in cells collected by aspiration of a cyst, but careful examination confirms that the individual nuclei have a normal distribution of chromatin and contain a single small nucleolus.

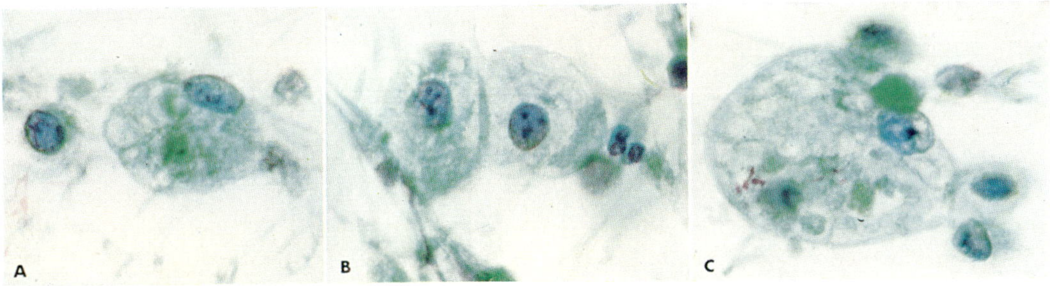

FIG. 125

Foamy Macrophages and Single Ductal Cells in a Discharge from the Nipple

A. A large foamy macrophage with an eccentric nucleus and vacuolated cytoplasm is seen to the right of a smaller ductal cell with opaque cytoplasm. Note the similarities between the two nuclei. (×800.)

B. Two vacuolated macrophages are present. (×800.)

C. A large foamy macrophage with intra-cytoplasmic debris is shown; note the prominent nucleolus. The two smaller cells to the right are ductal cells, but detail is not clearly illustrated in the upper cell. (×800.)

The nuclei of the macrophages and those of the ductal cells show similarities of size, shape and chromatin pattern, compatible with the fact that the foamy macrophages arise from the ductal lining. Although the macrophages show some variation in cell size their nuclei are fairly uniform in size and have normal chromatin patterns. Ductal cells are more commonly seen in groups, but even in these single cells their uniform small nuclei and normal chromatin patterns indicate their benign nature.

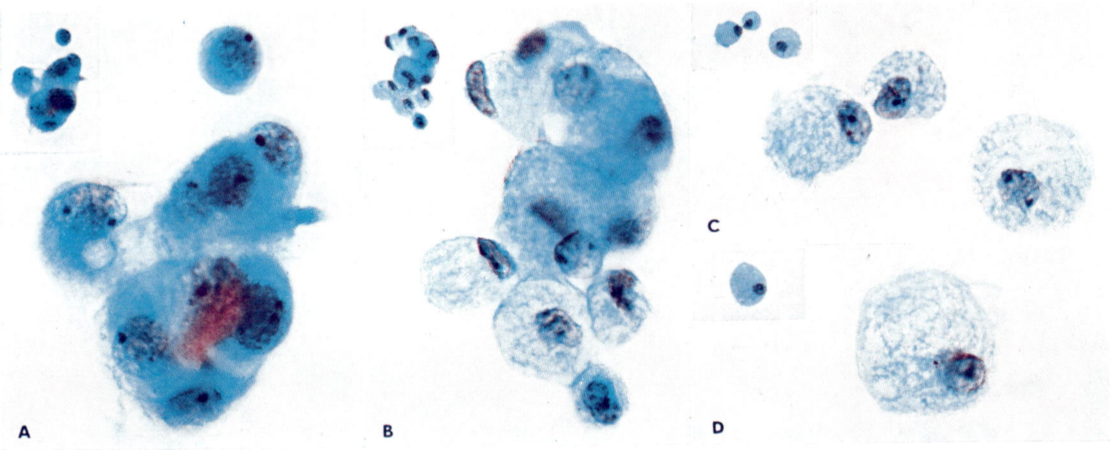

FIG. 126

Ductal Cells and Foamy Macrophages of Ductal Origin in an Aspirate of a Benign Breast Cyst

A. This group consists of ductal lining cells which show the compact grouping characteristic of this cell type. The individual cells have opaque cytoplasm and oval nuclei with finely granular chromatin patterns and small nucleoli. The good cellular cohesion and the uniformity of the nuclei indicate the benign nature of the cells. (Inset ×200.) (×800.)

B. In this compact group, the upper cells retain the opaque cytoplasm characteristic of the ductal epithelial cell, while the lower cells show a transition to the vacuolated cytoplasm of the macrophage. (Inset ×200.) (×800.)

C. and D. These foamy macrophages are readily identified by their vacuolated cytoplasm and eccentric nuclei. While showing variation in size, they are otherwise uniform in appearance and show the characteristic features of benign cells. (Insets ×200.) (×800.)

PRACTICAL APPLICATION

The cytological hazards of diagnosis, based on the examination of a discharge from the nipple, or of the content of a breast cyst, include the risk of missing an early malignant lesion and the equally grave risk of attributing too great a significance to atypical cellular features and thus producing a 'false' positive diagnosis of malignancy. It must be realised, therefore, that a negative cytological report cannot exclude a malignant lesion, so that the persistence of discharge, especially if blood-stained, or the recurrence of a cyst warrants further investigation by biopsy. Similarly a purely cytological diagnosis of malignancy should not be the sole basis for radical surgical or radiotherapeutic treatment; the cytological diagnosis should always be confirmed by biopsy.

CHAPTER 17

BREAST CANCER

CYTOLOGICAL methods can be applied to the diagnosis of cancer of the breast in three different ways; the examination of discharge from the nipple, the touch smear method of examination of the cut surface of an incisional or excisional biopsy and the technique of the aspiration smear. Of these three methods, the aspiration smear is the most controversial.

DISCHARGE FROM THE NIPPLE

The cytological examination of discharge from the nipple will occasionally lead to the early diagnosis of cancer, especially of intraduct carcinoma. In most cases, however, of persistent and particularly blood-stained discharge from the nipple, cytological examination is only a preface to biopsy.

When malignant cells are present they show the usual features of nuclear enlargement, producing abnormal nucleocytoplasmic ratios with abnormalities of chromatin pattern. Characteristically there is impairment of cellular cohesion in groups of malignant cells and, as a result, single abnormal cells detached from a group are often seen. These cells must be distinguished from ductal lining cells which are usually more uniform in size and chromatin pattern; in benign ductal cells their good cellular cohesion produces compact groups.

Biopsy is necessary to confirm the cytological findings, whether negative or positive, for malignant cells. It is impossible to exclude a malignant lesion by the examination of a nipple discharge, as the lesion will only give positive results if it lies within or encroaches on the lumen of a duct. Moreover, necrosis of a tumour may cause such severe cellular degeneration that diagnostic malignant cells are not apparent. Similarly it is advisable to confirm positive findings by biopsy before proceeding to radical therapy. Occasionally individual cells from a duct papilloma may show features suggestive of malignant change; it is then essential to see tissue relationships for definitive diagnosis.

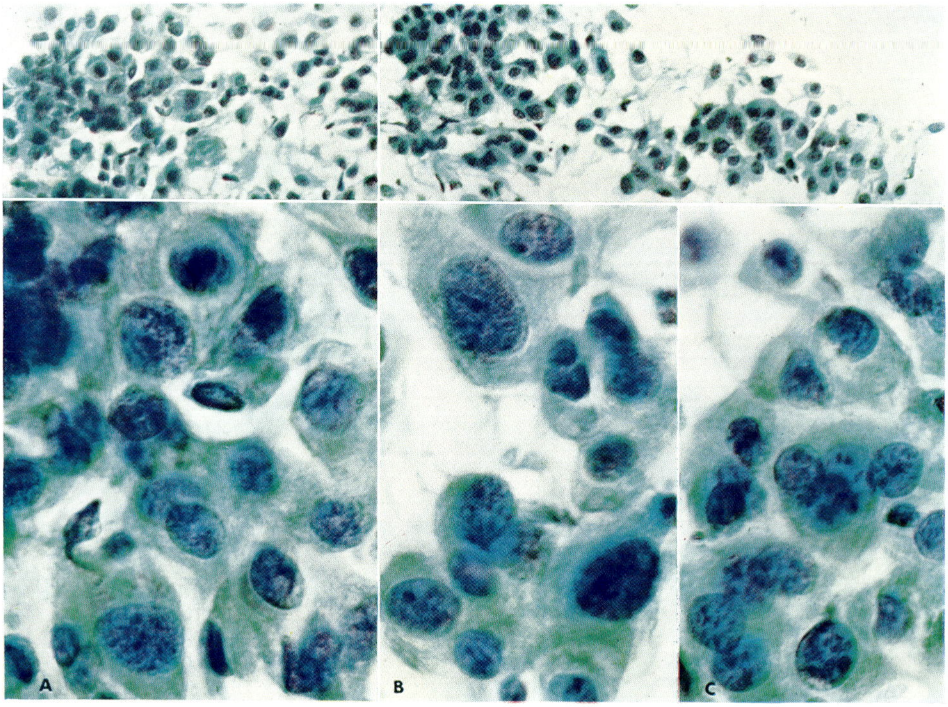

FIG. 127

Undifferentiated Malignant Cells in a Discharge from the Nipple

A. B. and C. In the two low-power fields at the top of the illustration, groups of large cells showing nuclear enlargement, hyperchromasia and pleomorphism are seen against a background of benign ductal cells with uniform small nuclei. Single malignant cells, detached from the groups, are readily recognised. In the high-power illustrations the large size of the cells, their abnormal nucleocytoplasmic ratios, with scanty cytoplasm, and the striking variation in nuclear size, shape and chromatin pattern are characteristic of malignant cells. A mitotic figure lies in the centre of field C. (× 200.) (× 800.)

Subsequent histological examination of a breast biopsy from this patient revealed an invasive anaplastic carcinoma.

THE TOUCH SMEAR METHOD

Touch smears can be prepared from the cut surface of an incisional or excisional biopsy. Once again it is advisable to confirm the findings obtained by this method with a frozen section technique, as these two methods complement each other. The touch smear gives excellent morphological detail in individual cells, which may not be as clearly defined in the frozen section, but the latter technique contributes more information on the relationships of the epithelial cells to the underlying structure of the breast tissue.

THE ASPIRATION SMEAR TECHNIQUE

The main advantage claimed for this method is its ease and simplicity from the point of view of the patient. For this reason it may, very occasionally, be of value in confirming the clinical diagnosis of an overt carcinoma where the patient's age and general condition preclude preliminary biopsy before radical surgery, or where it is hoped to avoid prolonging anaesthesia while awaiting the report of a frozen section. Counterbalancing this advantage, however, are the real hazards of this diagnostic method. Although the examination of fluid from a therapeutic aspiration of a single benign cyst is a useful procedure which may on rare occasions demonstrate an unexpected malignant lesion, the routine use of the aspiration smear technique as the main method of diagnosis of suspicious cystic or solid breast lesions is inadvisable. 'False' negative results may occur for a wide variety of reasons. The aspiration may miss the actual lesion and fail to obtain any diagnostic material. With a solid tumour it is difficult to sample the cellular components of the tumour adequately, while with a cystic tumour the fluid obtained may contain only necrotic non-diagnostic cells. Negative results, therefore, cannot provide any assurance of the benign nature of a breast lesion. 'False' positive results occur less often, but large foamy macrophages and ductal lining cells, especially those showing apocrine metaplasia, may be confused with malignant cells by the inexperienced observer. When malignant cells are present they show abnormal nucleocytoplasmic ratios, nuclear enlargement and pleomorphism, with abnormal chromatin patterns. It is extremely inadvisable to make a diagnosis of malignancy on an aspiration smear unless cells showing absolutely typical features are seen. In fact, even when this is the case, unless the observer has considerable experience of the method a confirmatory biopsy minimises the risk of exposing the patient to unnecessarily extensive surgery.

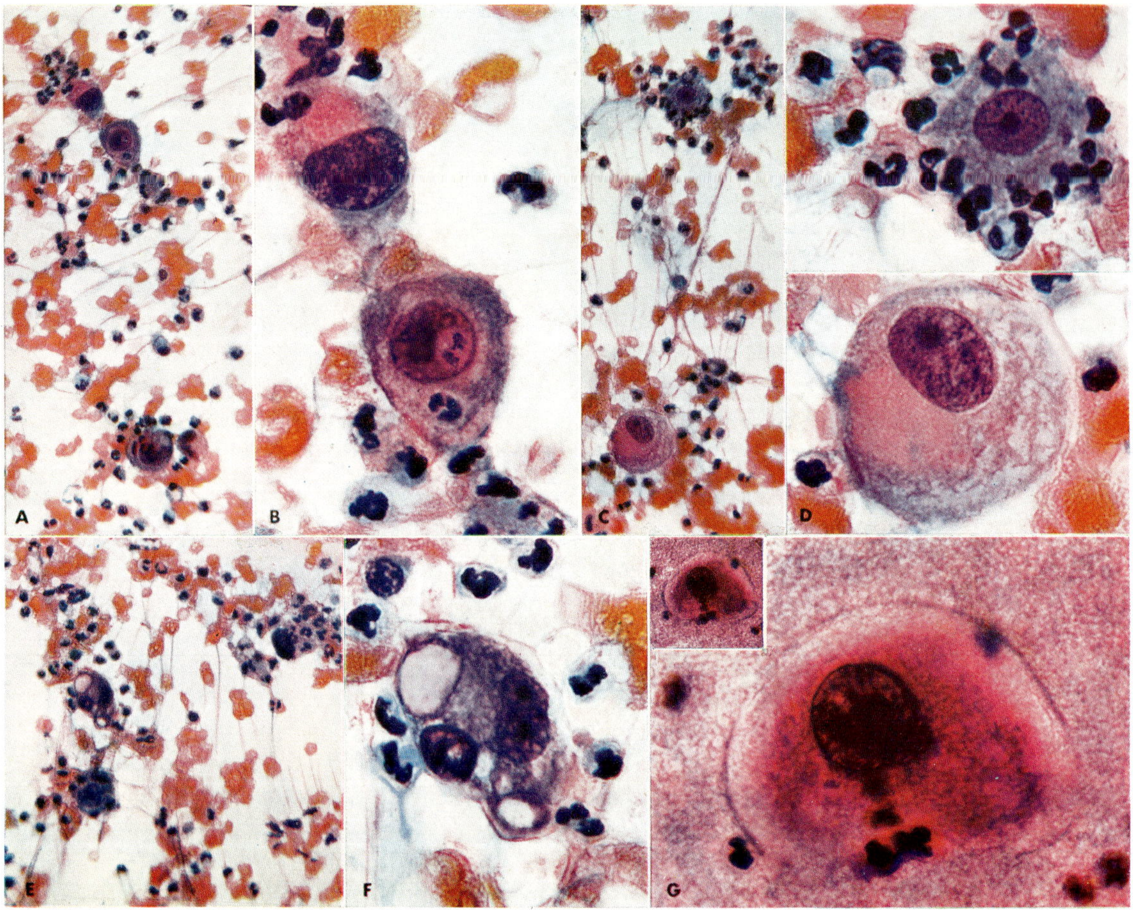

FIG. 128

An Aspirate from a Breast Cyst Showing Differentiated Malignant Cells from a Cystic Carcinoma of Breast

A. In this low-power field three large cells with hyperchromatic nuclei are seen against a background of erythrocytes, polymorphonuclear leucocytes and histiocytes. (×200.)

B. On high-power examination it is apparent that the two malignant cells have large nuclei and abnormal nucleo-cytoplasmic ratios. The two nuclei differ in shape, but both show chromatin clumping; the lower nucleus shows marginal condensation of chromatin. (×800.)

C. Two large malignant cells are present in this low-power field against a background of erythrocytes and polymorphonuclear leucocytes. (× 200.)

D. On high-power examination it is apparent that the nuclei of these two differentiated malignant cells show chromatin clumping and have prominent large nucleoli. Note the eccentric position of the nucleus in the lower cell. (× 800.)

E. Several large cells with hyperchromatic nuclei are present among the erythrocytes and polymorphonuclear leucocytes. (× 200.)

F. On high-power examination this malignant multinucleated giant cell shows cytoplasmic vacuolation and has nuclei which are eccentric in position. This appearance is compatible with an origin from an adenocarcinoma. The vacuolation, however, may only represent a degenerative change, as the nucleus on the left of this cell is densely hyperchromatic and degenerate. The two nuclei on the right show variation in size and shape, abnormal chromatin patterns and prominent nucleoli. (× 800.)

G. This huge malignant cell shows marked nuclear hyperchromasia with clumping and marginal condensation of chromatin. (Inset ×200.) (×800.)

A cystic swelling in the breast, with a provisional clinical diagnosis of a cyst or abscess, was aspirated and these large malignant cells were seen in the smear from the aspirate. Biopsy examination confirmed the diagnosis of an encephaloid carcinoma with an acute inflammatory reaction; unusually large malignant cells were seen in the section of the tumour. When a provisional clinical diagnosis of carcinoma of breast has been made, biopsy provides a safer and more reliable method of diagnosis than aspiration. Aspiration may be useful, however, in the recognition of unsuspected early malignant change in a 'simple' cyst or in an abscess.

SECTION VI

THE BODY FLUIDS

CHAPTER 18

NON-MALIGNANT DISORDERS

CYTOLOGICAL methods of diagnosis can be applied to a wide range of body fluids, but their practical value varies considerably with different types of fluid. The role of cytology in the diagnosis of lesions of the urinary tract, by examination of urine specimens, has previously been considered. Of the remaining body fluids, effusions into body cavities and aspirates from cysts are the fluids to which cytological examination is most readily applicable and in which the method is of the greatest diagnostic value. Cerebrospinal fluid and blood can be submitted to investigation by cytological methods for the diagnosis of malignant lesions, but there are considerable technical and diagnostic difficulties which make them unsuitable for routine investigation.

EFFUSIONS INTO BODY CAVITIES

In the healthy state only a thin film of fluid separates the parietal and visceral serosal surfaces. As a result specimens are not usually collected from the serosal spaces in healthy patients, although recently a technique for collection of aspirates from the Pouch of Douglas, through the posterior fornix, from normal female patients, in an attempt to make an early diagnosis of carcinoma of the ovary, has been described. In these normal patients the cellular content has been scanty and consisted mainly of serosal cells. In general, however, the presence of sufficient fluid, for a specimen to be collected by paracentesis and examined cytologically, is associated with a pathological process, whether benign or malignant, either in the serosa or in the underlying tissues. The main purpose of the cytological examination of this fluid is the recognition of malignant cells, but it is essential to be familiar with the 'normal' cellular content of the fluid and the changes seen in response to non-malignant disorders so as to establish a basis for comparison.

It is customary to subdivide effusions into transudates, with a low specific gravity and protein content, and exudates, with a high specific gravity and protein content. Cytologically, too, there are differences; the transudate, such as the effusion of congestive cardiac failure, has a relatively scanty cellular content consisting mainly of mesothelial cells of serosal origin with a few histiocytes and lymphocytes; the exudate, whether associated with the presence of inflammation or of tumour, is usually more densely cellular with numerous mesothelial cells, non-mesothelial inflammatory cells and, in the presence of a tumour, cells arising from the tumour.

MESOTHELIAL CELLS

Mesothelial cells appear singly or in small groups, usually with less than six cells in the group. The individual cells are round and approximately the same size as the smaller parabasal cells (10-20 μ). Their cytoplasm, which is usually basophilic, is opaque but sometimes finely vacuolated. The nuclei are large and vesicular, occupying about half the diameter of the cell. The nucleus is usually central in position, but occasionally a cuboidal cell is seen, in which the nucleus is eccentric in position. The shape of the nucleus is round or oval, with a smooth well-defined nuclear border; the chromatin pattern is finely granular, although an occasional small rounded clump of chromatin may be present. A single small nucleolus is often apparent. Multinucleation is a common feature; when it occurs there are usually two nuclei which are very similar to each other.

In non-specific inflammatory conditions considerable changes may be seen in these cells. Multinucleation increases in frequency and the number of nuclei in the multinucleated cells also increases. Mitotic figures are more frequently seen; in fluids the presence of normal mitotic figures is just as likely to be associated with benign as with malignant cells. Individual nuclei may show an increase in size and may have a more coarsely granular but still uniform chromatin pattern. Nucleoli may also enlarge and become more prominent and occasionally multiple nucleoli are seen. In benign cells, however, the nuclei retain a smooth outline, a uniform chromatin pattern with rounded nucleoli, and an adequate border of cytoplasm.

Degenerative changes in mesothelial cells can cause alterations in both cytoplasm and nuclei. The cytoplasm may be completely lost and the denuded nuclei may appear suspicious. However, where nuclear detail is retained the uniform round or oval outlines and normal chromatin patterns of the nuclei establish their benign nature. In addition these denuded nuclei usually appear singly or in small loosely arranged groups, contrasting with the compact arrangement characteristic of tumour cells. If degeneration of the nucleus occurs, it can take two forms; the degenerate nucleus may appear enlarged, pale and opaque or it may be densely hyperchromatic. In either case the significant feature is the complete loss of normal nuclear structure.

In view of the changes which can occur in mesothelial cells in response to inflammation or as a result of degeneration, caution should be exercised in identifying cells as malignant when the main features suggest a serosal origin, unless they show unequivocal abnormalities of chromatin pattern and nuclear shape. If cells whose features show transitional changes, ranging from a close approximation to normal mesothelial cells to markedly atypical forms, are present it is advisable to avoid a definitive diagnosis of malignancy.

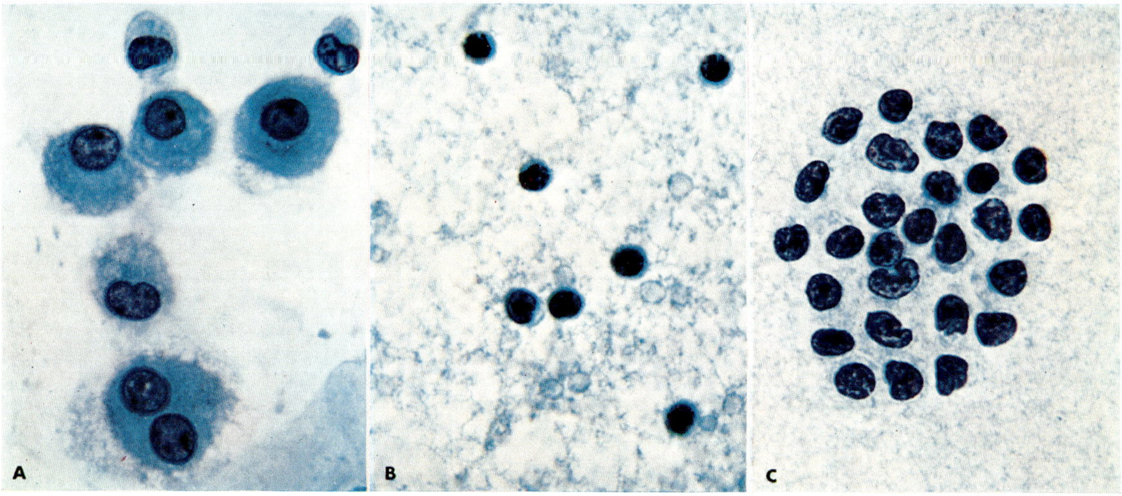

FIG. 129

Mesothelial Cells, Lymphocytes and Histiocytes in Ascitic Fluids

A. Well-preserved mesothelial cells and small histiocytes are seen. The mesothelial cells are round, with opaque, occasionally vacuolated cytoplasm and round central vesicular nuclei. One binucleate mesothelial cell is also seen. The nuclei of the mesothelial cells have finely granular chromatin patterns and nucleoli can be seen. The histiocytes are smaller and have foamy, finely vacuolated cytoplasm; their nuclei are eccentric and characteristically 'kidney-bean' shaped, but may be round or oval. (×800.)

B. These lymphocytes have a thin rim of basophilic cytoplasm. Their round nuclei are uniformly hyperchromatic and the nuclear borders frequently appear to be wrinkled. (×800.)

C. These small histiocytes show a loss of cytoplasm. Although there is considerable variation in nuclear shape, the nuclei are fairly uniform in size and chromatin pattern and several show the typical 'kidney-bean' appearance and serve to identify the group as histiocytes. (×800.)

Further illustrations of benign mesothelial cells and histiocytes are shown in Figs. 131, 132, 136, 138, 141, 142 and 163.

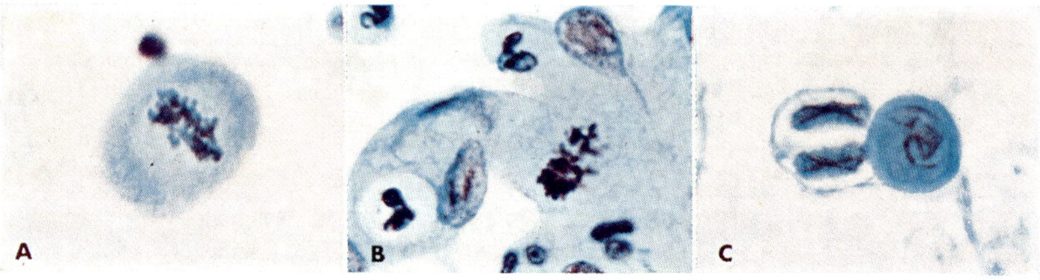

FIG. 130

Mitotic Figures in Benign Mesothelial Cells and Histiocytes from Non-Malignant Effusions

A. A benign mesothelial cell is shown in metaphase. This is the phase of mitosis most often seen. (×800.)

B. A group of histiocytes shows two cells in mitosis. One appears to be in prophase and the other in metaphase. Note the intracytoplasmic polymorphonuclear leucocytes in the adjacent histiocytes. (×800.)

C. A mesothelial cell is seen on the right of this field with another mesothelial cell in anaphase on the left. (×800.)

Mitoses are not often seen in cytological preparations. When they do occur they are just as likely to be seen in benign cells, especially histiocytes and mesothelial cells, as in malignant cells.

NON-MESOTHELIAL INFLAMMATORY CELLS

The non-mesothelial cells may show a mixture of polymorphonuclear leucocytes, lymphocytes, histiocytes and erythrocytes, but frequently there is a predominance of one or two cell types. In these cases the common combinations are of lymphocytes and histiocytes, or of lymphocytes and erythrocytes, although occasional polymorphonuclear leucocytes are usually seen as well.

The characteristically lobulated nucleus of the polymorphonuclear leucocyte is easily recognised. The lymphocyte has a narrow rim of palely basophilic cytoplasm surrounding a densely basophilic hyperchromatic nucleus, which often shows a crenated nuclear border; lymphocytes may show some size variation. Histiocytes have foamy vacuolated cytoplasm and an eccentric nucleus; the cytoplasm frequently contains phagocytosed debris and polymorphonuclear leucocytes. Large vacuoles in the cytoplasm may displace the nucleus to the periphery of the cell and produce a 'signet-ring' appearance. The nuclei vary in shape and may be round, oval or typically reniform, their chromatin patterns are more finely granular and their size often smaller than those of the nuclei of serosal cells. As in mesothelial cells nucleoli may be seen. Multinucleation is common and occasionally giant multinucleated histiocytes are present. Mitotic figures may be seen in histiocytes. Degeneration causes loss of cytoplasm; the denuded histiocyte nuclei then show considerable variation in shape, but are usually uniform in size and chromatin pattern. When erythrocytes are present they may appear as intact cells or they may have undergone cytolysis.

When there is a marked predominance of lymphocytes in a specimen, especially if these show considerable size variation, the possibility of lymphatic leukaemia or lymphosarcoma should be considered, although it is not usually possible to make this diagnosis solely on cytological grounds. If the specimen shows a combination of large numbers of lymphocytes and histiocytes further investigations to exclude tuberculosis should be performed.

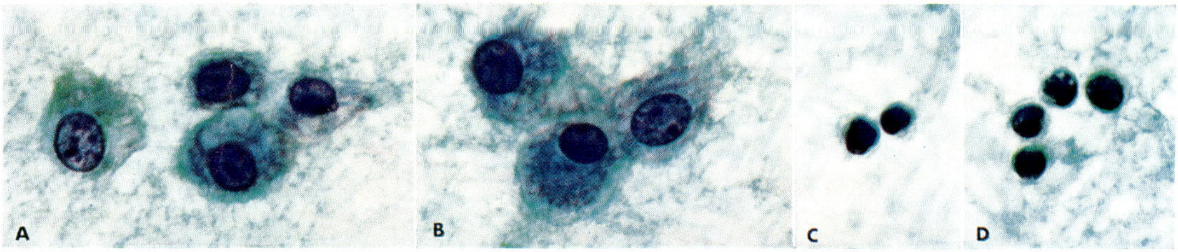

FIG. 131

Pleural Fluid showing Histiocytes, Mesothelial Cells and Lymphocytes

A. and B. The histiocytes, which predominate, show the characteristic vacuolated cytoplasm and eccentric nuclei. The mesothelial cells have rather larger central nuclei and the cytoplasm is more opaque. (×800.)

C. and D. These lymphocytes show densely staining basophilic nuclei which are surrounded by a narrow rim of paler staining basophilic cytoplasm. (×800.)

This pleural fluid was obtained from the patient whose ascitic fluid is illustrated in Fig. 138. Subsequent post-mortem examination showed a bronchopneumonia with no evidence of metastases in the lungs or pleura. It also revealed a poorly differentiated adenocarcinoma of ovary with peritoneal metastases, the source of the malignant cells shown in Fig. 138.

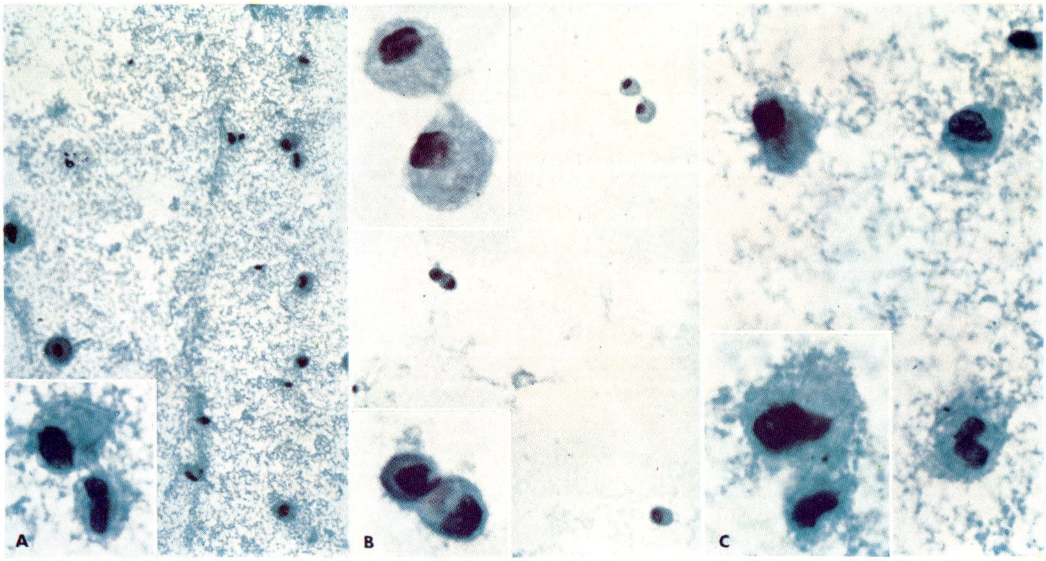

FIG. 132

Histiocytes and Mesothelial Cells in Fluid Aspirated from a Hydrocele

A. The cellular content illustrated in the low-power field consists mainly of histiocytes with only an occasional serosal cell with opaque cytoplasm and a central oval nucleus. High-power magnification (inset) shows two histiocytes with foamy cytoplasm. Although the nucleus of the smaller cell is central it is 'kidney-bean' shaped; the nucleus of the larger cell is typically eccentric in position. (×200.) (×800.)

B. Again histiocytes predominate in the low-power field and characteristic forms are shown in the upper high-power field (inset). The two cells in the lower high-power field (inset) are degenerate with opaque nuclei; the eccentric position and reniform outlines of the nuclei suggest that these cells are degenerate histiocytes. (×200.) (×800.)

C. Histiocytes of varying size are seen. The typical vacuolated cytoplasm and eccentric, often reniform, nuclei serve to identify them. The two cells (inset) with hyperchromatic nuclei are degenerate and show a complete loss of nuclear structure, but their characteristic cytoplasm and the remaining reniform nuclear shape suggest their histiocytic origin. (×800.)

ASPIRATES FROM CYSTS

Aspirates from cysts usually consist of epithelial and non-epithelial cells. However, where the cyst has been present for some time there may have been considerable flattening and distortion of the epithelial lining. As a result the epithelial cells may be outnumbered by the non-epithelial cells and, when seen, may appear only as round or cuboidal cells which are impossible to identify accurately. Usually it is possible to determine the type of epithelium from which the cyst has arisen by the presence of recognisable squamous or columnar cells which, in benign cysts, retain the characteristic features of the cells of the tissue of origin. For this reason it occasionally becomes apparent that a specimen, which is supposed to represent an effusion into a body cavity, is in fact a cyst aspirate. This is especially likely to occur in abdominal paracentesis, in the presence of an unsuspected ovarian cyst. Under these circumstances the absence of mesothelial cells and the presence of columnar or squamous cells in the specimen point to the correct diagnosis. Of course the issue may be confused by the fact that, occasionally, rupture of an ovarian cyst is the main source of an ascitic fluid.

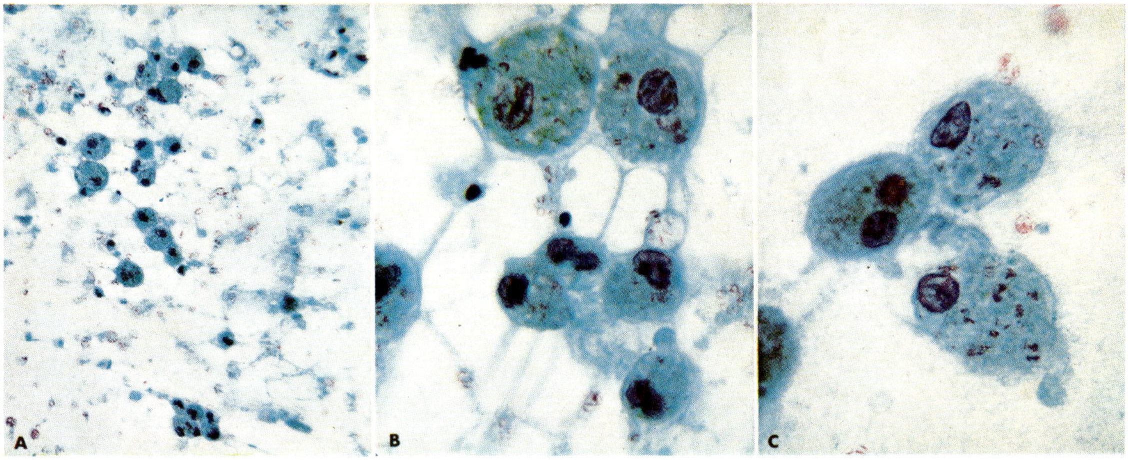

FIG. 133

Histiocytes in Fluid Aspirated from a Benign Ovarian Cyst

A. In this low-power illustration groups of cells with eccentric dark nuclei are seen against a background of erythrocytes. Although the cells show some variation in size, the nuclei are comparatively uniform in size. In the lowest group, the nuclei appear to be 'kidney-bean' shaped and eccentric in position. (×200.)

B. At high-power magnification the typical foamy cytoplasm of the histiocyte is now seen and the lower nuclei show the characteristic notched reniform shape. In the nucleus of the cell in the upper right quadrant the rather coarse but uniformly stippled chromatin pattern of an active histiocyte is seen. (×800.)

C. The nuclei in this group show variation in shape with a uniform chromatin pattern. Ingested cellular debris is seen in the cytoplasm of all these cells, but is especially prominent in the lowest cell in the group. (×800.)

These histiocytes contrast sharply with the malignant cells illustrated in Fig. 137, in which the eccentric dark nuclei show marked enlargement and variation, not only in shape, but also in size and in chromatin pattern.

THE CEREBROSPINAL FLUID

Normal cerebrospinal fluid is essentially acellular. In many non-malignant disorders, too, the cellular content is often insignificant, but in inflammation, lymphocytes and polymorphonuclear leucocytes can be readily recognised.

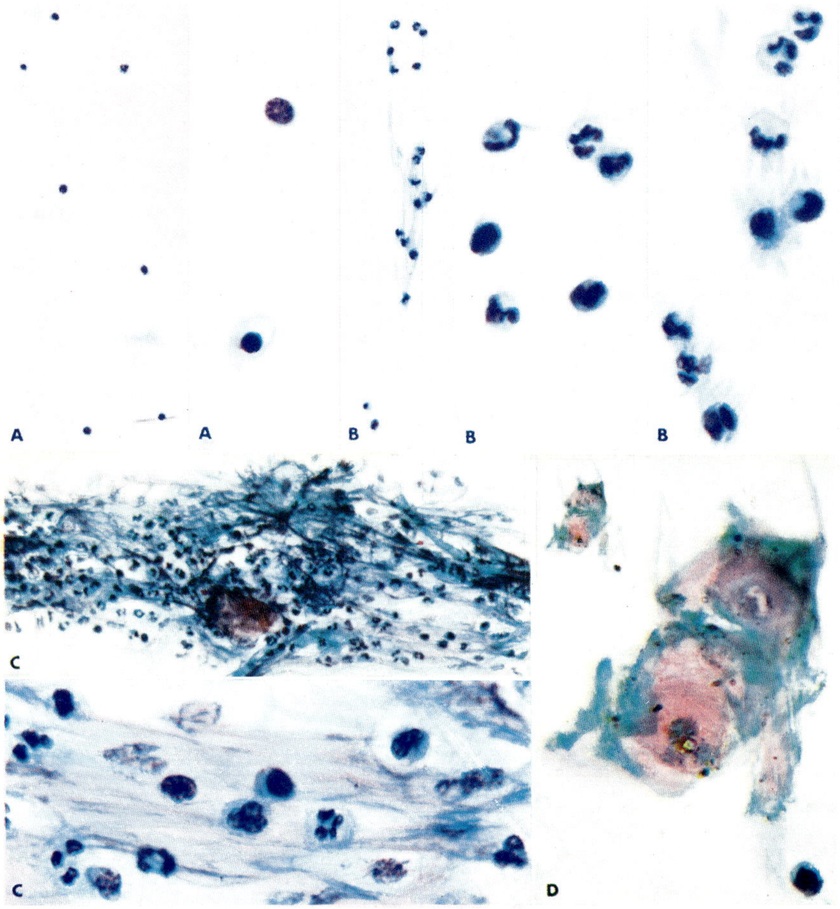

FIG. 134

Inflammatory Cells in Cerebrospinal Fluid

A. and B. These fields show lymphocytes, polymorphonuclear leucocytes and histiocytes at low and high-power magnifications. (×200.) (×800.)

C. In the presence of inflammation the inflammatory cells may be seen enmeshed in fibrin strands. (×200.) (×800.)

D. This field shows keratinised anucleated squames, which are present as contaminants. This type of cellular material and also superficial squamous cells of cutaneous origin are often present as contaminants in specimens collected by aspiration. (×200.) (×800.)

THE BLOOD

A detailed description of the appearance of blood films in non-malignant disorders will be found in any standard textbook on haematology.

CHAPTER 19

MALIGNANT CELLS IN THE BODY FLUIDS

MALIGNANT cells in the body fluids are most readily recognised in effusions into body cavities and aspirates from cysts. In cerebrospinal fluid and blood their identification presents considerable problems. Malignant cells in urine have been described previously (pp. 144-147).

EFFUSIONS INTO BODY CAVITIES

In the vast majority of cases the presence of malignant cells in effusions indicates the presence of a metastasis or of direct invasion of the serosa by an extrapleural or extraperitoneal primary tumour, since the primary tumour of mesothelium, the *Mesothelioma*, is extremely rare. The diagnosis of malignant cells, therefore, inevitably means a hopeless prognosis for the patient. It is apparent that this type of malignant diagnosis, which may lead to the discontinuation of certain types of therapy, should only be made in the presence of unequivocally malignant cells.

In practice making this diagnosis can be very difficult for two reasons. Mesothelial cells, in response to inflammation, can show markedly atypical features and their appearance may closely mimic the appearance of malignant cells. In contrast malignant cells, which have been present for some time in a body fluid, may lose many of their most strikingly abnormal characteristics. For these reasons a cell population which shows progressive gradations between recognisable mesothelial cells and abnormal cells should be interpreted with caution; this pattern should not be used as a basis for a definitive diagnosis of malignant cells. The presence of a second cell population distinct from the mesothelial cell population provides a more satisfactory basis for the diagnosis of malignant cells.

Malignant cells in effusions can be subdivided into differentiated and undifferentiated types. Undifferentiated malignant cells and differentiated malignant cells of the type originating in an adenocarcinoma are the two types most often seen. However, both differentiated and undifferentiated malignant cells may be identified in the same specimen.

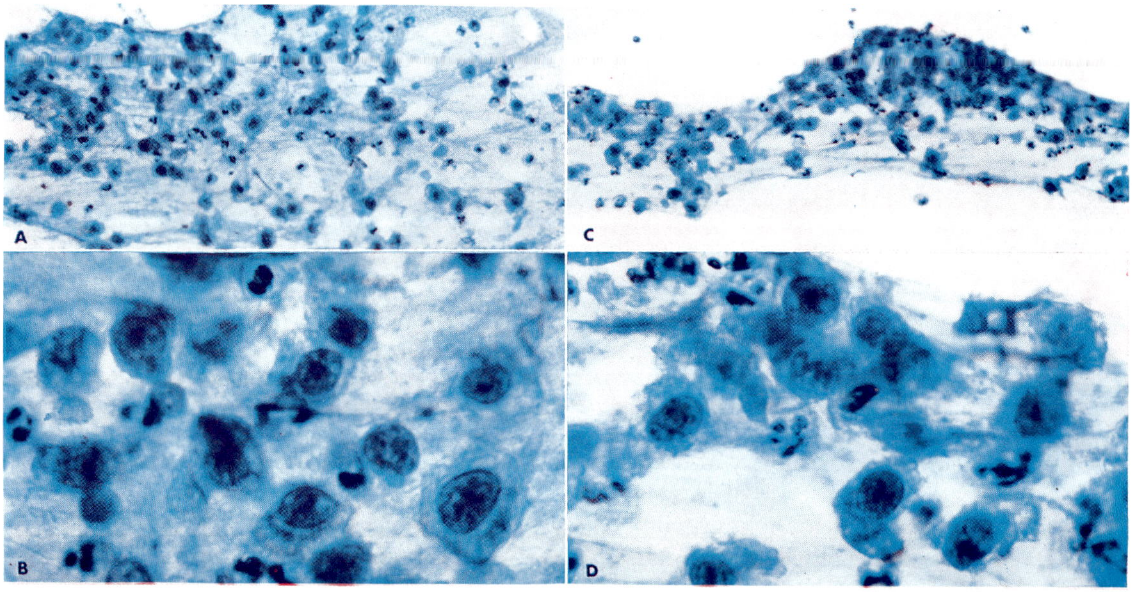

FIG. 135

A Hydrocele Fluid with Differentiated and Undifferentiated Malignant Cells from a Seminoma of Testis

A. and C. In these low-power fields numerous large cells are seen against a background of lymphocytes and polymorphonuclear leucocytes. Many of the larger cells have hyperchromatic nuclei; others show degenerative changes. (×200.)

B. At high-power magnification it is apparent that the larger cells are fairly uniform in size, but show variations in nuclear shape with abnormal chromatin patterns, chromatin clumping and uneven marginal condensation of chromatin. In many of the cells, the cytoplasm has partly or completely disappeared. Where it is present, as in the differentiated malignant cell in the lower right quadrant, the outline of the cell is polygonal and the nucleus is central. (×800.)

D. This high-power field shows a mixture of fairly well-preserved differentiated and undifferentiated malignant cells, essentially similar to those seen in plate B, and degenerate cells in which the nuclei show karyolysis. This admixture of degenerate cells was present in many areas of the smears. In the upper half of the field mitotic figures can be identified in several cells. (×800.)

Subsequent histological examination of the testis revealed a seminoma with frequent mitotic figures; the tumour showed numerous areas of necrosis.

The commonest type of differentiated malignant cell seen arises from an adenocarcinoma. Typically, malignant cells of this kind appear in sizeable compact groups of overlapping cells, although single cells may also be present. The cells have large eccentric nuclei of irregular size and shape, with abnormal clumped chromatin patterns. Often the presence of large multiple nucleoli is a striking feature. The cytoplasm may show vacuolation and contain mucus. These cells must be distinguished from groups of mesothelial cells and from histiocytes, especially those with 'signet-ring' forms. Mesothelial cells, however, rarely form large groups, although this can occur as a centrifugation artefact. Even when mesothelial cell nuclei are enlarged in response to inflammation, they are uniform in round or oval shape and also in chromatin pattern. Histiocyte nuclei are smaller, in relation to the size of the cell, than those of malignant cells and their chromatin patterns are uniform. Although varying in shape, histiocyte nuclei also tend to be of uniform size.

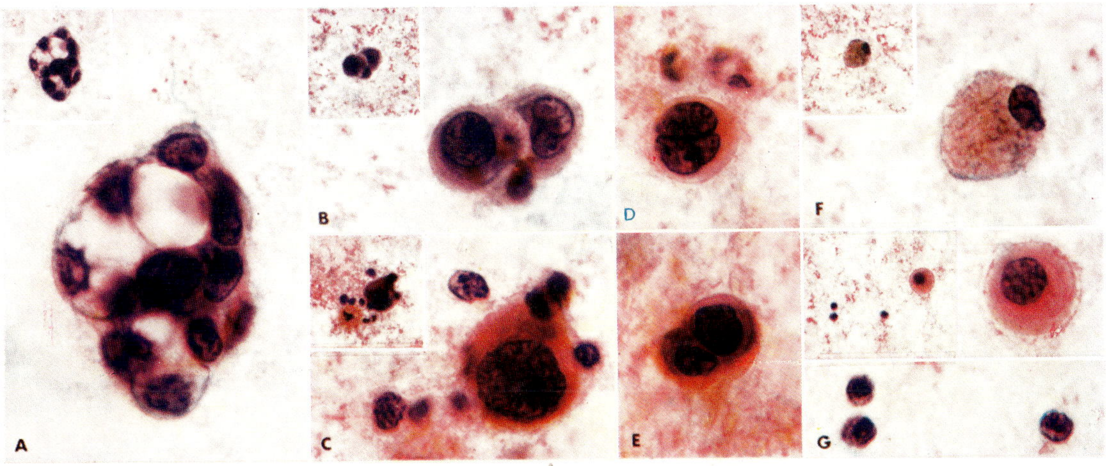

FIG. 136

An Ascitic Fluid showing Differentiated Malignant Cells from Metastases from a Cystadenocarcinoma of Ovary

A. This large group of malignant cells shows clearly the characteristic arrangement of malignant cells of this type. The nuclei tend to lie at the periphery of the group, producing a pseudo-acinar appearance. In the individual cells the nuclei are eccentric in position and vacuolation of the cytoplasm is a striking feature. The nuclei vary in size and shape and their abnormal clumped chromatin patterns can be identified. (Inset ×200.) (×800).

B. This group shows the compact arrangement of even a small group of malignant cells. Note the marginal chromatin condensation of the large nucleus of the cell on the left. (Inset ×200.) (×800.)

C. A large single malignant cell is seen against a background of inflammatory cells. The enlarged nucleus is hyperchromatic, with unevenly distributed chromatin and an irregular nuclear border. Contrast this single malignant cell with the histiocyte, seen in field F, and the mesothelial cell, shown in field G. (Inset ×200.) (×800.)

D. and E. The single multinucleated malignant cell shown in field D and the malignant cells in the small group in field E all display the characteristic features of nuclear enlargement, variation in nuclear size and shape, and hyperchromasia with abnormal chromatin patterns. Although these cells and those seen in fields B and C are readily recognised as malignant cells they do not show features which readily identify their origin from an adenocarcinoma. (×800.)

F. From the same specimen a typical histiocyte, with foamy cytoplasm and small eccentric nucleus, is shown for comparison. (Inset ×200.) (×800.)

G. A further low-power field (inset) from the same specimen shows lymphocytes and a single benign mesothelial cell. Note the smooth outline and evenly granular chromatin pattern of the round nucleus of the serosal cell which is clearly shown at high-power magnification in the upper right quadrant. (Inset ×200.) (×800.)

The eosinophilic staining of the cytoplasm of all the cells in this specimen is probably due to the fact that the specimen was heavily blood-stained. Further fields from the specimen are shown in Fig. 163; in these numerous erythrocytes are seen.

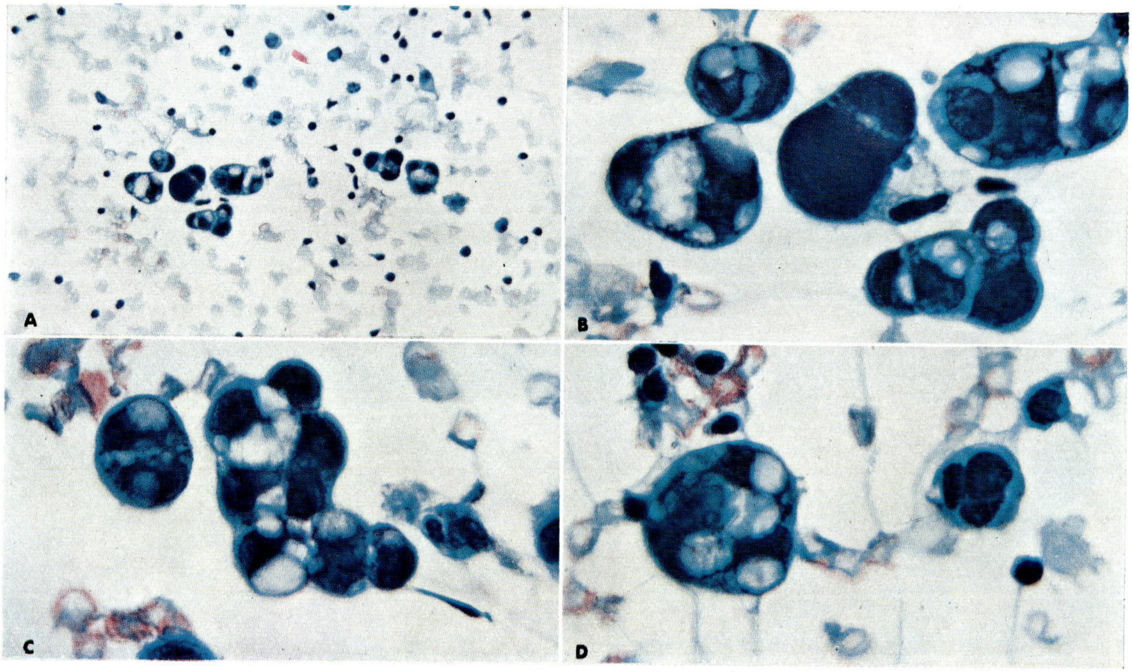

FIG. 137

Differentiated Malignant Cells, Originating from Metastatic Adenocarcinoma, in Pleural Fluid

A. Groups of differentiated malignant cells are seen against a background of erythrocytes and lymphocytes. The cells, originating from an adenocarcinoma, tend to form compact clumps, unlike mesothelial cells, which usually lie singly or in groups of only two or three cells. (×200.)

B. On high-power examination it is apparent that both the nuclear and cytoplasmic characteristics of cells arising from an adenocarcinoma are present. The cytoplasm shows coarse vacuolation and the nuclei, which are eccentrically situated within the cytoplasm, are peripherally located in each group. There is marked variation in nuclear size and shape. In some of the nuclei, as in the central group, degeneration is present. The presence of this type of degenerate cell group should always lead to further examination of the specimen, because of its large and disparate nuclei, but it should not be used for final diagnosis. In the other groups, however, diagnostic abnormal nuclear chromatin patterns and peripheral condensation of chromatin are seen. (×800.)

C. Coarse vacuolation of the cytoplasm is clearly seen in the centre group. The peripheral position of the nuclei is also a feature of this group. (×800.)

D. Even when many of the groups present show the features of differentiated malignant cells, this is not necessarily the case in all the malignant groups. The cells on the left of this field are similar to those already described in fields B and C and are compatible with an origin from an adenocarcinoma. The two cells on the right show the nuclear hyperchromasia and pleomorphism of malignant cells, but do not show any characteristic features to suggest the type of tumour from which they originate. This loss of differentiation is frequently seen in metastatic tumour cells in serous effusions. (×800.)

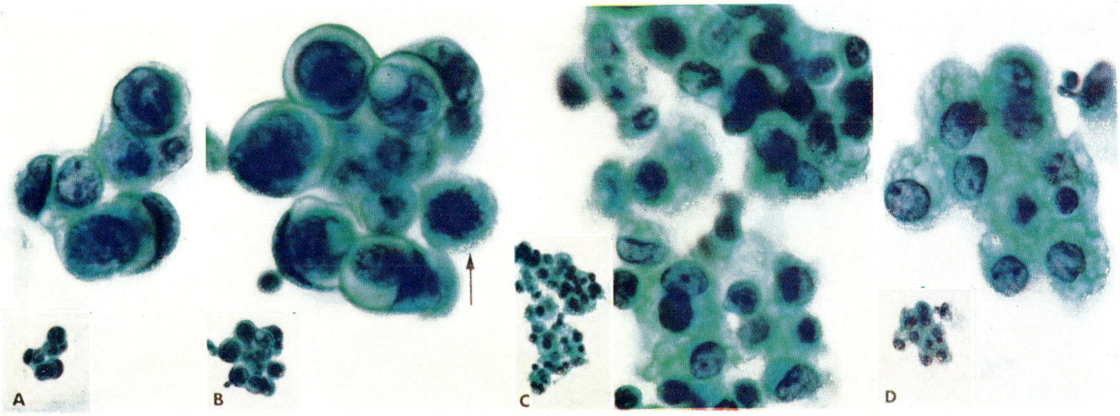

FIG. 138

An Ascitic Fluid showing Differentiated Malignant Cells, arising from an Adenocarcinoma of Ovary, and Histiocytes

A. and B. These fields show low- and high-power views of differentiated malignant cells arising from metastases from an adenocarcinoma of ovary. The groups show clusters of overlapping cells with moulding of adjacent nuclei, which are eccentric in position and vary in size, shape and chromatin pattern. The arrow in plate B indicates a cell showing a mitotic figure. The coarse vacuolation of the cytoplasm seen in several cells is compatible with an origin from an adenocarcinoma, but vacuolation may also occur as a degenerative change. (Insets ×200.) (×800.)

C. and D. In these plates groups of histiocytes are seen. The cytoplasm is finely vacuolated and the nuclei eccentric. Apart from a few degenerate pyknotic nuclei the other nuclei are fairly uniform in size and in chromatin pattern; they are mostly round or oval, but an occasional reniform nucleus is seen. The characteristic cytoplasm and eccentric uniform reniform nuclei identify these cells as histiocytes and distinguish them from the malignant cells shown in fields A and B. (Insets ×200.) (×800.)

Differentiated malignant cells of squamous type are only occasionally seen. When they are present they appear as cells with central large, irregular, often strikingly hyperchromatic nuclei, surrounded by opaque cytoplasm. The bizarre forms of differentiated malignant cells do not occur. The squamous type of malignant cell often appears singly or in small groups and must be distinguished from the mesothelial cell by careful examination of its abnormal nuclear characteristics.

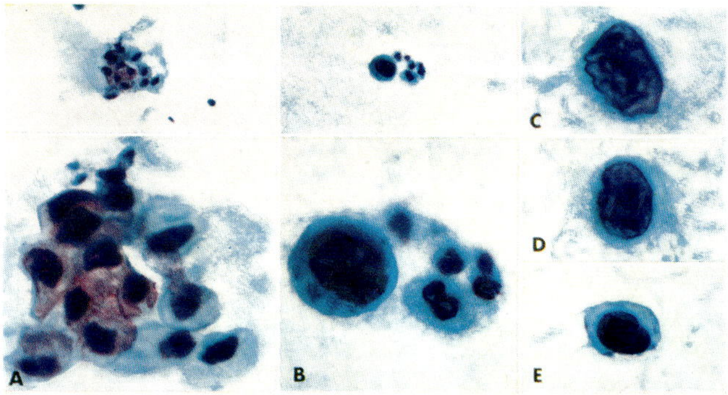

FIG. 139

Differentiated Malignant Cells of Squamous Type in a Pleural Fluid

A. This group of cells shows marked variation in nuclear size and shape, but the lower high-power field reveals that the hyperchromatic nuclei have completely lost their structure. These degenerate cells are unsuitable for definitive diagnosis, but their presence indicates a careful search for diagnostic cells. (Inset ×200.) (×800.)

B. The large differentiated malignant cell on the left has a hyperchromatic central nucleus with an irregular nuclear border and an abnormal chromatin pattern. A small histiocyte lies to the right of the malignant cell. (Inset ×200.) (×800.)

C. D. and E. These three single malignant cells all have large central nuclei with angular notching of the nuclear border. The cells in plates C and D also show marginal condensation of chromatin and irregular clumping of chromatin in linear streaks and angular masses. (×800.)

These cells were seen in a blood-stained pleural fluid aspirated from a patient with a squamous cell carcinoma of bronchus on the same side as the effusion. There is a loss of many of the characteristic features of differentiated malignant cells of squamous type when they are found in body fluids. Keratinisation is rarely seen. In this case the arrangement of the cells in the sheet of degenerate cells, the central position of the nuclei in the malignant cells and their opaque cytoplasm are compatible with their origin from a squamous cell carcinoma.

Undifferentiated malignant cells are most easily identified when they appear in groups, although rarely a single cell may show such bizarre features that it can be recognised as a malignant cell. The nuclei of undifferentiated malignant cells vary in size and shape; the nuclear borders are often irregular. Frequently the nuclei appear hyperchromatic, but even when this is not a striking feature, their chromatin patterns are abnormal with clumping and marginal condensation. Multiple or enlarged nucleoli may be a prominent feature. The cytoplasm of these cells may be seen only as a poorly defined background to the cell group, or may be present only over a part of the circumference of a malignant nucleus, or may not be identifiable at all. These undifferentiated malignant cells must be distinguished from denuded or degenerate histiocyte or mesothelial cell nuclei. By basing the diagnosis on groups of cells showing characteristic nuclear abnormalities confusion with benign nuclei can be avoided, as the benign nuclei tend to occur singly and show either normal chromatin patterns or the complete loss of nuclear structure associated with degeneration.

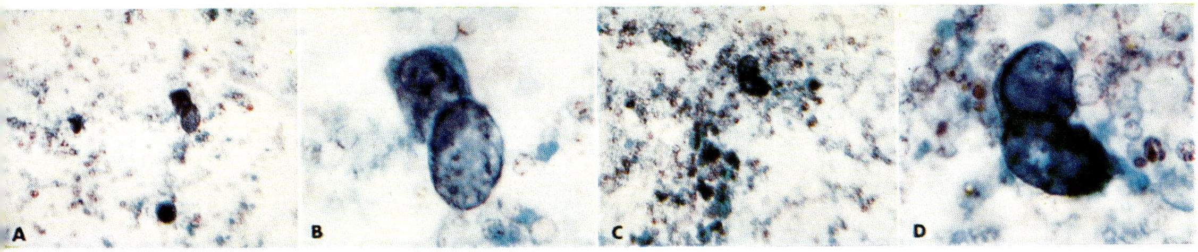

FIG. 140

Undifferentiated Malignant Cells in an Ascitic Fluid

A. In this low-power field two large dark undifferentiated malignant cells are seen against a background of erythrocytes. (×200.)

B. High-power magnification reveals a rim of cytoplasm incompletely surrounding a large notched nucleus in the upper cell, while the lower cell shows almost complete loss of cytoplasm. In both nuclei there is peripheral condensation and clumping of chromatin. (×800.)

C. In this low-power field two large sharply defined nuclei are seen lying above a group of smaller irregular, but ill-defined dark nuclei. (×200.)

D. High-power magnification shows a narrow rim of cytoplasm at the free border of the upper nucleus, which retains its abnormal chromatin pattern. The lower nucleus is almost completely denuded of cytoplasm, although the large size, irregular outline and hyperchromasia are compatible with a malignant cell. Degenerative changes, however, have produced a disappearance of diagnostic chromatin abnormalities in the lower nucleus, leaving only an opaque nucleus with unevenly distributed chromatin. (×800.)

At laparotomy an inoperable carcinoma of pancreas, with lymph node metastases, was found.

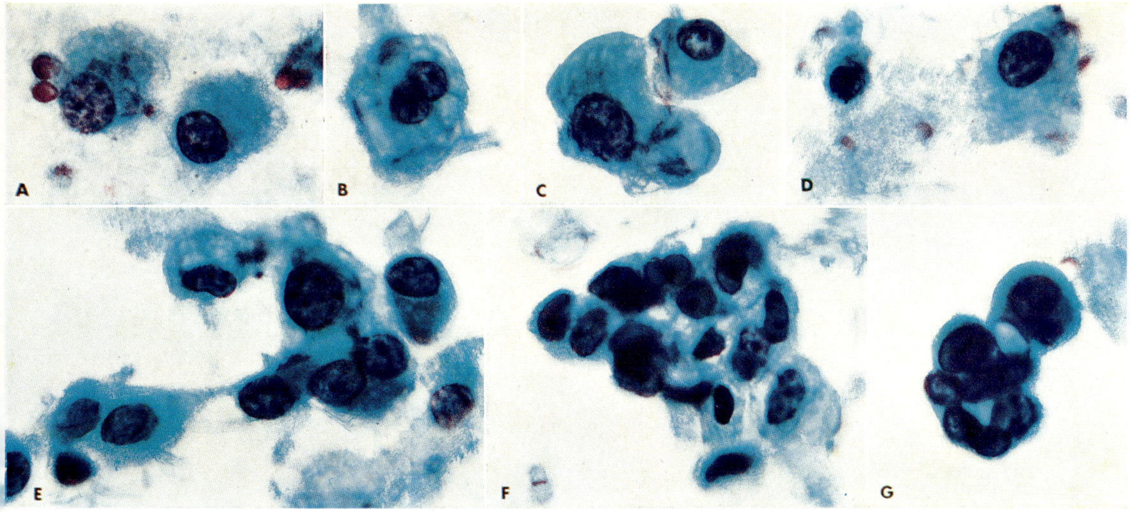

FIG. 141

Groups of Differentiated and Undifferentiated Malignant Cells together with Atypical Cells and Histiocytes in an Ascitic Fluid from a Patient with Metastatic Anaplastic Carcinoma

A. and B. These fields show histiocytes and erythrocytes which were present, in the background, throughout the specimen. Note the eccentric position of the histiocyte nuclei, their vacuolated cytoplasm, and the multinucleation of the histiocyte in field B. (×800.)

C. A histiocyte lies above an single abnormal cell. Although the larger cell has an eccentric nucleus its cytoplasm is opaque, its nucleus is enlarged and irregular in shape and its chromatin is unevenly distributed. Many of the features of this cell would arouse suspicion, but it is inadvisable to base a diagnosis on a single cell of this type. (×800.)

D. The larger cell on the right has a hyperchromatic nucleus with an abnormal chromatin pattern. Although its appearance is abnormal it is not sufficiently different from the histiocytes in field A to provide a basis for diagnosis on its own. (×800.)

E. F. and G. The examination of these cell groups readily provides cells showing characteristic features for a definitive diagnosis. The nuclei of these malignant cells display marked differences in size, shape and chromatin patterns. Abnormalities of chromatin pattern are seen in E and F, while prominent nucleoli can be identified in field F. In fields F and G the compact overlapping grouping of the cells is shown. All these cells have abnormal nucleocytoplasmic ratios; in some cells the nuclei are still surrounded by cytoplasm, but for other cells the cytoplasm only forms an indefinite background. (×800.)

Although single cells may arouse suspicion, groups of cells provide a more reliable basis for diagnosis. In this case subsequent pathological examination revealed an undifferentiated carcinoma of ovary with omental and peritoneal metastases.

It is only rarely possible to determine the site of the primary tumour on purely cytological grounds, but deductions based on a combination of the cytological findings with the site of the effusion and the clinical probabilities may suggest the most likely areas for further investigation to establish the site of the primary lesion.

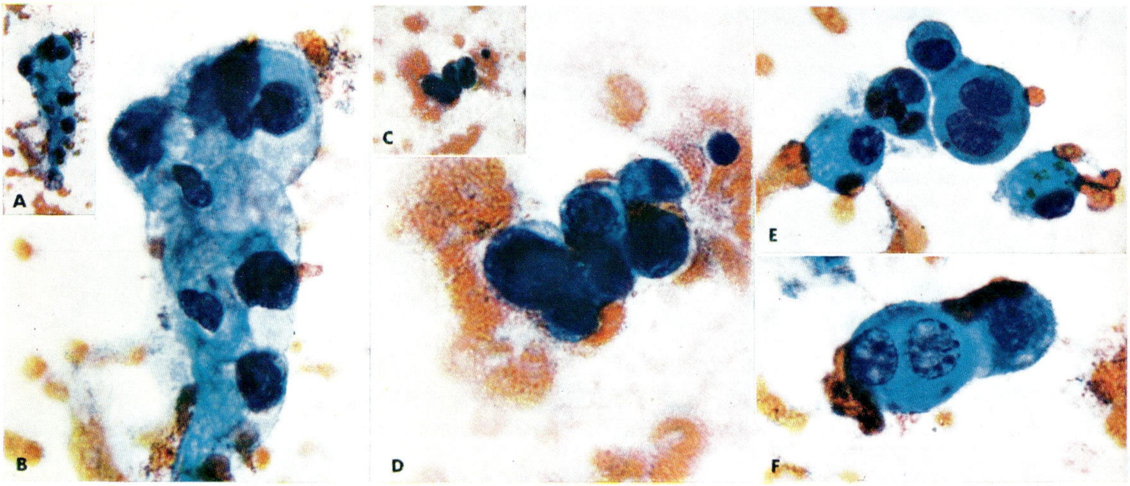

FIG. 142

Differentiated and Undifferentiated Malignant Cells, together with Mesothelial Cells and Histiocytes, in Pleural Fluid from a Case of Metastatic Renal Carcinoma

A. A large amount of clear cytoplasm is the most striking feature of this low-power view of a group of differentiated malignant cells. (×200.)

B. On high-power examination it is now seen that the group is composed of large round cells with abundant clear cytoplasm. The nuclei show marked variation in size and shape, with irregular nuclear outlines, and abnormal chromatin patterns. (×800.)

C. In contrast to the cells in field A this group of dark cells appears to be devoid of cytoplasm at low-power magnification. (×200.)

D. Higher magnification reveals that the dark nuclei of these undifferentiated malignant cells are similar to those of the differentiated malignant cells. Only a small amount of cytoplasm is visible and it fails to surround the nuclei. (×800.)

E. Mesothelial cells and histiocytes are shown for comparison with the malignant cells in fields A to D. Two mesothelial cells with round outlines and central nuclei are seen; the lower one is binucleate, with evenly matched nuclei. Three histiocytes with eccentric nuclei and foamy cytoplasm are present. One contains ingested nuclear debris, probably from a polymorphonuclear leucocyte. (×800.)

F. Mesothelial cells are again seen in this field; the binucleate one is large as are its nuclei. The nuclei, however, are evenly matched in size and shape and have smooth outlines. One shows a prominent nucleolus and both have rather coarse chromatin patterns. This type of 'active' mesothelial cell appears as part of a non-specific response to many stimuli and it retains sufficient of the essential characteristics of a mesothelial cell for identification. It does **not** have those specific qualitative differences from a mesothelial cell which malignant cells show and which make it possible to recognise that a second species of cell is present. (×800.)

Malignant cells in serous fluids frequently become de-differentiated. The degree of differentiation seen is not necessarily related to the differentiation of the primary tumour. In the case illustrated, however, different areas of the tumour did show two histological patterns. One pattern was highly differentiated and composed of large malignant cells with abundant clear cytoplasm. In other areas the tumour appeared to be anaplastic. The two histological patterns were unusually accurately mirrored in these cells seen in the pleural effusion.

ASPIRATES FROM CYSTS

When a cyst is aspirated, as a therapeutic procedure, the fluid should be examined cytologically as this may occasionally reveal unsuspected malignant change in a clinically benign lesion, but the examination of cyst fluid cannot exclude a malignant lesion, as the specimen may not be representative of the whole of the cyst epithelium. For this reason the recurring or rapidly enlarging cyst is more safely investigated by excision biopsy.

Malignant cells in cyst fluids are usually of adenocarcinomatous type and the appearance of the cells is similar to that of differentiated malignant cells of adenocarcinomatous type in effusions (see Fig. 128). Occasionally a cyst fluid may contain cells of adenomatous type as well as those of adenocarcinomatous type. When this occurs it is important to recognise the characteristic nuclear abnormalities before identifying cells as malignant.

THE CEREBROSPINAL FLUID AND THE BLOOD

The recognition of malignant cells in these two types of specimen demands technical skill in the collection and processing of specimens and considerable experience for the observer. Even when these conditions are fulfilled experienced observers may differ significantly in their interpretation of the cells seen, especially in the case of malignant cells in blood. For these reasons and because of the relatively limited clinical application of cytological methods in these fields, the reader is referred to the literature on these subjects.

SECTION VII

THE EFFECTS OF RADIATION

CHAPTER 20

RADIATION EFFECTS

An awareness of the effects of radiation on benign and malignant cells and of the changes produced by radiation in the overall pattern of smears is an essential part of diagnostic cytology for a variety of reasons. The first and the most important reason is the possibility that a failure to recognise these changes, so that benign cells with radiation effects are misinterpreted as malignant cells, can lead to a wrong diagnosis. The second reason is the necessity, in specimens from patients who have had radiotherapy for malignant disease, to distinguish the characteristic features of malignant cells from those of benign cells, showing radiation effects, and from those of malignant cells, destroyed by radiation, in order to diagnose residual or recurrent carcinoma. Lastly, changes described both in individual cells and smear patterns can be used in the assessment of the response of the patient to radiotherapy.

THE EFFECTS OF RADIATION ON CELLS

The first effect of radiation is the destruction of cells, both benign and malignant; this can be considered as a non-specific effect. The cellular features include nuclear pyknosis, with a densely hyperchromatic nucleus; karyorrhexis, with disruption of the nucleus into fragments; and karyolysis, in which the nucleus becomes opaque and structureless, then ultimately disappears. These changes occur similarly in benign and malignant cells and it is frequently impossible to determine the nature of a cell prior to destruction. This, however, is unimportant as cells showing these nuclear degenerative changes are all dead cells. Moreover, these changes are unlikely to be confused with the features of malignant cells, provided that the preservation of nuclear structure is retained as an essential criterion in the recognition of a malignant cell.

In cytological diagnosis the other responses to radiation present greater practical problems, as they are seen in cells which retain their nuclear and cytoplasmic structure. Following radiation one of the most striking features is the enlargement of benign cells and nuclei. Often this occurs together in both nucleus and cell, so that in benign cells a normal nucleocytoplasmic ratio is retained. In some benign cells, however, cellular enlargement is not accompanied by nuclear enlargement, so that the nucleus appears disproportionately small. In either case, a normal nuclear structure is retained. As a result of nuclear enlargement, wrinkling of the nuclear membrane may occur. Initially this may create the impression that a benign nucleus is irregular in shape and that strands of chromatin are present, but careful focusing will identify the folds and wrinkles. Multinucleation may be seen in benign cells after radiation, but the individual nuclei retain their normal chromatin patterns, although they may vary in size. Cytoplasmic changes are also striking following radiation, so benign cells of bizarre shape may occur. These can cause diagnostic difficulty if benign radiated cells of tadpole or fibre shape are confused with malignant cells of similar shapes, but this will not occur if the nuclei of the cells are carefully examined. The benign radiated cells retain normal nuclear structures, which contrast sharply with the abnormal features of the nuclei of malignant cells. Another cytoplasmic change is the appearance of vacuolation, which may be coarse or fine.

Although these effects are most striking and significant when they occur in benign cells, since these may then be confused with malignant cells, similar changes do occur in some malignant cells, which may show marked enlargement of cell and nucleus, multinucleation and cytoplasmic vacuolation. In those cells, however, which retain the diagnostic characteristics of malignant cells, the incidental presence of features which may be attributable to radiation is insignificant.

VAGINAL SMEAR PATTERNS AFTER RADIATION

In addition to recording detailed observations on individual cell responses during and after radiotherapy, Graham has stressed the different types of smear pattern which may occur. The recognition of these patterns is valuable in the diagnosis of residual and recurrent malignant lesions. Moreover, observations of cellular and smear pattern changes in gynaecological patients have been successfully used to assess the best method of treatment and to estimate the prognosis with alternative methods of treatment.

After radiotherapy the smear, which ultimately proves to be completely negative for malignant cells, may appear extremely abnormal and 'suspicious' on superficial examination. The epithelial cells show both the degenerative features, previously described, and also the response to radiation, with occasional bizarre cell shapes, cytoplasmic vacuolation, cellular and nuclear enlargement and multinucleation. The nuclei, however, retain their normal structure and chromatin pattern. The predominant epithelial cells are usually parabasal and intermediate cells, but complete atrophy is also seen, in which the smear consists of parabasal cells with an admixture of basal cells. The background to the epithelial cells consists of polymorphonuclear leucocytes and histiocytes, with a few erythrocytes. Histiocytes are usually present in large numbers and multinucleated giant forms are frequently seen. This marked histiocyte response is one of the most striking features of the post-radiation smear, which is negative for malignant cells.

It should be noted that some of these features may be seen in patients who have received comparatively small amounts of radiation, as, for example, in those cases where radiotherapy has been employed to procure an artificial menopause. In addition there appears to be wide individual variation both in the severity of the epithelial and non-epithelial cell changes seen following radiotherapy and in the persistence of these effects. Graham reports persistent changes thirty-five years after radiotherapy.

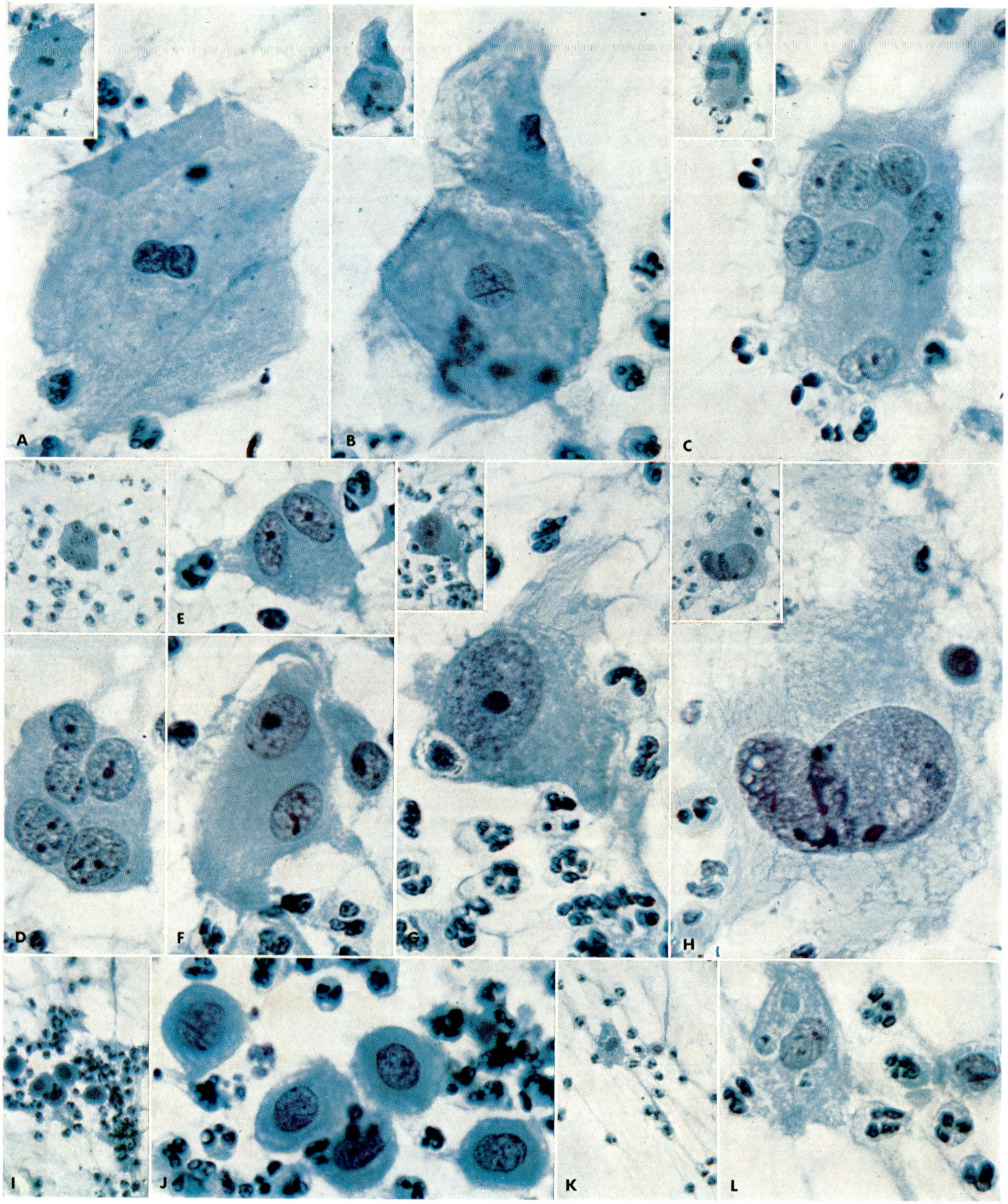

FIG. 143

The Effects of Radiation on Benign Cells in Cervical Smears

For detailed description see facing page.

FIG. 143

The Effects of Radiation on Benign Cells in Cervical Smears

A. The benign squamous cell, illustrated at low-power magnification (inset) is enlarged but its nuclei are of normal size. The high-power field shows the similarity between the two nuclei and the absence of abnormalities of chromatin pattern. (Inset ×200.) (×800.)

B. The nuclei of these two benign squamous cells are wrinkled. In the upper cell this gives the impression of an irregular nuclear outline, while in the lower cell a fold across the nucleus mimics the appearance of a strand of chromatin. These features, due to wrinkling and folding, can be distinguished from chromatin abnormalities by careful focusing. (Inset ×200.) (×800.)

C. D. E. F. K. and L. These fields illustrate the features of histiocytes seen in cervical smears after radiotherapy. Giant forms with high multinucleation are shown in fields C and D. Multinucleation is also present, but is less striking, in the cells in fields E and F. Although the nuclei of these histiocytes display considerable variation in size, especially in field D, they retain uniformly smooth round or oval outlines. Enlarged nucleoli are a prominent feature in the cells in fields D, E and F. Although the chromatin patterns of the nuclei are granular no significant abnormalities of chromatin arrangement are seen. Note the phagocytic activity of the histiocyte shown at low- and high-power magnifications in fields K and L. (Insets ×200.) (800.)

G. This benign squamous cell shows gross nuclear enlargement with a strikingly prominent nucleolus. The nuclear outline, however, is smoothly oval and the chromatin pattern normal. There is a loss of definition of the cytoplasmic border; this is probably a degenerative change, as a polymorphonuclear leucocyte is also seen within a vacuole in the cytoplasm. (Inset ×200.) (×800.)

H. This bizarre, but benign, squamous cell shows multinucleation, in which the two nuclei are enormously enlarged. The cytoplasm is both finely and coarsely vacuolated. The nuclei also show vacuolation, which is especially marked in the nucleus on the left of the cell. (Inset ×200.) (×800.)

I. On low-power examination the nuclei of this group of squamous cells appear to be enlarged, irregular and hyperchromatic. (×200.)

J. On high-power examination, however, it is apparent that these benign cells show the effects of radiation. Their cytoplasm is vacuolated and their nuclei enlarged. The irregular nuclear outlines and the impression of an abnormal chromatin pattern are the results of wrinkling and folding of the nuclei. (×800.)

These cells were seen in smears from two patients who had been treated by radiotherapy for carcinoma of the cervix, three years and one year previously. At the time the smears were taken there was no evidence of a recurrent or residual lesion. Both patients have remained well in the two years since these smears were taken.

In contrast, smears which contain malignant cells, persisting after radiotherapy, may appear remarkably 'normal' on first examination. In this type of smear the epithelial cells may be predominantly superficial and intermediate cells, which appear to be unaffected by radiation. Polymorphonuclear leucocytes may be seen, but histiocytes are not usually present in significant numbers. It is essential to screen this type of smear carefully, searching for small undifferentiated malignant cells; as these may occur singly and be obscured by polymorphonuclear leucocytes, high-power screening may be necessary for their identification. These undifferentiated malignant cells can be distinguished from the denuded nuclei of histiocytes or benign basal cells by their chromatin patterns. In practice it is usually found that these two types of benign cell, with which undifferentiated malignant cells might be confused, occur mainly in the other type of post-radiation smear which is actually negative for malignant cells, although it appears abnormal on first examination, because the benign epithelial cells show the effects of radiation.

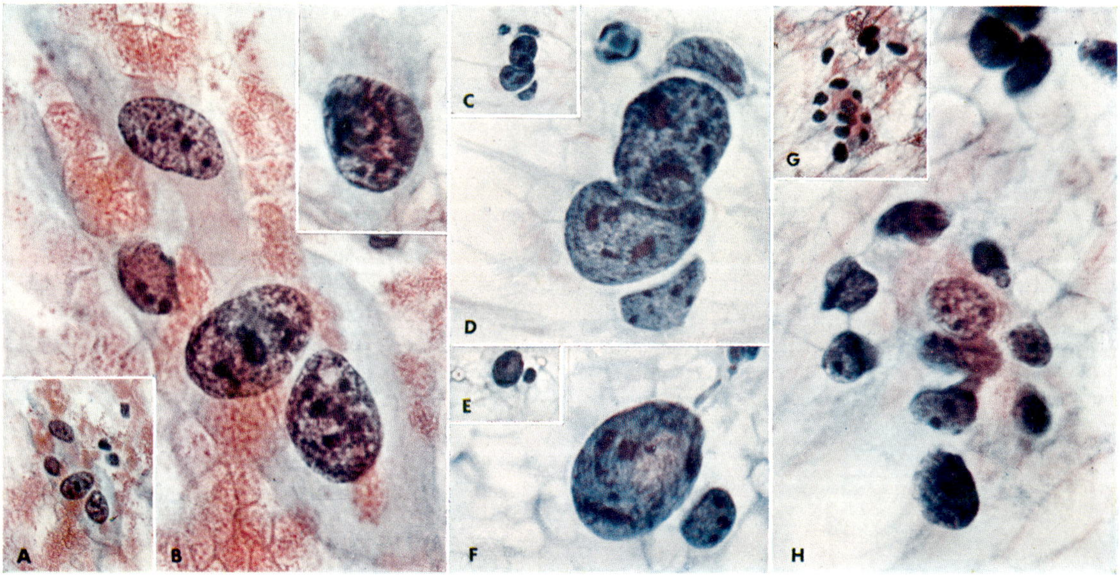

FIG. 144

Undifferentiated Malignant Cells in a Vaginal Smear from a Patient with a Recurrent Carcinoma of Cervix following Radiotherapy

A. Large dark nuclei are seen against a background of erythrocytes. (×200.)

B. The nuclei vary in size and have abnormal chromatin patterns with marked clumping of chromatin. The single nucleus in the inset shows similar features. Frequently this type of malignant nucleus occurs singly and requires high-power examination for recognition. (×800.)

C. A compact group of hyperchromatic nuclei is seen. (×200.)

D. These malignant nuclei show marked variation in shape and size, with abnormal chromatin patterns. Multiple angular eosinophilic nucleoli are present. There is striking moulding of adjacent nuclei. (×800.)

E. and F. These two malignant nuclei, seen at low-power magnification in field E, show marked variation in size and abnormal chromatin patterns. Multiple angular eosinophilic nucleoli are again present. (×200.) (×800.)

G. A large group of hyperchromatic nuclei is shown. (×200.)

H. In many of these nuclei, degenerative changes have occurred, which make them unsuitable for final diagnosis. In the centre of the group, however, a nucleus retains its abnormal chromatin pattern. The presence of degenerate hyperchromatic nuclei should lead to a further search for diagnostic cells. (×800.)

Smears from patients with a recurrent malignant lesion present rather a different picture. In addition to undifferentiated malignant cells these smears often contain differentiated malignant cells which may, occasionally, be the predominant malignant cells. Once again, the benign epithelial cells present usually show little evidence of the effects of radiation. Although differentiated malignant cells can be identified with certainty in this type of smear, it is inadvisable to use these cells alone as a basis for diagnosis, as benign squamous cells showing marked effects of radiation can readily be confused with differentiated malignant cells of squamous type. It is, therefore, wiser to search for the presence of undifferentiated malignant cells as a basis for final diagnosis.

THE PROGNOSTIC VALUE OF SMEARS

The significance of the finding of malignant cells in smears depends on the particular type of radiotherapy employed, on the time interval and on the type of malignant cell. The presence of malignant cells on completion of treatment is considered by Graham to be significant only when radium has been the primary method of treatment. Graham considers that the subsequent persistence of small undifferentiated malignant cells after radiotherapy signifies a complete failure to respond to radiotherapy, and, therefore, an indication for an alternative method of treatment. The reappearance of differentiated malignant cells and undifferentiated malignant cells after an interval following treatment is associated by Graham with an eventual deterioration in the condition of the patient. However, this deterioration may not occur for a considerable period of time.

In addition to observations on malignant cells in smears Graham has used counts on benign epithelial cells showing the effects of radiation (the radiation response or R.R.) and counts of untreated benign epithelial cells showing similar changes (the sensitisation response or S.R.) to assess the probable effects of radiotherapy on individual patients, and thus select the most suitable method of treatment. However, the application of these methods by other workers has not yielded such consistent results as those obtained by the Grahams.

An alternative method of assessment of the smear, as evidence of the response to radiation, has been suggested by Wachtel, who employs the karyopyknotic index. It is suggested that a smear with a high karyopyknotic index in a patient who has received radiotherapy (or radical surgery) for a genital tract neoplasm is an unfavourable prognostic sign (in the absence of complicating exogenous oestrogen or other therapy). A low post-therapy karyopyknotic index is considered a good prognostic feature in those cases where initially the karyopyknotic index was high. The value of the low karyopyknotic index is limited by the fact that it is not applicable to pre-menopausal women, because of their pre-treatment cyclical variation in karyopyknotic indcx, nor is it applicable to those post-menopausal women who do not show a high karyopyknotic index prior to treatment.

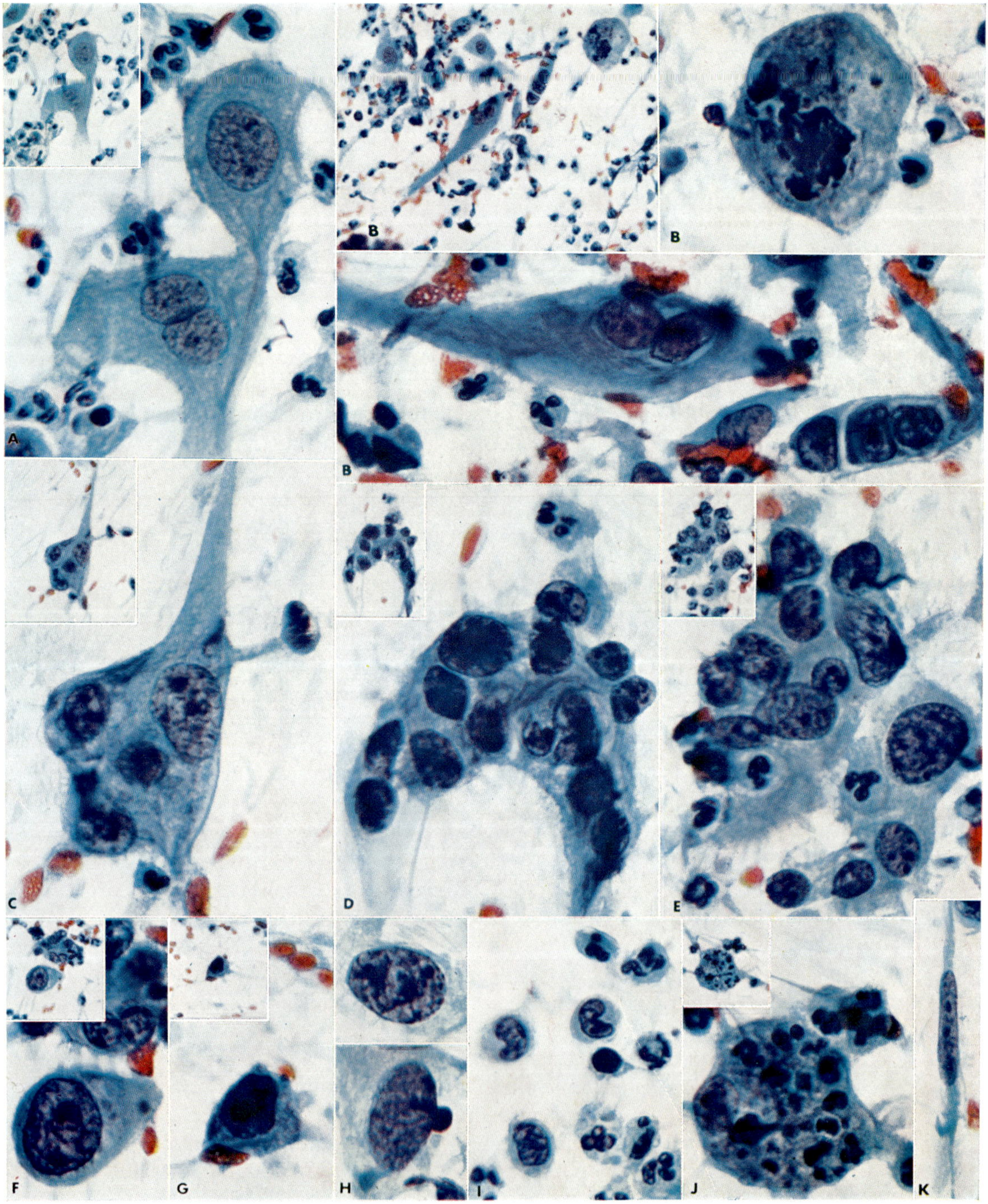

FIG. 145
The Effects of Radiation on Benign and Malignant Cells in a Cervical Smear
For detailed description see facing page.

FIG. 145

The Effects of Radiation on Benign and Malignant Cells in a Cervical Smear

A. These benign squamous cells show the characteristic effects of radiation. Their cytoplasm is vacuolated and the lower cell is abnormal in shape. The nucleus of the upper cell is enlarged and the lower cell shows multinucleation. All three nuclei, however, retain smooth oval outlines and normal chromatin patterns. (Inset ×200.) (×800.)

B. The low-power field at the upper left-hand side shows bizarre forms of benign and malignant cells, including two tadpole cells. In the lower high-power field the benign tadpole cell on the left has two nuclei surrounded by a vacuole. The nuclei are similar in size and shape, although there is some wrinkling of the border of the nucleus on the right. Although there are prominent nucleoli, no abnormality of chromatin pattern is seen in either of these nuclei. In contrast the tadpole cell on the right has three nuclei, which show the features of malignant nuclei; they vary in size and shape, have abnormal chromatin patterns and demonstrate moulding of adjacent nuclei. The single cell illustrated in the upper high-power field has undergone complete disruption of the nuclear structure. When this occurs it is not possible to determine whether this is a benign or a malignant cell, but this is not of any practical importance. (Inset ×200.) (×800.)

C. In this bizarre cell the individual nuclei vary in size and shape, multiple eosinophilic nucleoli are present and angular masses of chromatin are seen. Its appearance is compatible with a multinucleated differentiated malignant cell. (Inset ×200.) (×800.)

D. and E. In these groups of differentiated and undifferentiated malignant cells the cytoplasm is present mainly as a poorly defined background. The individual nuclei show striking variation in size and shape; their chromatin patterns are abnormal with irregular angular clumping of chromatin. (Inset ×200.) (×800.)

F. A single differentiated malignant cell is seen in the centre of the field. The high-power view below shows the large nucleus with abnormal nucleocytoplasmic ratio, the prominent nucleolus and the abnormal chromatin pattern with irregular strands and clumps of chromatin. (Inset ×200.) (×800.)

G. This single cell is similar to the cell shown in field F, but its nucleus is completely degenerate. It is, therefore, impossible to determine whether this is a benign or malignant cell; in practice this is unimportant in a completely degenerate cell. (Inset ×200.) (×800.)

H. These two fields contrast the appearance of benign and malignant nuclei. The upper field shows an undifferentiated malignant cell, in which there is a strikingly abnormal chromatin pattern with irregular angular clumps and strands of chromatin. In the benign enlarged nucleus in the lower field the chromatin is evenly distributed; the nipple-like protrusion of chromatin suggests that this is the nucleus of a columnar cell. (×800.)

I. This field illustrates the characteristic background of the smear with small histiocytes and polymorphonuclear leucocytes. (×800.)

J. In this degenerate cell numerous polymorphonuclear leucocytes are present in the cytoplasm. (Inset ×200.) (×800.)

K. A fibroblast is seen in this field with typical elongated cellular and nuclear form. Note the finely granular chromatin pattern which distinguishes this cell from the differentiated malignant cell of 'fibre-cell' type, which it resembles on superficial examination. (×800.)

This specimen was collected twelve weeks after radiotherapy for a vault recurrence of an adenocarcinoma of the uterine body, originally treated by total hysterectomy and bilateral salpingo-oophorectomy two years previously. When this specimen was obtained there was no clinical evidence of a malignant lesion. Three months later, however, smears again showed numerous undifferentiated malignant cells and a few differentiated malignant cells. By this time a friable mass of tumour was present in the upper vagina. During the subsequent four months the patient has required treatment for pain associated with extensive pelvic tumour spread. An interesting feature of these smears is the 'squamoid' appearance of some of the malignant cells, despite the fact that the original tumour was an adenocarcinoma. These bizarre cell shapes and tadpole forms represent a response to radiation.

SECTION VIII

PRACTICAL PROBLEMS IN DIFFERENTIAL DIAGNOSIS

CHAPTER 21

PRACTICAL PROBLEMS IN DIFFERENTIAL DIAGNOSIS

DIAGNOSTIC difficulties and errors can be subdivided into two main groups, although there is some degree of overlapping of the two categories. The first can be considered as extrinsic in origin, including artefacts produced during collection, preparation or staining of the specimen, and contaminants. The problems presented by this group are usually fairly readily resolved. The second group, which presents greater diagnostic difficulty, consists of those problems which can be considered to be intrinsic in the specimen. This includes non-cellular material, non-epithelial cells and, most important, those changes in benign epithelial cells which may cause confusion with malignant cells.

PROBLEMS OF EXTRINSIC ORIGIN

Drying Distortion

One of the commonest causes of diagnostic difficulty is distortion due to drying of smears before fixation. It is always essential to fix smears as rapidly as possible, to ensure fixation while the cellular material is still wet. Drying distortion is characterised by three features:—

(*a*) an appearance of 'ground glass' opacity. Initially the inexperienced observer may merely think that it is very difficult to get the specimen 'in focus';

(*b*) eosinophilic staining;

(*c*) apparent nuclear enlargement.

It is this last feature which is most likely to lead to diagnostic error by creating a false impression of an abnormal nucleocytoplasmic ratio. Any smear which shows extensive areas of drying distortion should, therefore, be regarded as unsatisfactory.

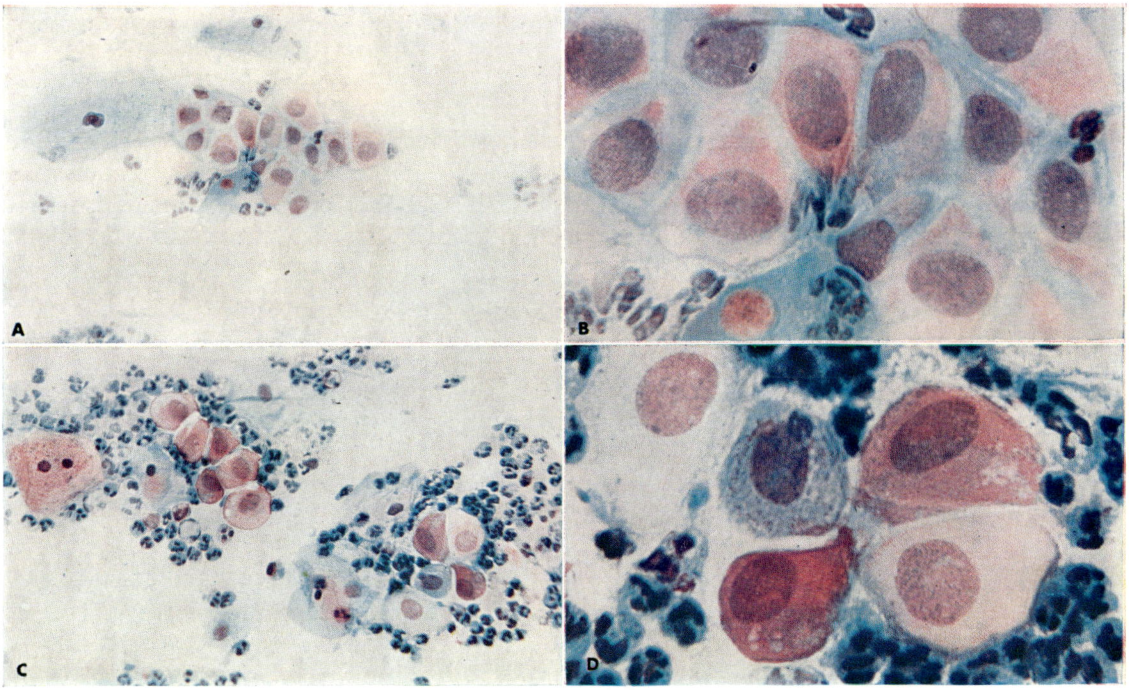

FIG. 146

Drying Distortion

A. Severe drying distortion. The field appears indistinct, as though seen through ground-glass. (× 200.)

B. At high-power magnification it is apparent that there is a loss of definition of nuclear and cellular borders. The nuclei appear to be enlarged, and the nuclear structure cannot be identified. The dried squamous cells show eosinophilic staining. Note the smudged appearance of the adjacent polymorphonuclear leucocytes. (× 800.)

C. and D. In this field from the same smear the drying distortion is less marked, but the eosinophilic staining and nuclear enlargement are still present. (× 200.) (× 800.)

When a smear shows drying distortion, i.e. a 'ground-glass' opacity with nuclear enlargement and eosinophilic staining, *the specimen is not suitable for diagnosis.*

Extrinsic Contaminants

Contamination of a specimen can cause difficulty in a variety of ways. Opaque material diffusely distributed on a smear may obscure cellular detail. This occurs when a cervical smear is contaminated by lubricant, used to facilitate the introduction of the speculum. A different type of difficulty arises when a contaminant is present only in small foci on the smear. This occurs when a smear contains glove powder, although its characteristic appearance is readily recognised. A greater problem is presented when the contaminant appears hyperchromatic on superficial examination. This can occur if small precipitates of stain are deposited on top of cellular material, but careful examination soon reveals the lack of structure of the hyperchromatic areas (see Fig. 152).

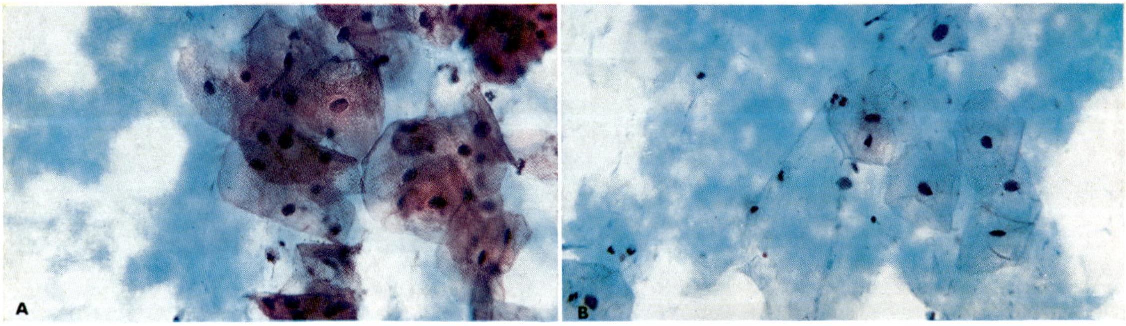

FIG. 147

Contamination of a Cervical Smear by Lubricant

A. The background is masked by amorphous basophilic material. Although eosinophilic staining of cytoplasm defines the cellular borders of the squamous cells, the nuclear detail is obscured. (× 200.)

B. These squamous cells, with basophilic cytoplasm, show poor definition of cellular borders and loss of nuclear detail due to contamination by a lubricant cream. (× 200.)

In view of the loss of cellular and nuclear detail caused by contamination by a lubricant, it is essential that smears should be taken prior to bi-manual examination and that the speculum should be inserted without the use of a lubricant.

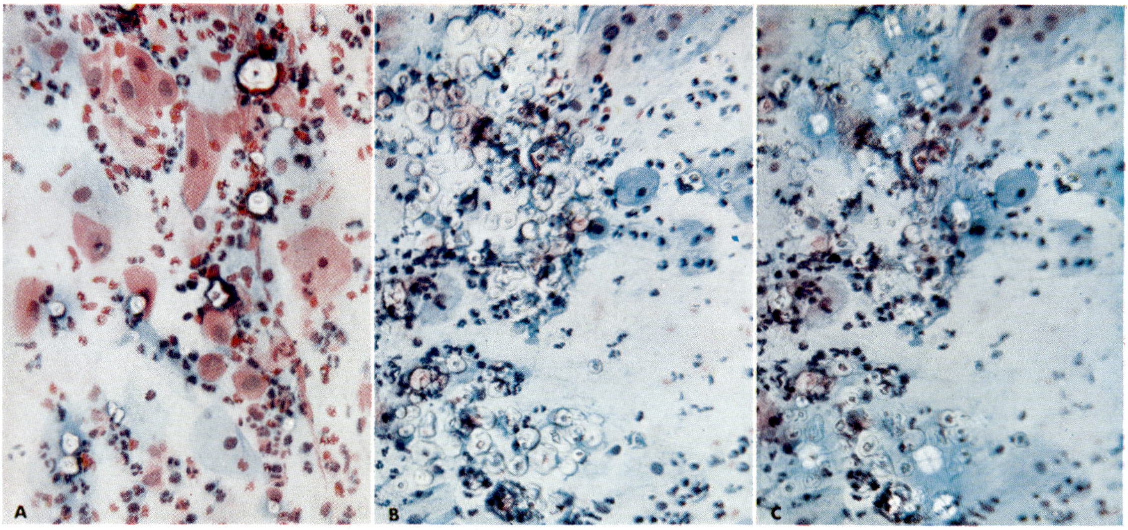

FIG. 148

Contamination of a Cervical Smear by Glove Powder

A. This area of a cervical smear shows two technical faults. During collection of the specimen it has been contaminated by glove powder and the specimen has also been allowed to dry prior to fixation. Starch granules are recognisable as refractile round or hexagonal structures, of variable size, in which the hilum is seen as a dot. The cells present show the characteristic features of drying distortion with an opaque ground-glass appearance, eosinophilic staining of the cytoplasm and apparent nuclear enlargement. (× 200.)

B. Another field from the same specimen again shows contamination with glove powder. Typical starch granules are present in large numbers, especially in the left half of the field. (× 200.)

C. On examination with polarised light the characteristic 'Maltese Cross' appearance of the birefringent starch granules shown in field B is demonstrated. (× 200.)

'Floaters'

This term is used to describe a particular type of contaminant. A 'floater' consists of cellular material which has been transferred from one smear to another, usually during the staining process. The recognition of a 'floater' is most important when it consists of malignant cells, which might lead to a 'false positive' diagnosis. The exact appearance of a 'floater' will vary according to the type of smear from which it originates and the type of smear to which it has been inadvertently transferred. In general, however, it will display some or all of the following features:—

(*a*) 'the floater' is in a different plane of focus from the majority of the cells of the smear;

(*b*) the 'floater' often shows a different staining reaction from the rest of the smear, because it has been exposed to a particular stain for a period of time before becoming attached to the next set of smears;

(*c*) the 'floater' may consist of cells which are quite alien to the rest of the smear. For example, if cells from a serous fluid contaminate a cervical smear they will be strikingly different from the rest of the smear, whether they are benign or malignant.

The possibility of a 'floater' should always be considered if only one group of abnormal or malignant cells is seen in the whole of a specimen. The risk of 'floaters' is greatest from fluid specimens; for this reason such specimens should be stained quite separately from cervical smears. In addition routine precautions with stain filtration help to minimise the risk.

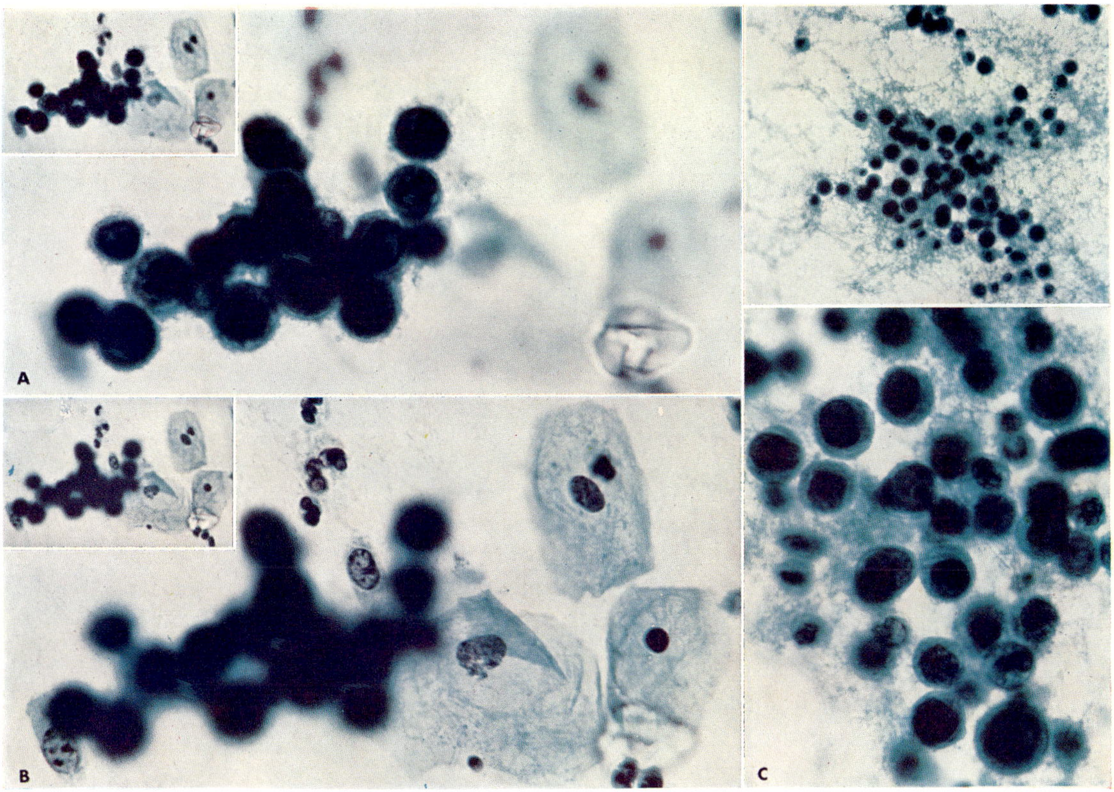

FIG. 149

Contamination of a Benign Cervical Smear by a 'Floater' of Malignant Cells from an Ascitic Fluid

A. and B. The low-power views (insets) show the striking appearance of a group of small malignant cells, with nuclear hyperchromasia and pleomorphism, against a field of benign superficial and intermediate squamous cells. In the high-power view in plate A the abnormal features of the malignant cells are confirmed, but it is now apparent that they are sharply in focus only when the squamous cells are out of focus. Examination of the high-power view in plate B confirms the impression that the malignant cells are in a different plane from the squamous cells by showing only their blurred outlines when the squamous cells are clearly in focus. (Insets × 200.) (× 800.)

C. Illustrates low- and high-power views of malignant cells from an ascitic fluid processed shortly before the cervical smear. It can be seen that these malignant cells closely resemble the cells in the malignant 'floater'. This patient had previously had a histologically proven carcinoma of ovary with multiple peritoneal metastases. (Inset × 200.) (× 800.)

This type of 'floater' in which the contaminant is obviously in a different plane from the specimen, shows over-staining and consists of material totally alien to the rest of the specimen and compatible with an origin from another contemporaneous specimen, is relatively easily diagnosed. However, where there is any similarity between the 'floater' and the basic specimen there may be difficulty in diagnosis. It is, therefore, inadvisable to make a definitive diagnosis of malignancy on a single cell or small group of cells without confirmatory evidence. On the other hand any patient whose specimens have shown the presence of malignant cells in what is apparently a 'floater' should be carefully investigated.

PROBLEMS OF INTRINSIC ORIGIN

Intrinsic Contaminants

The difficulties presented by this group are similar in many ways to those arising from the extrinsic contaminants. Thus in a heavily blood-stained cervical smear, the diffusely spread erythrocytes may hide the details of the epithelial cells. Again focal problems may occur from the presence of foreign cellular material in the actual specimen, as is the case with food contamination of a gastric lavage specimen or a specimen of sputum (see Fig. 87). Non-cellular material, too, as with the extrinsic contaminants, can cause difficulty. Examples of this are seen in the suspicious appearance, which may be produced by plugs of mucus or aggregates of micro-organisms in sputum (see Figs 84 and 85). Most problems of this type can be resolved by careful high-power examination which will reveal the typical features of foreign cellular material and the structurelessness of non-cellular material.

Non-epithelial Cells

The non-epithelial cells, which are most often confused with malignant cells, are histiocytes. However, on occasion lymphocytes and plasma cells may give rise to diagnostic difficulty.

Histiocytes may be misdiagnosed as malignant cells, arising from an adenocarcinoma. This occurs because both histiocytes and malignant cells from an adenocarcinoma have eccentric nuclei with granular chromatin patterns and nucleoli. Careful scrutiny, however, shows that the histiocyte chromatin pattern is normal, without any clumping or condensation of chromatin. Frequently, too, examination of a group of cells will reveal some histiocytes with the typical reniform nuclei (see Figs. 11 and 52. Compare Figs. 102 D and E with Fig. 104.). Occasionally when the cytoplasm of the histiocytes shows eosinophilic staining, it may also appear more opaque and less strikingly vacuolated. When this occurs, histiocytes may mimic differentiated malignant cells of squamous type, but the absence of chromatin pattern abnormalities distinguishes the histiocytes from the malignant cells.

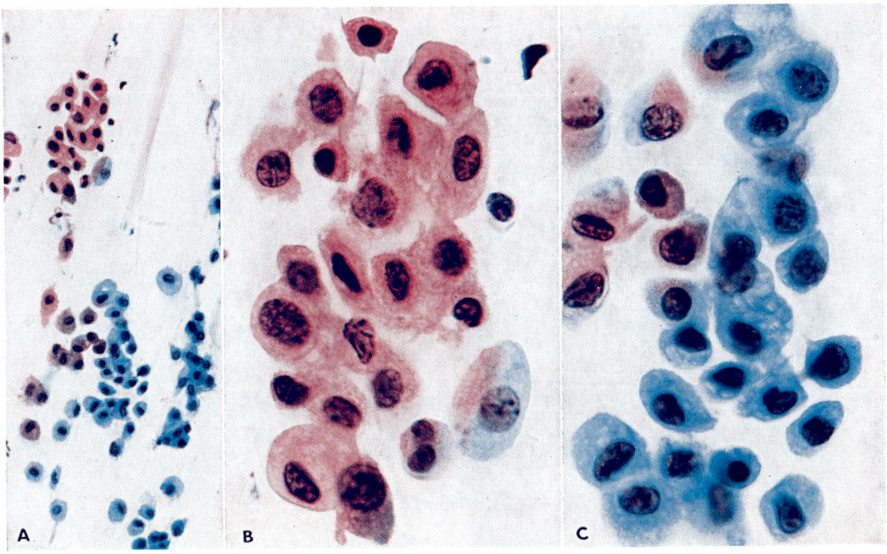

FIG. 150

Histiocytes with Eosinophilic Cytoplasmic Staining in a Benign Cervical Smear

A. On low-power examination, the group of eosinophilic cells in the upper half of the field shows marked nuclear pleomorphism and hyperchromasia. There is a superficial similarity to parabasal cells showing dyskaryosis, or even remotely to small differentiated malignant cells of squamous type. (× 200.)

B. and C. High-power examination, however, reveals the essential similarity between these histiocytes with eosinophilic cytoplasm and those with basophilic cytoplasm. Foamy, finely vacuolated cytoplasm and occasional, typically reniform nuclei, with rather coarse chromatin patterns, can be identified in both groups. In the eosinophilic group, several degenerate histiocytes with pyknotic nuclei are seen; on low-power examination these contribute to the abnormal appearance of the group. (× 800.)

When only the nuclei of histiocytes, denuded of cytoplasm, are seen, they may superficially resemble undifferentiated malignant cells, but again the occasional reniform outline and the absence of chromatin pattern abnormalities serve to identify the histiocyte nuclei (compare Fig. 129 with Fig. 103). Degenerate histiocytes with densely hyperchromatic nuclei will only cause diagnostic difficulty if insufficient attention is paid to the necessity to see a recognisable nuclear chromatin pattern before making the diagnosis that a cell is malignant.

Lymphocytes and plasma cells cause diagnostic confusion only in comparatively limited areas. Lymphocytes may be a problem in sputum specimens and in serous fluids. In sputum the main hazard is that small undifferentiated malignant cells (oat cells) may be overlooked if they are thought to be lymphocytes. Although the main distinction lies in the abnormal chromatin patterns of the malignant cells, there are other points which may be helpful. The undifferentiated malignant cells have sharply defined nuclear borders, which tend to give them a punctate appearance, even on low-power examination. Moreover groups of small undifferentiated malignant cells usually show little admixture of inflammatory cells, while the lymphocytes are frequently found in groups with small histiocytes (compare Fig. 83 with Fig. 102). In fluids lymphocytes are again likely to be confused mainly with undifferentiated malignant cells, from which they can be distinguished by the uniformity of their round nuclear outlines and their lack of an abnormal chromatin pattern. Plasma cells are only likely to give difficulty, if the possibility of their presence is overlooked, as their characteristic eccentric 'cart-wheel' nuclei are readily recognised (see Fig. 14). If they are misinterpreted it is likely to be in a gastric lavage specimen from a case of chronic gastritis, in which they may be mistaken for malignant cells, or in a cervical smear, in which a group of plasma cells is most likely to be misinterpreted as a group of endometrial cells, to which undue significance may then be attached in a post-menopausal patient.

Epithelial Cells

Confusion between benign and malignant epithelial cells arises in three ways. The first occurs when insufficient attention is paid to the essential nuclear characteristics for the diagnosis of a benign or malignant cell.

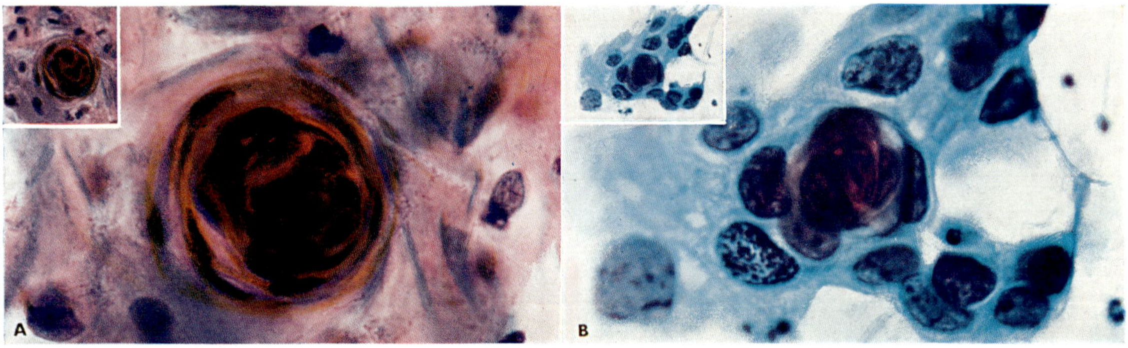

FIG. 151

Epithelial 'Pearl' Formation by Benign and Malignant Squamous Cells in Cervical Smears

A. The low-power illustration (inset) shows a group of benign intermediate squamous cells with 'pearl' formation in the centre. In the high-power field it can be seen that the nuclei of the cells composing the pearl, although degenerate, are small and fairly uniform in size. The vesicular nuclei of the squamous cells in the background have normal chromatin patterns. (Inset × 200.) (× 800.)

B. Even on low-power examination (inset) the irregular arrangement of this epithelial 'pearl' of malignant cells can be seen. The adjacent cells have large hyperchromatic nuclei. On high-power examination the 'pearl' consists almost entirely of nuclei, which vary in size and shape and show considerable moulding of adjacent nuclei. The cells in the background show the variation in nuclear size, shape and chromatin pattern characteristic of differentiated malignant cells of squamous type. The nucleus immediately to the left of the 'pearl' is degenerate and shows karyorrhexis. (Inset × 200.) (× 800.)

As a result of this, cells may be identified as malignant cells of squamous type because they have bizarre shapes (e.g., tadpole forms, see Fig. 31) or unusual staining reactions (e.g., brightly eosinophilic inflammatory parabasals, see Fig. 18). Alternatively, the loss of cytoplasm may cause the observer to identify benign cells as undifferentiated malignant cells (e.g., denuded endocervical cell nuclei, see Fig. 20. Compare these nuclei with those of the undifferentiated malignant cells seen in Fig. 34. See also the denuded benign and malignant nuclei in Fig. 49.). The second difficulty arises when nuclei exhibit degenerative changes. Once these are recognised in a nucleus the cell is unsuitable for diagnosis. However, often the dense structureless hyperchromatic appearance of a benign, but pyknotic nucleus may be incorrectly identified as a malignant change (see Fig. 45 and 46). Conversely, degenerative changes in malignant cells with obviously bizarre nuclei may make them unsuitable for definitive diagnosis (see Fig. 44).

The last group of diagnostically difficult epithelial cells are those in which the nuclei are 'active', often enlarged, with prominent nucleoli and coarsely but uniformly granular chromatin patterns (see Fig. 90). Usually these benign cells can be distinguished from malignant cells quite readily by the absence of chromatin pattern abnormalities in the benign cells. Occasionally, however, if doubt arises, it can only be resolved by repeating the investigation after treatment or time have remedied the condition which gave rise to the 'active' appearance of the nuclei. Such repetition, although often serving only to demonstrate a progressive return of the cellular pattern within normal limits, will occasionally reveal progression to a malignant lesion.

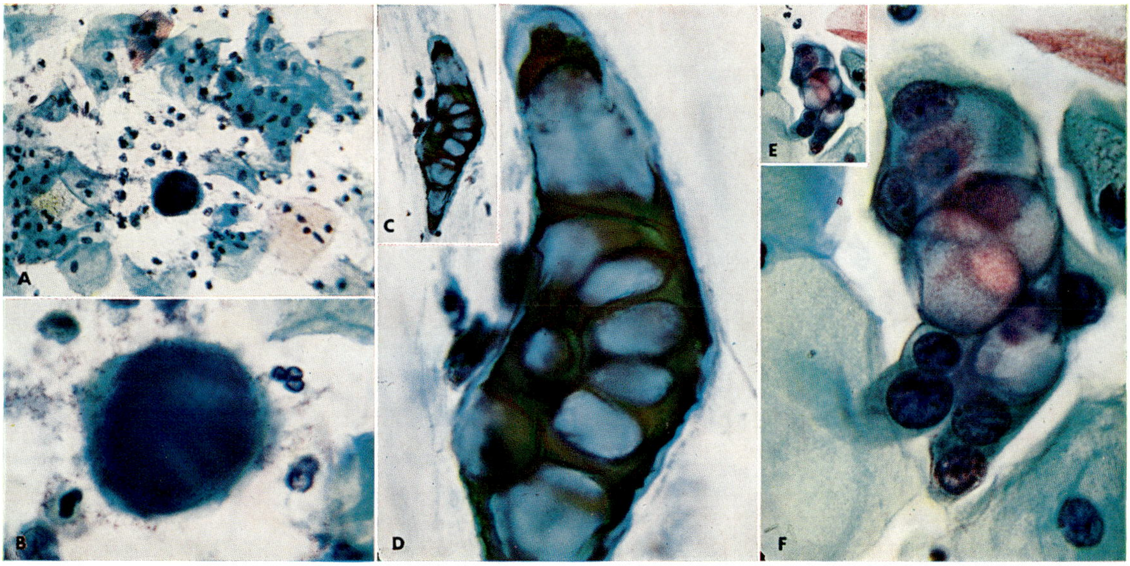

FIG. 152

Three possible sources of a false 'Positive' Diagnosis of Malignant Cells

A On low-power examination a large 'cell' with a hyperchromatic nucleus and a narrow incomplete rim of cytoplasm appears to be present among the squamous cells and polymorphonuclear leucocytes in this cervical smear. This 'cell' is similar to a large undifferentiated malignant cell. (× 200.)

B. Closer scrutiny, at high-power magnification, reveals that the dark material is not cellular. The dense basophilic 'nucleus' shows no internal structure. The whole of this material is amorphous. This is an artefact formed by a deposit of incompletely dissolved stain. (× 800.)

C. On low-power examination this specimen of sputum appears to contain a large group of cells with eccentric nuclei and vacuolated cytoplasm. The appearance resembles a group of malignant cells from an adenocarcinoma. (× 200.)

D. High-power magnification, however, again fails to show any cellular structure. The material consists of mucus, which has absorbed the stain. (× 800.)

E. On low-power examination of this cervical smear, a group of cells is seen in which the nuclei are eccentric in position and variable in size, while the cytoplasm is vacuolated. Superficially it resembles a group of cells originating in an adenocarcinoma. (× 200.)

F. Careful examination at high-power magnification confirm that this is cellular material. The nuclei do vary in size but are uniformly oval in shape. The hyperchromatic appearance of some nuclei is in fact, due to degeneration. The other nuclei have finely stippled uniform chromatin; the arrangement of the nuclei follows a linear pattern. This is a group of benign endocervical cells showing degenerative changes associated with inflammation. (× 200.)

There are many possible sources of a false 'positive' diagnosis. Three common types are illustrated; contamination by hyperchromatic non-cellular material from an outside source, hyperchromatic staining of non-cellular material within the specimen and hyperchromasia due to degeneration of cellular material. In every case the risk of error can be minimised by insisting on the recognition of well-preserved cellular and nuclear characteristics of malignant cells before accepting cells as a basis for diagnosis.

APPENDIX IX

THE ORGANISATION OF A CYTOLOGY SERVICE

CHAPTER 22

GENERAL PRINCIPLES

DIAGNOSTIC cytology depends on the fact that cells exfoliated or collected from a tissue surface reflect changes occurring in the underlying tissue. However, the correct interpretation of changes seen in these cells requires a knowledge of the relevant clinical features and pathological processes. Unless diagnostic cytology is carried out against this background of information, the method may prove not only unreliable but also positively misleading. It is, therefore, essential that clinical medicine and pathology form part of the basic training of the cytopathologist and that his work should be closely integrated with the findings of colleagues in these other disciplines, so as to provide the greatest benefit to each individual patient.

A valid cytological diagnosis can only be obtained if the following conditions are fulfilled:—

(*a*) The lesion must be accessible, e.g. the direct collection of cellular material must be possible, or it must communicate with a hollow tube which extends on to the body surface, or it must encroach on a body cavity, or it must be explored by aspiration technique. The fact that the lesion is inaccessible accounts, for example, for the frequent negative results in small peripheral carcinomata of lung, and for the occurrence of negative cervical and vaginal smears in the presence of ovarian carcinomata.

(*b*) The lesion must exfoliate cells or be accessible to an abrasion technique. An example of false negative results for this reason may be encountered in some early cases of carcinoma of stomach where comparatively little exfoliation of abnormal cellular material occurs.

(*c*) The cells must exhibit diagnostically abnormal features. When the tumour cells are so highly differentiated that the pathologist's diagnosis of malignancy is essentially a histological one, as may be the case in carcinoma of prostate and carcinoma of endometrium, the individual cells may show insufficient abnormality for cytological diagnosis.

(*d*) The method and site of specimen collection selected should be those most likely to yield diagnostic cellular material. For example, cervical scrape smears will yield better results in the diagnosis of carcinoma of the cervix than vaginal smears.

(*e*) The specimen must be suitably fixed and stained, as the cytopathologist depends on morphological detail for diagnosis. Drying, prior to fixation, produces cellular distortion, which can make a specimen completely valueless for diagnosis. The staining method chosen should be one which clearly demonstrates morphological detail, especially in the nucleus.

(*f*) The stained preparation must be critically examined in its entirety. Although it may be possible, with an occasional specimen, to make a positive diagnosis after a rapid scrutiny, a negative report cannot be given until the whole specimen has been examined. Even with a specimen which shows malignant cells, further examination may yield additional information on tumour type and differentiation.

(*g*) A final assessment of a specimen should only be made in the light of all available relevant clinical and pathological information.

If these criteria are to be applied successfully the cytopathologist must co-operate with the clinician and pathologist to establish satisfactory patterns for the following stages in the investigation and ultimate treatment of patients:—

(*a*) the selection of patients for cytological investigation;
(*b*) the provision of relevant clinical information;
(*c*) the collection of specimens and preparation of smears;
(*d*) the fixation and staining of specimens;
(*e*) the subdivision of work in the cytological laboratory;
(*f*) the provision of complete 'screening' of all specimens;
(*g*) the evaluation of the cytological findings in the light of clinical information;
(*h*) reporting the specimen in a way which contributes to the clinical management of the patient;
(*i*) planning with the clinician the further investigation of the patient;
(*j*) comparing the cytological findings with the pathological findings when a biopsy is performed;
(*k*) assessing with the clinician the extent to which cytological investigation could play a part in the follow-up of the patient.

CHAPTER 23

THE COLLECTION AND PREPARATION OF SPECIMENS

THE SELECTION OF PATIENTS FOR INVESTIGATION

PATIENTS can be subdivided into two groups for investigation by cytological methods. The first group, the symptomatic patients, consists mainly of those in whom suspicious symptoms, signs or laboratory tests warrant further investigation in the search for a malignant lesion. However, this symptomatic group also includes patients with benign disorders, in which cytological investigation can contribute additional information. Examples in this second category include patients with endocrine disorders and patients with allergic conditions. The selection of symptomatic patients lies with the clinician; the cytopathologist can only help by suggesting additional approaches to the problem, which may become apparent, after an initial specimen has been examined. It should not be forgotten that the cytopathologist may assist the clinician even when it is known that the patient has a carcinoma. In the untreated case cytology may help in localising and in typing the tumour. In the treated case it may be used to assess the adequacy of surgical excision or the response to radiotherapy or to cytotoxic drugs. In the case of the patient who has completed treatment, cytology can play a valuable part in follow-up and may lead to the early recognition of residual or recurrent disease. In all these symptomatic patients, cytological investigation is a diagnostic procedure.

In the second group, the asymptomatic patients, the role of cytology is somewhat different. This group is, of course, far larger and careful selection of lesions and patients is essential if available laboratory facilities are to be used to the greatest advantage. In general, this means the selection of lesions which are readily accessible for investigation, such as carcinoma of cervix, and of high-risk groups of patients, such as workers in the dye and rubber industry for investigation for carcinoma of the bladder. In these asymptomatic patients cytology is essentially a detection technique.

THE COLLECTION OF SPECIMENS AND PREPARATION OF SMEARS

Although techniques may require modification for special problems, it is advisable to use standardised methods for common procedures. In all of them the same general principles apply: collection from as near the site of the suspected lesion as possible and rapid fixation.

THE FEMALE GENITAL TRACT

The techniques of specimen collection available include the vaginal pool aspirate (collected either by a doctor or by the patient herself), the lateral vaginal wall smear, the cervical scrape smear, the endocervical smear and the endometrial aspirate. The most important practical application of cytology in this field is the early diagnosis of carcinoma of the cervix. For this purpose the method of choice is the cervical scrape smear taken from the squamo-columnar junction.

Before any specimen is collected the patient should be advised to avoid intercourse and douching for twenty-four hours. When, in well women, the timing of the smear is a matter of choice, it is preferable to collect the specimen at or around the middle of the cycle and to avoid the days of menstruation. In symptomatic patients, however, bleeding is *not* a contra-indication to the collection of specimens.

An adequate clinical history must accompany each specimen and should include the following information about the patient:—age, civil status, occupation of patient and husband, parity, date of first and last pregnancies (including, if there has been a recent pregnancy, information on lactation), menstrual history including cycle and date of last menstrual period or menopause, presenting symptom and clinical diagnosis (with especial care to record suspicious symptoms such as intermenstrual, post-coital or post-menopausal bleeding), and a note of previous or current treatment, especially hormonal therapy, including the contraceptive 'pill', radio-therapy and gynaecological surgery. The results of any previous relevant cytological or pathological investigation should also be recorded.

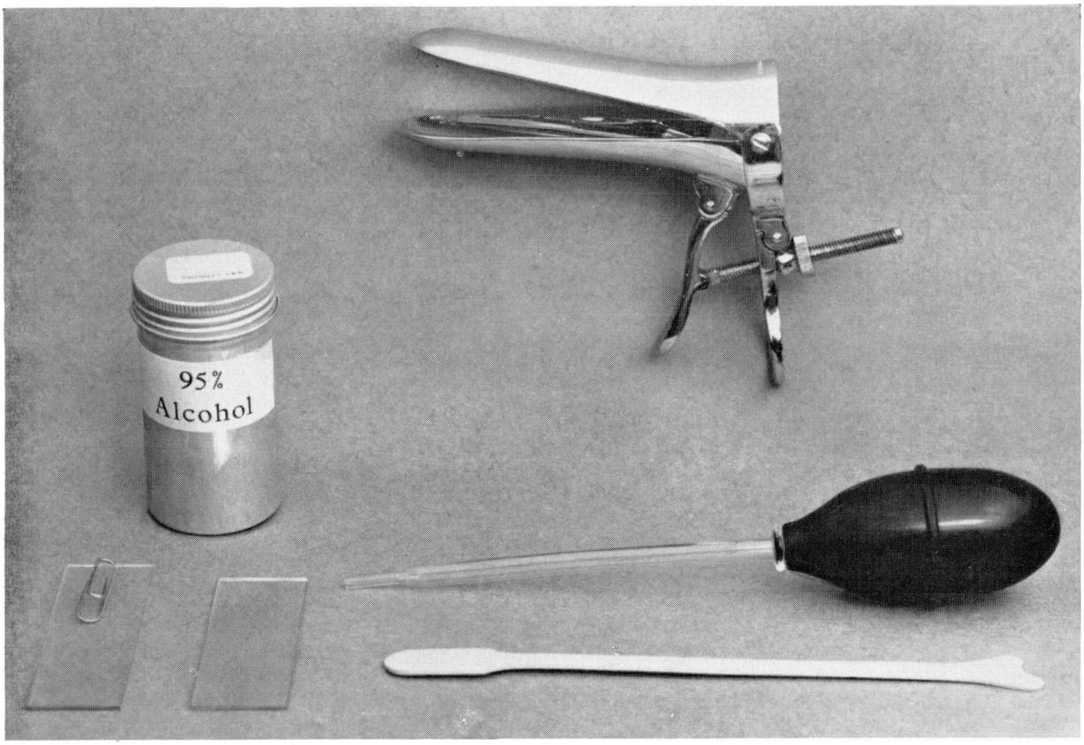

FIG. 153

Equipment for the Collection and Fixation of a Vaginal Pool Aspirate and Cervical or Vaginal Scrape Smears

The equipment includes a bivalve speculum for visualisation of the cervix and/or posterior fornix, a pipette and suction bulb for aspiration of the pool in the posterior fornix, an Ayre's spatula for the cervical scrape smear technique (its rounded end may also be used for the collection of lateral vaginal wall smears), glass slides, which may be kept apart in the fixative by a paper clip, and a container of fixative.

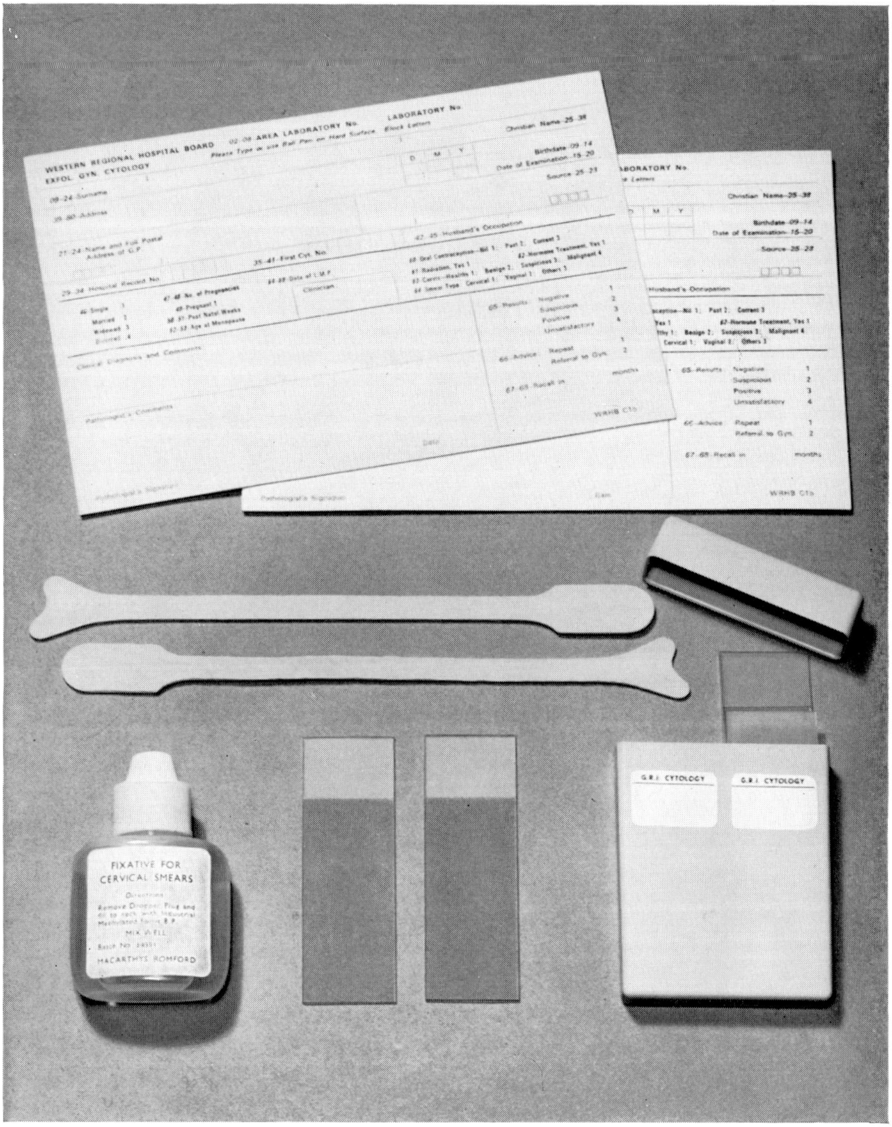

FIG. 154

Equipment for the Collection and Fixation of Cervical Smears for despatch by post

This standard kit can be sent by post to a medical practitioner by the laboratory. It includes Ayre's spatulae for the collection of cervical or vaginal smears, frosted-end glass slides, which can be labelled with a lead pencil, a plastic bottle containing fixative concentrate which will provide a protective film over the smears (to be diluted with 95 per cent alcohol), a plastic container to hold two slides from each of two patients, and appropriate laboratory request forms. The doctor provides a speculum for the visualisation of the cervix.

ASPIRATION OF THE POOL IN THE POSTERIOR FORNIX OF THE VAGINA

Advantages

(*a*) This method yields better results in the diagnosis of carcinoma of the uterine body and carcinoma of the vagina than cervical smears, since the pool contains cellular material exfoliated from the endometrial cavity or vaginal surface over the preceding twenty-four to forty-eight hours, as well as cervical cellular material;

(*b*) an assessment of endocrine function can be made on this type of specimen;

(*c*) the method can be used in those patients in whom it is impossible to pass a speculum to visualise the cervix;

(*d*) since collection is possible without passing a speculum, less highly trained staff can carry out the procedure. The method has also been modified to permit collection of the specimen by the patient (q.v.).

Disadvantages

(*a*) When a cervical lesion is present the vaginal pool aspirate contains a far smaller number of abnormal cells than the cervical smear and may give 'false negative' results in carcinoma of the cervix;

(*b*) some of the cellular material present may show degenerative changes;

(*c*) for the assessment of endocrine function the specimen is less satisfactory than the lateral vaginal wall smear, because of contamination with cervical cellular material.

Technique

Aspiration can be performed after the introduction of a speculum or as a 'blind' procedure, but it should be done before, not after, bimanual examination. The preliminary introduction of an unlubricated speculum allows collection, under direct vision, of cells from the posterior fornix pool. The cellular material is obtained by aspiration with a glass or plastic pipette and a suction bulb. When a speculum is not employed the pipette is gently introduced into the vagina, until resistance is encountered. It is important to compress the suction bulb during the introduction of the pipette to avoid collecting cellular material of lower vaginal origin. After the specimen has been obtained the contents of the pipette are blown on to a clean glass slide and the pipette can then be used to spread the cellular material by gentle rolling. *Immediate* fixation of the smear, while the material is still moist, is essential. The patient's name, which has been confirmed at the time of specimen collection, is recorded on the slide, either with a lead pencil on frosted-end slides or with a diamond pencil on plain slides.

PATIENT COLLECTION OF VAGINAL SMEARS

The vaginal aspiration technique has been modified so that the patient herself can collect the specimen.

Advantages

(*a*) Patient acceptance of this method of investigation in population surveys has been far higher than with other methods;

(*b*) the method is suitable where distance or other local factors make it impossible for the patient to attend a doctor.

Disadvantages

(*a*) The method shares the disadvantages of any vaginal aspiration technique;

(*b*) failure to collect a satisfactory specimen occurs more often than with other methods;

(*c*) more technical work is required in the laboratory to process the specimen;

(*d*) even when the specimen is correctly collected and processed the number of 'false negative' results in the diagnosis of carcinoma of the cervix is larger than with other methods.

Technique

The patient is supplied with a plastic pipette and bulb containing fluid*, and a set of detailed instructions, indicating the time in the cycle most suitable for specimen collection, the importance of avoiding intercourse and douching in the previous twenty-four hours and also a form for clinical particulars, which must be completed and returned with the specimen. The patient is instructed that she should introduce the pipette into the upper vagina and then compress and release the bulb, so that cellular material suspended in the solution is finally aspirated into the bulb. On arrival at the laboratory, the specimen is centrifuged, smears are then prepared from the deposit and fixed. Assessment of each specimen must include an estimation of whether sufficient material is present for diagnostic purposes.

THE LATERAL VAGINAL WALL SMEAR

Advantage

This specimen is the most suitable one for the assessment of hormonal function. It is less likely to be contaminated by cervical, endometrial and non-epithelial cells than the posterior fornix aspirate.

Disadvantages

(*a*) The method is *not* suitable for the diagnosis of malignant lesions of the female genital tract;

(*b*) trained staff are necessary to pass a speculum to visualise the appropriate area; experience is necessary to obtain a standardised technique for the collection of serial specimens.

Technique

After an unlubricated speculum has been introduced, the lateral vaginal wall, at the level of the cervix (or in the upper third of the vagina, if there is cervical descent) is scraped with a wooden spatula. The cellular material is rapidly but gently smeared on clean glass slides and the smears are fixed and identified in the routine way.

THE CERVICAL SCRAPE SMEAR

Advantages

(*a*) This is the *method of choice for the diagnosis of carcinoma of the cervix*, since cells are collected directly from the squamo-columnar junction;

(*b*) as the cellular material is fixed immediately after it is removed from the underlying epithelium there is good preservation of cellular and nuclear morphological detail.

*Physiologic salt solution containing 20% ethyl alcohol, adjusted to slightly alkaline pH against phenol red. Davis, H. J., Acta cytol. 6. 459, 1962.

Disadvantages

(*a*) The cervix must be visualised. This means that trained staff are necessary;

(*b*) the cellular material obtained comes from a limited area and, therefore, this method is not so likely to yield good results in the diagnosis of carcinoma of the vagina, endometrium, tube or ovary;

(*c*) the specimen is unsuitable for the assessment of endocrine function.

Technique

The specimen should be collected prior to bimanual examination. An unlubricated speculum is passed to visualise the cervix (if a lubricant is necessary water may be used). A wooden spatula is used to scrape material from the squamo-columnar junction. Ayre's spatula is designed for this purpose with a bifid end for the small round nulliparous cervical os and a broad rounded end for the parous cervix. In collecting the specimen, care should be taken to obtain material from the whole circumference of the cervix. If an excessive amount of mucus or discharge obscures the external os, it may be removed with the spatula before collecting the specimen, but the cervix should not be dried or cleaned with swabs, as this often means that the subsequent smears are dry and scanty. The cellular material should be rapidly spread on two clean glass slides, taking care that material from each side of the spatula is present on each slide. The smears should be fixed *immediately* and identified in the way described for vaginal smears.

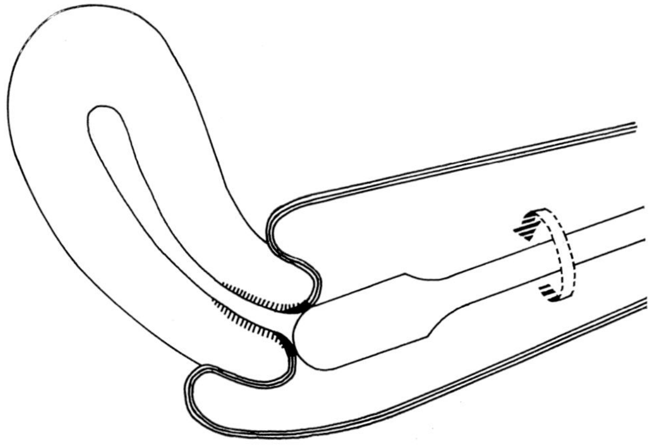

Fig. 155

The Collection of a Cervical Scrape Smear

The essential points illustrated in the collection of this specimen are:—

(*a*) the spatula is introduced into the cervical os to ensure collection of cellular material from the squamo-columnar junction;

(*b*) material is collected from the whole circumference of the os.

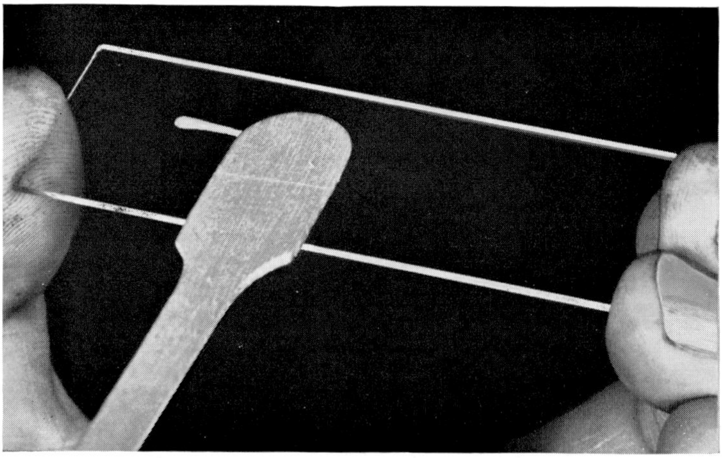

FIG. 156

The Technique of Smear Preparation—Incorrect

If the spatula is held at a right angle to the slide during smear preparation, the cellular material is distributed in streaks of considerable density rather than as an evenly spread smear.

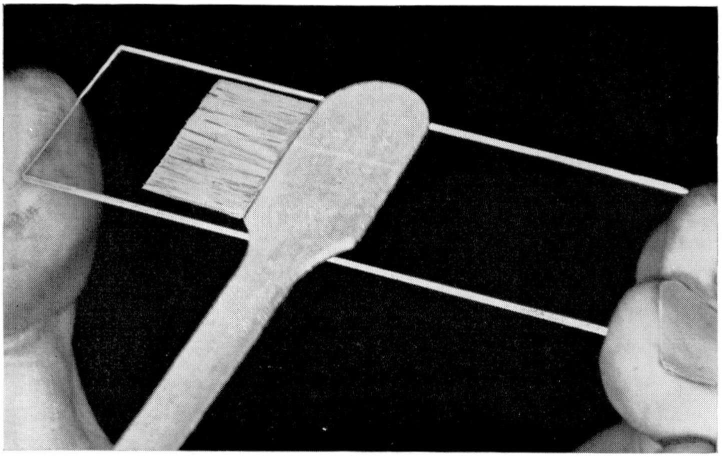

FIG. 157

The Technique of Smear Preparation—Correct

If the spatula is held parallel to the slide during the smear preparation the cellular material is evenly distributed over a wide area of the slide.

THE ENDOCERVICAL SMEAR

Advantage

This method can be used to collect cellular material in patients in whom previous surgery (especially cone biopsy or partial amputation of the cervix) has produced some degree of stenosis of the cervical os. This may be especially valuable in the follow-up of patients in whom carcinoma-in-situ or dysplasia have been demonstrated by previous biopsy.

Disadvantages

(*a*) The technique requires trained staff for visualisation of the cervix and collection of the specimen;

(*b*) a smaller amount of cellular material is obtained from the squamo-columnar junction than with the cervical scrape smear as this area is not completely investigated throughout its circumference. This method should, therefore, always be combined with a cervical scrape smear.

TECHNIQUE

After preliminary visualisation, with an unlubricated speculum, and thorough cleansing of the cervix, a moistened sterile cotton wool tipped applicator is introduced into the cervical canal and rotated. The swab is then rolled along a clean slide. The smear is fixed *immediately*, as this type of specimen is especially likely to suffer from drying distortion if there is any delay. The specimen is identified in the routine manner.

THE ENDOMETRIAL ASPIRATION SMEAR

Advantage

This method permits sampling of endometrial cellular material, which would otherwise be inaccessible unless exfoliation of endometrial cells occurred shortly before or during specimen collection.

Disadvantages

(*a*) The method requires a high degree of technical skill;

(*b*) unless suitable precautions are taken, the patient is subjected to the risk of infection and the remote possibility of injury;

(*c*) for these reasons this method is unsuitable for routine application to all patients. When the investigation is indicated because of a suspicious symptom, diagnostic curettage offers more fully representative sampling of the endometrium.

TECHNIQUE

After preliminary visualisation and cleansing of the cervix, a sterile cannula or polythene tube is introduced into the uterine cavity and aspiration is then carried out with a syringe. The specimen is squirted on to clean glass slides, gently spread, rapidly fixed and identified in the routine manner.

RESPIRATORY TRACT

The specimens which may be submitted from the respiratory tract include nasal secretion, the laryngeal swab, sputum, bronchial aspirates and washings.

NASAL SECRETION

This specimen is usually submitted for examination for eosinophil polymorphonuclear leucocytes in allergic disorders. Nasal mucus is collected with a moistened cotton wool tipped applicator and smeared on a clean glass slide. Fixation and identification are carried out in the routine manner.

THE LARYNGEAL SWAB

If a swab is taken from a suspicious area of the larynx, during clinical examination, it is essential to smear the cellular material on clean glass slides, while it is still moist, and to fix the smears immediately. Additional information about laryngeal lesions may be obtained by the examination of sputum specimens.

SPUTUM AND BRONCHIAL ASPIRATES OR WASHINGS

The main importance of these specimens is in the diagnosis of malignant lesions, but specimens may also be submitted for examination for eosinophil polymorphonuclear leucocytes in allergic disorders and for the diagnosis of various industrial diseases. However, these specimens are principally considered in relation to the diagnosis of malignant disease.

Specimens should be accompanied by a history which includes the patient's age, sex, occupation, smoking history, presenting symptoms, clinical diagnosis, radiological findings and relevant previous medical history.

SPUTUM

Advantages

(*a*) The specimen is easily collected, so that the examination can be repeated as required;

(*b*) the specimen is widely representative of the cellular changes occurring in the respiratory tract.

Disadvantages

(*a*) The specimen does not offer any information which helps to localise a lesion;

(*b*) occasional patients fail to produce sputum; only saliva is submitted for investigation.

Technique

The specimen should be collected before the patient's breakfast to minimise food contamination. The patient should understand that it is *sputum* produced by deep coughing and not saliva, which is required. Collection into a fresh container, preferably of transparent plastic, permits direct preparation of the smears from the collection container without intermediate transfer to a Petri dish. Fixative *should not* be added to the specimen, which should be sent to the laboratory the same morning.

It is advisable to obtain specimens of sputum (not saliva) on three separate mornings, before accepting the results of the investigation as negative for malignant cells. If positive findings are obtained in one specimen, the investigation should be continued until a further confirmatory specimen is positive for malignant cells.

In the laboratory the specimen is examined and suitable areas for smear preparation selected. These include blood-stained flecks and small white fragments. The selected small areas of sputum are transferred to four labelled clean slides with forceps or a spatula. The sputum is most satisfactorily spread by a 'crush-smear' technique. A second glass slide is used to cover and compress the sputum on the first slide, then the two slides are drawn apart and the sputum smears fixed.

Modifications in the processing of sputum, designed to increase the concentration of abnormal cells, include digestion techniques and the preparation of cell blocks.

BRONCHIAL ASPIRATES AND WASHINGS

Advantages

(*a*) A higher concentration of abnormal cells may be obtained if the specimen is collected from the appropriate bronchus;

(*b*) evidence from a bronchial aspirate or washing may confirm or even identify the site of a lesion. Great care must be taken in evaluating this evidence, however, as abnormal cellular material may be transferred to another site during bronchoscopy.

Disadvantages

(*a*) The cellular material obtained is not representative of the lower respiratory tract as a whole;

(*b*) the patient must be subjected to bronchoscopy;

(*c*) surgical skill is essential for satisfactory specimen collection.

Technique

The actual details of the method of collection will depend on the choice of the individual surgeon, but it is desirable that the tube for collection of the aspirate, during bronchoscopy, should be sited as close to the lesion as possible. Many methods involve collection of the specimen by aspiration into a tube at some intermediate or proximal point; these methods waste a considerable amount of the aspirated material. When there is difficulty in aspirating a specimen, 2-3 ml. of sterile normal saline may be instilled into the bronchus and immediately reaspirated. This procedure may be repeated once or twice during the bronchoscopy. Bronchial aspirates and washings should be transferred to the laboratory as rapidly as possible. If this is impracticable smears should be prepared from aspirates in the theatre and fixed immediately.

In the laboratory smears are prepared directly from bronchial aspirates, but bronchial washings are centrifuged in a bench centrifuge at 2,000 r.p.m. for twenty minutes. The sediment is transferred to labelled clean glass slides, spread and fixed. As with sputum, cell blocks may be prepared from any residue, after making the smears for cytological examination.

THE ALIMENTARY TRACT

The specimens which may be collected include oral specimens, oesophageal, gastric, colonic and rectal washings. All specimens should be accompanied by a full clinical history.

ORAL SPECIMENS

Oral specimens may be collected either by direct scraping of suspicious areas of buccal or lingual mucosa with a moistened wooden spatula or by the use of a mouth wash technique. The direct scrape smear is the most suitable for the investigation of a known lesion, but the mouth wash technique offers a more representative sample in any type of screening project. The spatula smear should be rapidly spread on a glass slide and immediately fixed. The mouth wash fluid is centrifuged, in a bench centrifuge, at 2,000 r.p.m. for twenty minutes and smears are then prepared from the sediment.

OESOPHAGEAL AND GASTRIC WASHINGS

Although oesophageal or gastric washing may be carried out alone, oesophageal washing can immediately precede gastric washing when both investigations are indicated.

The patient should fast for twelve hours prior to the investigation. This is most conveniently arranged overnight.

Oesophageal Washing Technique

A Levine tube (No. 14 or 15 F) is introduced, if possible, as far as the distal end of the oesophagus, connected to a syringe and aspirated. Any material obtained in this way is likely to consist of food debris, accumulated as a result of obstruction, and should be discarded. While suction is constantly maintained on the tube with the syringe, the patient is asked to swallow 200 ml. of saline, and the entire volume is aspirated. As soon as possible after collection the specimen is centrifuged at 2,000 r.p.m. for twenty minutes. Smears are then prepared from the sediment and immediately fixed.

Gastric Washing Technique

A Levine tube (No. 14 or 15 F) is passed through the nose into the stomach. A few mouthfuls of tap water may be swallowed while the tube is being passed. The stomach is then emptied, using a 50 ml. syringe, and this aspirate is labelled as specimen 1. 200-300 ml. of normal saline are then injected down the tube and a spigot is inserted in the end of the tube. The patient is then asked to lie prone, supine and in both lateral positions, to sit up, and, if ambulatory, to walk. The fluid is then aspirated, reinjected several times, and finally the total volume is withdrawn, and this aspirate is labelled as specimen 2. The two specimens should be sent to the laboratory immediately and centrifuged separately at 2,000 r.p.m. for twenty minutes. Smears are prepared from the whole of the sediment and immediately fixed. Rapid transfer to the laboratory and immediate centrifugation are essential, as the specimen readily undergoes cytolysis. It should be noted that simple gastric aspirates are not suitable for the diagnosis of malignant lesions of stomach.

Abrasive Techniques

A variety of abrasive techniques are described for the collection of gastric cellular material. These methods have the disadvantages of causing the patient greater discomfort and exposing him to the risk of trauma, while equally satisfactory specimens can be obtained by the simpler and safer mechanical action of a gastric lavage technique.

Colonic and Rectal Washing Technique

The disadvantage of these investigations is that they require considerable preparation of the patient by purgation and enemata.

If a suitable specimen is to be obtained, however, enemata must be given until a clear return is obtained, before commencing specimen collection. After this has been achieved, 500-1,000 ml. of saline is slowly introduced by catheter into the colon, with the patient in the left lateral position, and the patient is then turned to the supine and right lateral positions. If possible the fluid should be retained for five minutes. The total volume is then collected and immediately sent to the laboratory. After centrifugation at 2,000 r.p.m. for twenty minutes, smears are prepared from the sediment and rapidly fixed. When only rectal washings are required, 100-200 ml. of saline are introduced by catheter into the rectum, collected after five minutes and processed by a similar method.

THE URINARY TRACT

Specimens of urine may be obtained by collection of a mid-stream or total specimen, or by catheterisation of bladder or ureter. All specimens should be accompanied by the relevant clinical history. The type of specimen selected depends largely on the purpose of the investigation. When screening asymptomatic patients from industrial high-risk groups the most suitable specimen is the whole volume of an early morning specimen of urine, as the cellular content is likely to be scanty. In symptomatic patients a midstream morning specimen is likely to yield sufficient cellular material for diagnosis, but the whole volume of a specimen should be examined if this is not the case. Catheterisation of the bladder is not necessary and should be avoided because of the risks of infection. Ureteric catheterisation, however, is of value in the detection of lesions of ureter or renal pelvis, and may help to localise the site of a lesion. Care must be taken in interpreting the findings of ureteric catheter specimens, as malignant cells of vesical origin may be carried into the ureter by the catheter tip.

All urine specimens should be sent to the laboratory as rapidly as possible, as degenerative changes soon appear in transitional epithelial cells. When this is not possible an equal volume of 50 per cent or 70 per cent ethyl alcohol should be added. The lower proof should be used if the urine contains significant amounts of protein, as the higher proof alcohol will produce protein precipitation.

In the laboratory the whole specimen is centrifuged in a bench centrifuge for twenty minutes at 2,000 r.p.m. The sediment is smeared on previously albuminised labelled slides and fixed.

FLUIDS

The preparation of all fluids, whether aspirated from body cavities or cysts, is essentially similar. If the specimen cannot be sent directly to the laboratory an equal volume of 50 per cent ethyl alcohol is added. All specimens should be accompanied by a full clinical history. The whole volume collected should be dispatched to the laboratory, but when the volume is very large approximately 200 ml. should be sent.

In the laboratory the specimen is centrifuged at 2,000 r.p.m. in a bench centrifuge for twenty minutes and smears are then prepared from the sediment. Occasionally it may be necessary to use albuminised slides (e.g., for C.S.F.), but in most fluids collected from serous cavities or cysts, there is sufficient protein present for this to be unnecessary. The smears are fixed while still moist. If desired cell blocks may be prepared from any residue after making smears for cytological examination.

Millipore Filter Technique

This method is designed to concentrate the cellular content of a fluid specimen on to a filter, which can then be fixed and stained. It is, therefore, especially suitable for the examination of urine and other body fluids, but it can also be used for the examination of sputum specimens after previous digestion. The main disadvantages of the method are the fact that the membrane may become blocked if the cell density is high, and that numerous non-epithelial cells, if present, may obscure the details of the epithelial cells. Details of filtration techniques are available from the filter manufacturers.

FIXATION

Adequate fixation is essential for the preservation of morphological detail. The best results are achieved when smears are prepared from freshly collected material and immediately fixed while still moist. If this is not possible, fluid specimens should be fixed by the addition of an equal volume of 50 per cent alcohol. Higher concentrations of alcohol produce protein precipitation in the fluid.

The essential component in most cytological fixatives is 95 per cent ethyl alcohol. Originally this was combined with an equal volume of ether, but this entails considerable explosion and fire risks. It has been found possible to omit the ether completely without loss of morphological detail. Smears of all types can be satisfactorily fixed by immersion in 95 per cent ethyl alcohol for twenty to thirty minutes. One advantage of this fixative is that smears can be left in it for several days if necessary.

Modifications of the method can be used for different circumstances or specimens. To avoid immersion fixation or if it is necessary to post the specimen, a protective film over the smear can be obtained by employing a fixative containing polyethylene glycol in ethyl alcohol. A rapid drying version of this type of fixative is commercially available as an aerosol spray. The alcohol fixative can be modified for sputum examination by the addition of acetic acid.

Alternative methods of fixation, which may be employed if ethyl alcohol is not available, include fixation with glycol, with acetone, or with Schaudinn's fixative. Most satisfactory results, however, are obtained by adhering to 95 per cent alcohol as the fixative of choice.

ROUTINE FIXATION

(1) **Fixative** 95 per cent ethyl alcohol.

Method Immersion of the moist smear in the fixative for thirty minutes.

OR

(2) **Fixative:**—STOCK DILUTE SOLUTION OF POLYETHYLENE GLYCOL (CARBOWAX 1500)

Absolute alcohol (74 O.P.)	100 ml.
Carbowax 1500	3 gm.
Glacial acetic acid	0.2 ml.

The carbowax is dissolved in warm alcohol. After cooling the acetic acid is added and the solution mixed.

Method The fixative is dropped on to the smear while it is still moist. The slide is left in the horizontal position for ten to fifteen minutes, until the fixative dries. (The smear can then be placed in a plastic bag or box for posting.)

FIXATION FOR POSTING

(1) **Fixative** POLYETHYLENE GLYCOL (CARBOWAX) CONCENTRATE FIXATIVE

(*To be diluted before use*)

Carbowax 1500	60 gm.
Distilled water	100 ml.
Glacial acetic acid	4 ml.

The carbowax is dissolved in warm water. After cooling the acetic acid is added and the solution mixed. 1.5 ml. of this concentrate is placed in the plastic dropper bottle (capacity 25 ml.). *Before use the plastic dropper bottle is filled with 95 per cent alcohol.*

Method The fixative is used in the way described above for the stock dilute solution of polyethylene glycol fixative.

OR

(2) Smears may be fixed by the routine method of immersion in 95 per cent ethyl alcohol, dried and covered with a protective film of glycerine.

SPUTUM FIXATION

Fixative 3 per cent acetic acid in 95 per cent ethyl alcohol.

Method Immersion of the moist smear in the fixative for thirty minutes.

CHAPTER 24

STAINING TECHNIQUES

A wide variety of staining techniques is available for cytological preparations. As with any other technique, individual preference and experience may determine the choice of a particular method. However, routine use of the standard Papanicolaou technique yields consistently satisfactory results in the field of diagnostic cytology. Unless otherwise indicated, all preparations illustrated in this volume were prepared by staining with the Papanicolaou technique after wet fixation in 95 per cent ethyl alcohol.

THE PAPANICOLAOU STAINING TECHNIQUE

Advantages

(*a*) Excellent demonstration of nuclear morphological detail.
(*b*) Cytoplasmic translucency.

Disadvantages

(*a*) An elaborate technique.
(*b*) The stability of differential staining is not satisfactory, if this criterion is going to be employed for endocrine assessment.

Stains Required

EA 50, EA 65, Orange G (O.G.6) and Harris's haematoxylin are all commercially available or may be prepared in the laboratory from the following formulae:—

EA 50

Light green S.F. (yellowish) 0.1 per cent solution in 95 per cent ethyl alcohol	45 ml
Bismark brown 0.5 per cent solution in 95 per cent ethyl alcohol	10 ml
Eosin (yellowish) 0.5 per cent solution in 95 per cent ethyl alcohol	45 ml
Phosphotungstic acid	0.2 g
Lithium carbonate (saturated aqueous solution)	1 drop

EA 65

The formula is similar to the formula for EA 50 but the amount of light green S.F. (yellowish) is halved.

Orange G (O.G.6)

Orange G stock solution (0.5 per cent in 95 per cent ethyl alcohol)	100 ml
Phosphotungstic acid	0.015 g

Harris's haematoxylin

Haematoxylin	5 g
Absolute ethyl alcohol	50 ml
Potassium alum	100 g
Mercuric oxide (red)	2.5 g
Water	1000 ml

Dissolve haematoxylin in alcohol and the alum in water. Mix the two solutions together in a flask (2 litre). Rapidly bring to boil and, withdrawing the flame immediately, add the mercuric oxide. Cool at once by placing the flask in a dish of cold water. When cool the solution is ready for use. 4 per cent glacial acetic acid is added just before use to increase the precision of nuclear staining.

The preparations illustrating this book were stained using commercially prepared EA 50, EA 65 and Orange G (O.G.6). Harris's haematoxylin was prepared in the laboratory.

PAPANICOLAOU STAINING METHOD

(*1)	95 per cent ethyl alcohol	1 min to 24 hours
	Rinse in distilled water (to rid slides of fixative)	1 min
	Harris's haematoxylin	2 to 4 min
	Rinse in distilled water	2 min
(*2)	Acid alcohol (to differentiate)	1 or 2 dips
	Rinse in distilled water	5 to 10 min
(*3)	Scott's tap water (to 'blue')	1 min
	Wash in running tap water	5 min
	95 per cent ethyl alcohol	2 min
	O.G.6	1½ to 2 min
	95 per cent ethyl alcohol	4 dips (quick)
	95 per cent ethyl alcohol	4 dips (quick)
(*4)	EA 50	2 to 3 min
	95 per cent ethyl alcohol	4 dips (quick)
	100 per cent ethyl alcohol (absolute)	4 dips (quick)
	100 per cent ethyl alcohol (absolute)	4 dips (quick)
	Xylol	1 min
	Xylol	1 min
	Examine and mount	

*1. Slides may be left in 95 per cent ethyl alcohol until a complete rack is ready for staining.

*2. Acid alcohol: 1 per cent hydrochloric acid in 95 per cent ethyl alcohol.

*3. Scott's tap water:

Sodium bicarbonate	3.5 g
Magnesium sulphate	20 g
Water	1000 ml

*4. For non-gynaecological specimens EA 65 is substituted for EA 50.

Results

The squamous epithelial cells range from basophilic (or cyanophilic) cytoplasmic staining in the deepest, least mature cells to eosinophilic (or acidophilic) staining in the most superficial, mature cells. The colour reactions, however, are not stable and are affected by alterations in pH, the presence of erythrocytes, inflammatory changes and even the thickness of the cell groups.

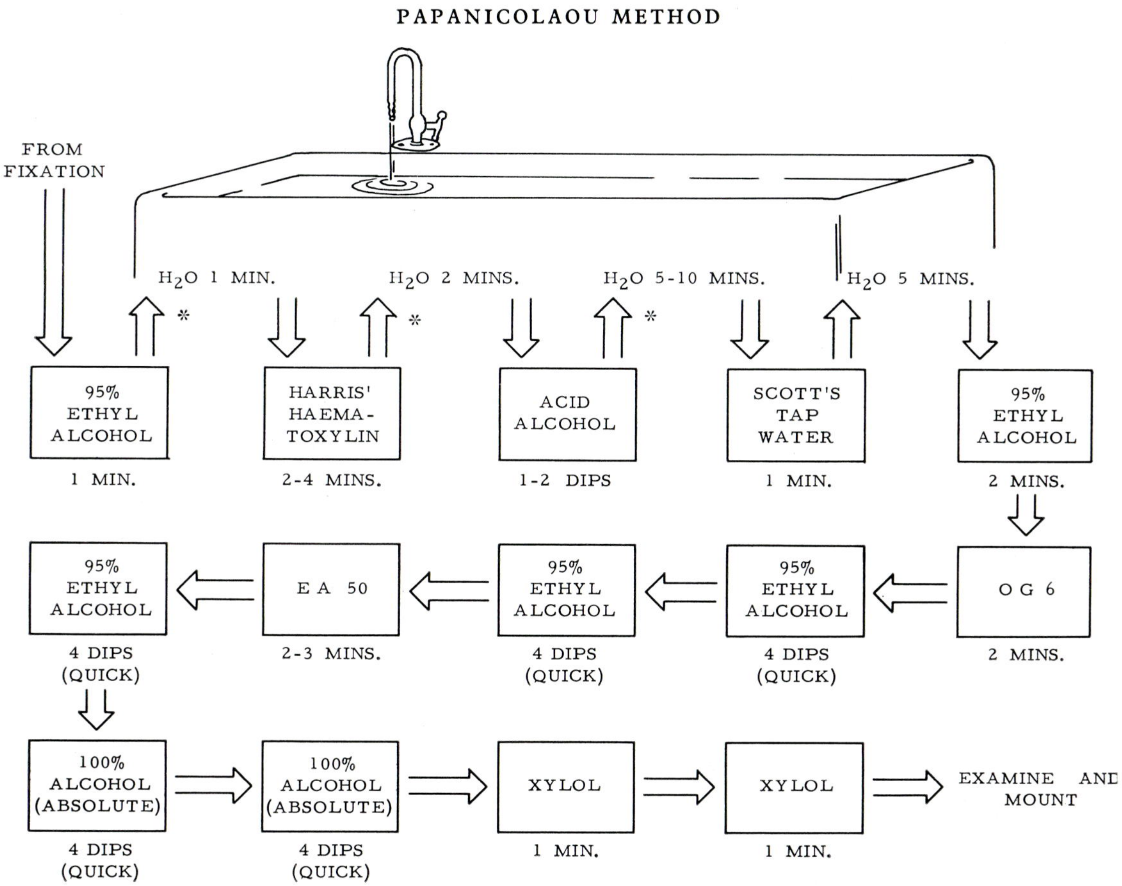

* Distilled water is generally recommended for these rinses but in certain areas, where the tap water is very 'soft', tap water can be used instead of distilled water.

FIG. 158
The Papanicolaou Staining Method

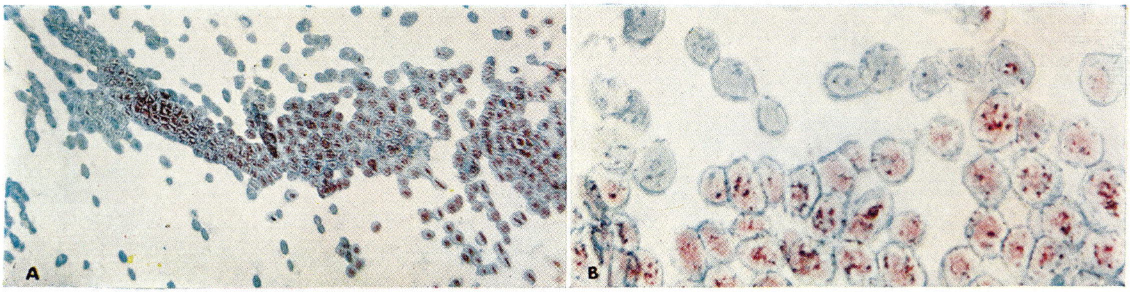

FIG. 159
Variable Staining of Erythrocytes with the Papanicolaou Technique

A. This low-power illustration shows a wide range of colour variation in a single field of erythrocytes, with predominantly basophilic staining of the cells on the left and predominantly eosinophilic staining of the cells on the right. (×200).

B. The high-power illustration again shows variable staining of erythrocytes. (×800.)
Compare these cells with the erythrocytes shown in Figure 10.

THE ACRIDINE ORANGE FLUORESCENT TECHNIQUE (BERTALANFFY)

Advantages

(*a*) The method is designed to provide a simple, rapid technique for easy recognition of abnormal malignant cells, by producing brilliant red cytoplasmic fluorescence of ribonucleic acid (R.N.A.) at pH6. Modification of pH (3.8) can be used to demonstrate brilliant yellow nuclear deoxyribonucleic acid (D.N.A.).

(*b*) The staining technique is simpler than the Papanicolaou method.

Disadvantages

(*a*) Since other active cells may produce red cytoplasmic fluorescence, as may protozoa, morphological criteria are still necessary for accurate diagnosis. Morphological features are not so readily recognised as with the Papanicolaou technique and considerable observer training and skill are still required.

(*b*) Undifferentiated malignant cells are not readily recognised.

STAIN

Stock solution of acridine orange: 0.1 per cent in distilled water

Phosphate buffer pH6 M/15 KH_2PO_4 2 parts
M/15 Na_2HPO_4 5 parts

Working solution acridine orange: 1 part stock solution of A.O. in 9 parts phosphate buffer. Both stain and buffer must be kept in the refrigerator. The A.O. solution should be kept in a dark bottle.

METHOD

N.B. For this method the smears are air dried prior to fixation in the standard manner in 95 per cent ethyl alcohol for 20-30 minutes.

80 per cent ethyl alcohol	4 dips
70 per cent ethyl alcohol	4 dips
50 per cent ethyl alcohol	4 dips
Distilled water	4 dips
1 per cent acetic acid	4 dips
Distilled water	1 dip
Working solution acridine orange (0.01 per cent)	3 min
Phosphate buffer (pH6)	1 min
M/10 calcium chloride	30 sec
Phosphate buffer (pH6)	wash
M/10 calcium chloride	30 sec
Phosphate buffer (pH6)	dip

Allow slides to dry.

Directly prior to screening mount in buffer.

Results

The cytoplasm of malignant cells shows brilliant orange-red fluorescence when the smears are examined in ultra-violet light. (Modification of the method using McIlvaine's buffer pH 3.8 yields brilliant yellow nuclear fluorescence.) Nucleoli also show orange fluorescence. In contrast benign cells show green cytoplasmic staining with yellow nuclei. Since any active cells may show some red cytoplasmic fluorescence or yellow nuclear fluorescence, while protozoa or micro-organisms show red fluorescence, morphological criteria in addition to the fluorescent phenomena must be employed for the definitive diagnosis of malignant cells.

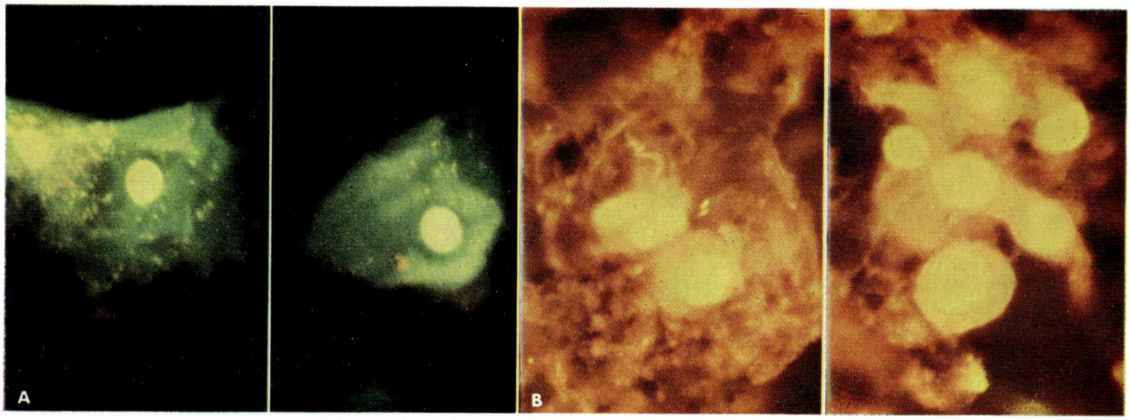

FIG. 160

Benign Squamous Cells and Differentiated Malignant Cells of Squamous Type in Cervical Smears Stained by the Acridine Orange Fluorescent Technique (pH6)

A. The cytoplasm of these benign squamous cells appears green and their nuclei are yellow. (×800.)

B. The cytoplasm of these malignant cells shows brilliant reddish-orange fluorescence. The nuclei show yellow fluorescence. (×800.)

HAEMATOXYLIN AND EOSIN TECHNIQUE

Advantage

The method produces a cellular appearance similar to the one which is already familiar to the histopathologist from studying tissue sections.

Disadvantages

(*a*) Nuclear detail is not as clear as with the Papanicolaou technique.

(*b*) Cytoplasm is less translucent than with the Papanicolaou technique.

STAINS REQUIRED

Harris's haematoxylin (q.v.)

Eosin.

Eosin 7	16 g
Potassium dichromate	8 g
Picric acid (saturated aqueous)	160 ml
95 per cent ethyl alcohol	160 ml
Distilled water	1280 ml

The eosin and potassium dichromate are dissolved in warm water: the picric acid and alcohol are added.

METHOD

Absolute ethyl alcohol	10 dips
Absolute ethyl alcohol	10 dips
95 per cent ethyl alcohol	10 dips
95 per cent ethyl alcohol	10 dips
Tap water	10 dips
Harris's haematoxylin	3-7 min
Tap water	10 dips
Acid alcohol (q.v.) to differentiate	4-5 dips (until only nuclei are stained)
Wash in running tap water	5 min
Scott's tap water (q.v.) to 'blue'	1 min
Tap water	10 dips
Eosin	1-2 min
Tap water	10 dips
95 per cent ethyl alcohol	10 dips
95 per cent ethyl alcohol	10 dips
Absolute ethyl alcohol	10 dips
Absolute ethyl alcohol	10 dips
Xylol	10 dips
Xylol	10 dips

Results

The cytoplasm is stained pink and the nuclei are stained blue.

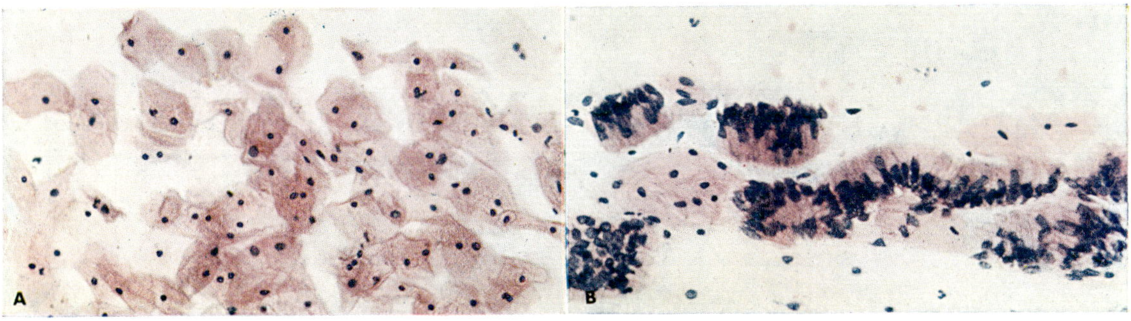

FIG. 161

Benign Squamous and Columnar Cells from a Cervical Smear Stained by Haematoxylin and Eosin

A. These superficial and intermediate squamous cells show haematoxylin nuclear staining while the cytoplasm is counterstained with eosin. (×200.)

B. Haematoxylin staining of the nuclei of these columnar endocervical cells is shown against the eosinophilic staining of the cytoplasm. (×200.)

STAINS FOR SPECIAL PURPOSES

SHORR'S STAINING TECHNIQUE (FOR ENDOCRINE CYTOLOGY)

Advantages

(*a*) A simple technique.
(*b*) Gives good differential staining of deep and superficial squamous cells.

Disadvantage

The nuclear detail is not as clear nor the cytoplasm as translucent as with the Papanicolaou technique. Although nuclear detail can be improved by preliminary staining with Harris's haematoxylin the results are still inferior to those obtained by the Papanicolaou method for the diagnosis of malignant cells.

SHORR SINGLE STAIN

50 per cent ethyl alcohol	100 ml
Biebrich scarlet (water soluble)	0.5 g
Orange G.	0.25 g
Fast green F.C.F.	0.075g
Phosphotungstic acid	0.5 g
Phosphomolybdic acid	0.5 g
Glacial acetic acid	1 ml

METHOD

Shorr single stain	1-2 min
70 per cent ethyl alcohol	10 dips
90 per cent ethyl alcohol	10 dips
Absolute ethyl alcohol	10 dips
Xylol	to clear

Results

The cytoplasm of superficial squamous cells stains orange-red. The cytoplasm of intermediate and parabasal cells stains green-blue; the deeper cells show darker staining than the more mature intermediate cells. The nuclei stain red.

METHYLENE BLUE TECHNIQUE (FOR SPUTUM OR BRONCHIAL ASPIRATES)

Advantages

(*a*) A rapid and simple staining method for preliminary screening.

(*b*) As the preparation is still wet, it is possible, by gentle pressure on the slide, to 'spread' during screening, any thick or 'suspicious' areas of sputum.

Disadvantage

The method does not produce a permanent preparation. However, the specimen can, subsequently, be decolourised in ethyl alcohol, fixed and stained by any standard technique.

METHOD

A small portion of sputum is selected and placed on a clean glass slide. An approximately equal volume (2 or 3 drops) of 0.5 per cent aqueous methylene blue solution is added to the sputum and rapidly mixed with it. The specimen is cover-slipped while still moist and read immediately.

Results

Both cytoplasm and nucleus stain blue. The hyperchromatic dark blue nuclei of malignant cells are prominent against the paler blue background of cytoplasm.

MAY-GRÜNWALD-GIEMSA TECHNIQUE (FOR FLUIDS AND FOR NON-EPITHELIAL CELLS IN SPUTUM, Etc.)

Advantages

(*a*) This is essentially a haematological stain, which demonstrates detail in non-epithelial cells. For this reason it is often used for serous fluids in which non-epithelial components are likely to predominate.

(*b*) Used to identify particular types of non-epithelial cell, e.g., to identify eosinophil polymorphonuclear leucocytes in allergic disorders (see Fig. 82).

Disadvantages

(*a*) The method does not give as clear detail, especially of nuclear morphology, in epithelial cells as the Papanicolaou technique and is, therefore, less suitable for the diagnosis of malignant epithelial cells.

(*b*) The heavy staining of non-epithelial cells may obscure the detail of epithelial cells.

Stains Required

May-Grünwald	0.3 per cent powder stain in methyl alcohol diluted with equal volume buffered water (pH 6.8).
Giemsa	diluted 1/10 with buffered water (pH 6.8).

METHOD

N.B. An air-dried smear is used for this method.

Methyl alcohol	10 min
May-Grünwald stain	8 min
Giemsa stain	17 min

Differentiate in buffered water (pH 6.8).

Results

The cytoplasm of epithelial cells stains light blue and their nuclei dark blue. Polymorphonuclear leucocytes have blue nuclei with eosinophilic (red), basophilic (dark blue) or neutrophilic granules in the cytoplasm. Erythrocytes stain pale reddish-brown.

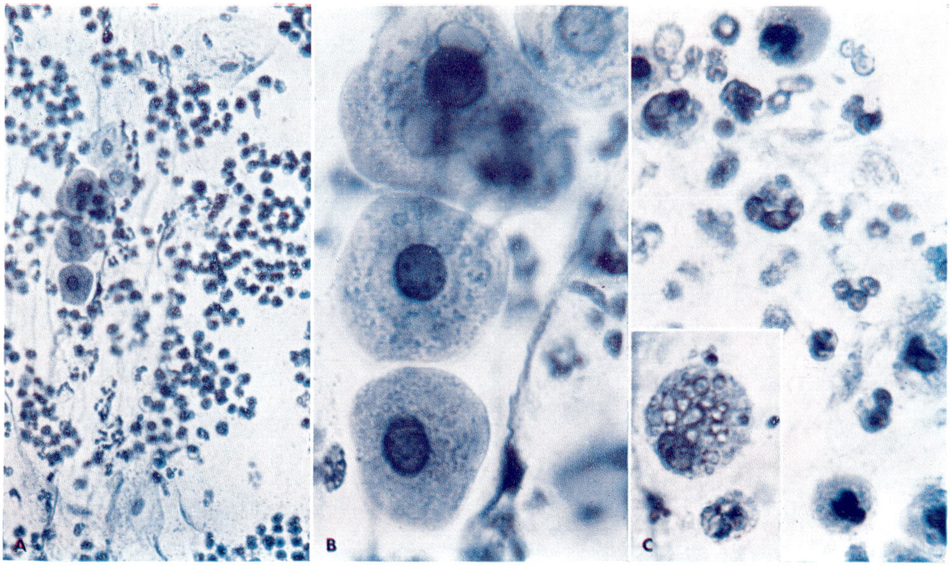

FIG. 162

Sputum Stained with Methylene Blue

A. This low-power field shows benign squamous cells and histiocytes against a background of polymorphonuclear leucocytes. (×200.)

B. Benign squamous cells, with deeply stained, uniform round or oval central vesicular nuclei and pale stained cytoplasm are shown at high-power magnification. (×800.)

C. A large histiocyte and numerous polymorphonuclear leucocytes are illustrated at high-power magnification. Note the small eccentric nucleus and vacuolated cytoplasm of the large histiocyte in the inset. (×800.)

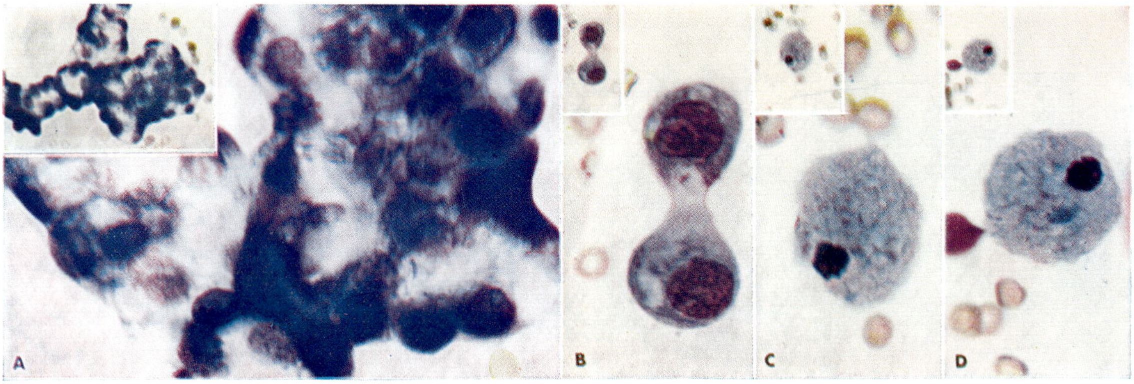

FIG. 163

An Ascitic Fluid Containing Differentiated Malignant Cells from Metastases from a Cystadenocarcinoma of Ovary, Stained by the May-Grünwald-Giemsa Method

A. The low-power view (inset) shows a group of cells in which the large hyperchromatic nuclei are peripherally situated. At high-power magnification it is apparent that the nuclei vary in size and shape, and that the individual nuclei are eccentrically placed in the cells and tend to form the outer border of the cell group. Little detail of nuclear structure can be seen. (Inset ×200.) (×800.)

B. Two mesothelial cells are shown at low-power magnification in the inset. At high-power magnification the rounded nuclear outlines, the similarity between the two nuclei and the adequate cytoplasmic borders identify these as benign cells. (Inset ×200.) (×800.)

C. and D. Single histiocytes are shown against a background of erythrocytes. Note the finely vacuolated cytoplasm of the histiocytes. Again there is little nuclear detail. (Insets ×200.) (×800.)

Compare these cells with the similar benign and malignant cells from the same specimen, stained by Papanicolaou technique, shown in Figure 136. Note the superior demonstration of nuclear detail by the Papanicolaou method. A further example of the May-Grünwald-Giemsa technique is shown in Figure 82.

THE PERIODIC ACID-SCHIFF REACTION (FOR SEROUS FLUIDS)

Advantages

(*a*) Mesothelial cells can be identified by the presence of magenta cytoplasmic granules.

(*b*) Some differentiated malignant cells (mucus-secreting) of adenocarcinomatous type may contain PAS-positive material in their cytoplasm.

Disadvantage

The fact that some malignant cells may contain PAS-positive material impairs the specificity of the method for the demonstration of mesothelial cells.

REAGENTS

Schiff reagent. Sulphite rinses—freshly prepared.

METHOD

Water	1 min
Periodic acid (1 per cent aqueous)	5-10 min
Running water	5 min
Distilled water	10 dips
Schiff reagent	10 min
Sulphite rinse	1 min
Sulphite rinse	2 min
Sulphite rinse	2 min
Running water	10 min
95 per cent ethyl alcohol	10 dips
Absolute ethyl alcohol	10 dips
Absolute ethyl alcohol	10 dips
Xylol	10 dips
Xylol	10 dips

If desired a counterstain may be used, prior to dehydration and clearing.

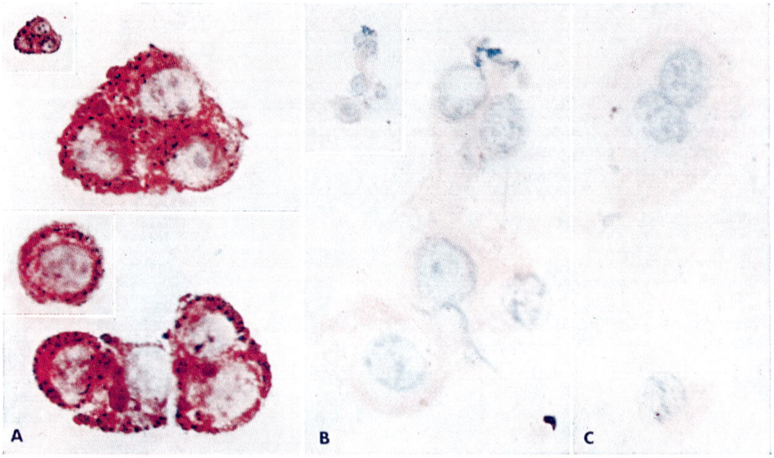

FIG. 164

Benign Mesothelial Cells and Histiocytes in an Ascitic Fluid Stained by the Periodic Acid-Schiff Technique

A. The small group of mesothelial cells inset in the upper left corner shows the characteristic peripheral distribution of the magenta, PAS-positive, staining of the cytoplasm. The composite high-power field shows the typical intracytoplasmic granules in mesothelial cells. (Inset ×200.) (×800.)

B. and C. These single and multinucleated histiocytes do not show PAS-positive staining. (Inset ×200.) (×800.)

Results

PAS-positive material stains magenta. In mesothelial cells it appears as granular masses peripherally located in the cytoplasm.

THE PRUSSIAN BLUE REACTION (FOR THE DEMONSTRATION OF IRON)

Advantages

(*a*) The method can be used directly to demonstrate the presence of iron, e.g., haemosiderin in histiocytes in sputum.

(*b*) The presence of iron may be a helpful pointer in the search for asbestos bodies.

REAGENT FOR THE PRUSSIAN BLUE REACTION

Equal parts 2 per cent aqueous potassium ferrocyanide and 2 per cent hydrochloric acid.

METHOD

Water	1 min
Reagent (see above)	30 min
(If result doubtful carry out reaction at 60°C.)	
Distilled water (×3)	5 min, 5 min, 5 min
1 per cent neutral red	10-15 sec
Running water	3 min
95 per cent ethyl alcohol	10 dips
Absolute ethyl alcohol	10 dips
Absolute ethyl alcohol	10 dips
Xylol	10 dips
Xylol	10 dips

Result

Pigments containing ferric ion appear dark blue (haemosiderin).

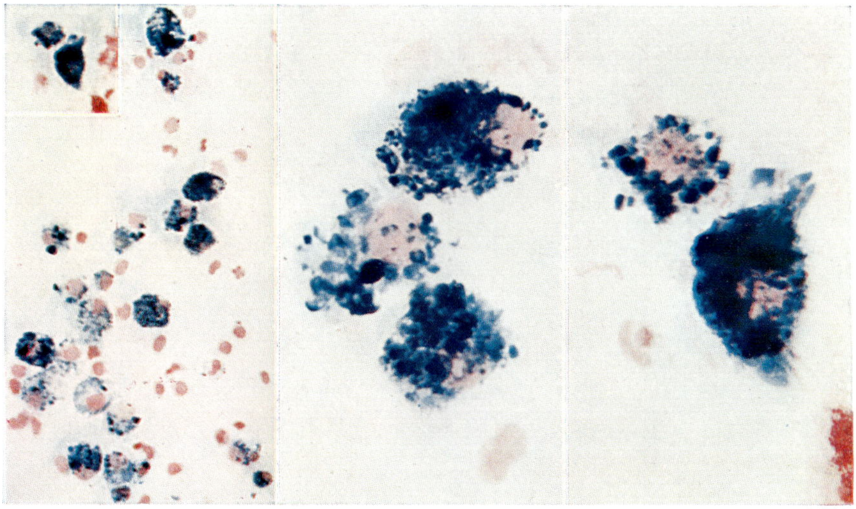

FIG. 165

Histiocytes in Sputum Stained by the Prussian Blue Reaction

This specimen of sputum from a case of pulmonary haemosiderosis shows, in the low-power fields on the left, a positive Prussian blue reaction in many histiocytes. The high-power fields (centre and right) show the granular cytoplasmic distribution of the haemosiderin. Its loosely bound ferric ion is responsible for the positive Prussian blue reaction. (×200.) (×800.)

THE FEULGEN REACTION (FOR THE DEMONSTRATION OF D.N.A.)

Advantage

The method gives specific staining of deoxyribonucleic acid. With accurate standardisation the method can be used for the measurement of nuclear D.N.A. on the integrating microdensitometer.

STAINS REQUIRED

Schiff reagent
Fast green F.C.F. 0.01 per cent in 95 per cent ethyl alcohol.

METHOD

N/1 hydrochloric acid (at 60°C.)	10 min
Rinse in distilled water	1 min
Schiff reagent	5-10 min
Sulphite rinses (×3)	1 min, 2 min, 2 min
Wash in running water	5 min.
Fast green F.C.F.	1-2 dips (excess may be removed with water)
Absolute ethyl alcohol	10 dips
Absolute ethyl alcohol	10 dips
Absolute ethyl alcohol/Xylol (50:50)	10 dips
Xylol	10 dips
Xylol	10 dips

Results

Nuclear D.N.A. stains bright reddish-purple. Cytoplasm stains pale translucent green.

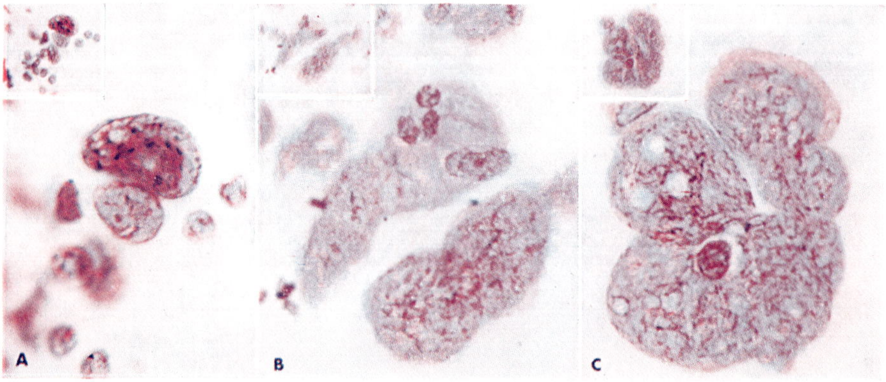

FIG. 166

Undifferentiated Malignant Cells in a Cervical Smear Stained by the Feulgen Reaction

A. B. and C. In each field large nuclei, which vary in size and shape, are present. Since these are undifferentiated malignant cells only an occasional fragment of pale green cytoplasm is noted (field B). (Insets ×200.) (×800.)

Compare these cells with similar cells from the same specimen, stained by the Papanicolaou technique, illustrated in Figure 35.

CHAPTER 25

PHASE CONTRAST MICROSCOPY

PRINCIPLE

THE components of cells and organisms have slight differences in refractive index. The phase contrast microscope converts these differences into changes of amplitude (or brightness) thus producing, in the unstained specimen, an image which can be resolved.

Advantages

(*a*) Provides a rapid method for preliminary screening of gynaecological smears, in which completely clean normal smears can be identified and discarded.

(*b*) A partial assessment of hormonal function can be made by this method.

Disadvantages

(*a*) Since all the details of nuclear structure, and especially evidence of hyperchromasia, are not defined by the method, any smear which shows anything but *completely normal* cells must be rescreened using the Papanicolaou technique.

(*b*) If the method is used directly on a wet unfixed preparation, no permanent preparation is available. This can be obviated by using fixed, unstained smears.

TECHNIQUE

A Fresh specimen examination

After routine visualisation of the cervix or vaginal pool, a specimen is collected with a spatula and a drop of the material is placed on a clean slide. If the specimen is insufficiently fluid in consistency a drop of physiologic saline is added, before covering the specimen with a cover-slip.

B Fixed specimen examination

The cervical or vaginal smear is collected, prepared and fixed in 95 per cent ethyl alcohol in the usual way. Prior to reading a drop of immersion oil or glycerine is placed on the smear, which is then cover-slipped.

CRITERIA

A For the diagnosis of malignant cells

Nuclear and cytoplasmic outlines and nucleocytoplasmic ratios are clearly demonstrated. Any cell which shows an abnormal nucleocytoplasmic ratio, variation in nuclear size or irregularity of nuclear shape should be regarded as abnormal. The specimen should then be investigated by the Papanicolaou technique.

B For the assessment of hormonal function

The recognition of nuclear and cellular size and shape and of nucleocytoplasmic ratios permits identification of the superficial cells and parabasal cells, but details of intermediate cell types are not so well demonstrated. It is, however, possible to make some assessment of the degree of epithelial maturation which is present. The method is, therefore, useful in assessing the effect of unopposed oestrogen activity, but does not permit full hormonal assessment.

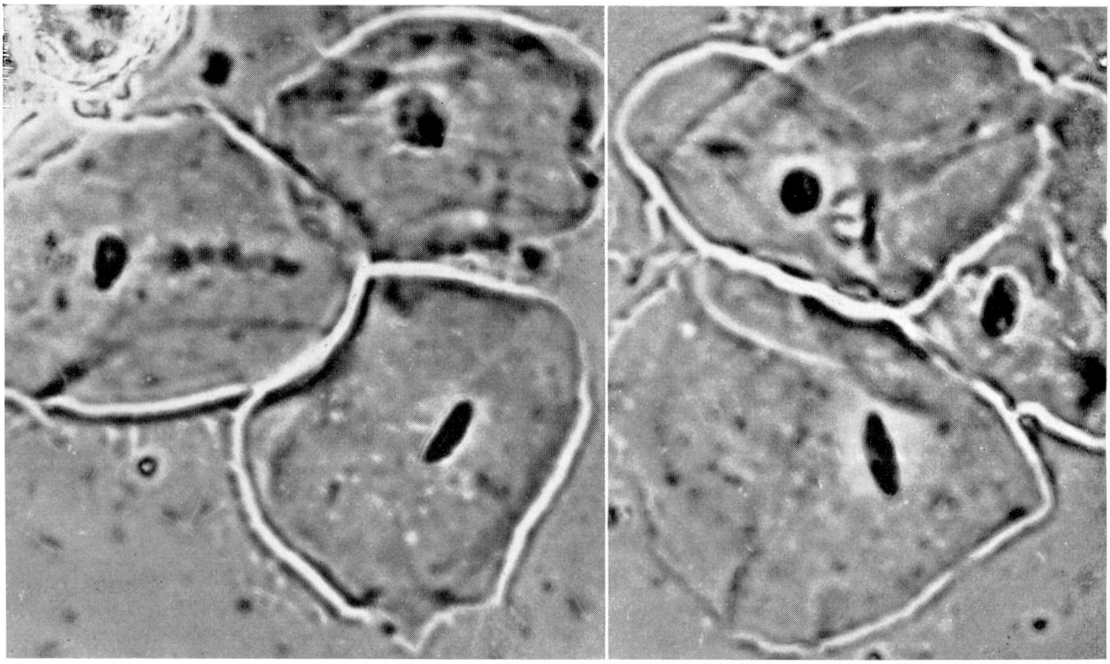

FIG. 167

Benign Superficial and Intermediate Squamous Cells in a Cervical Smear

Both fields show the appearance of benign superficial and intermediate squamous cells on phase contrast microscopy. The small round central nuclei of the superficial cells can be distinguished from the larger oval nuclei of the intermediate cells, but it is impossible to identify any detail of nuclear structure. Note the normal nucleocytoplasmic ratios of all these benign cells. (×800.)

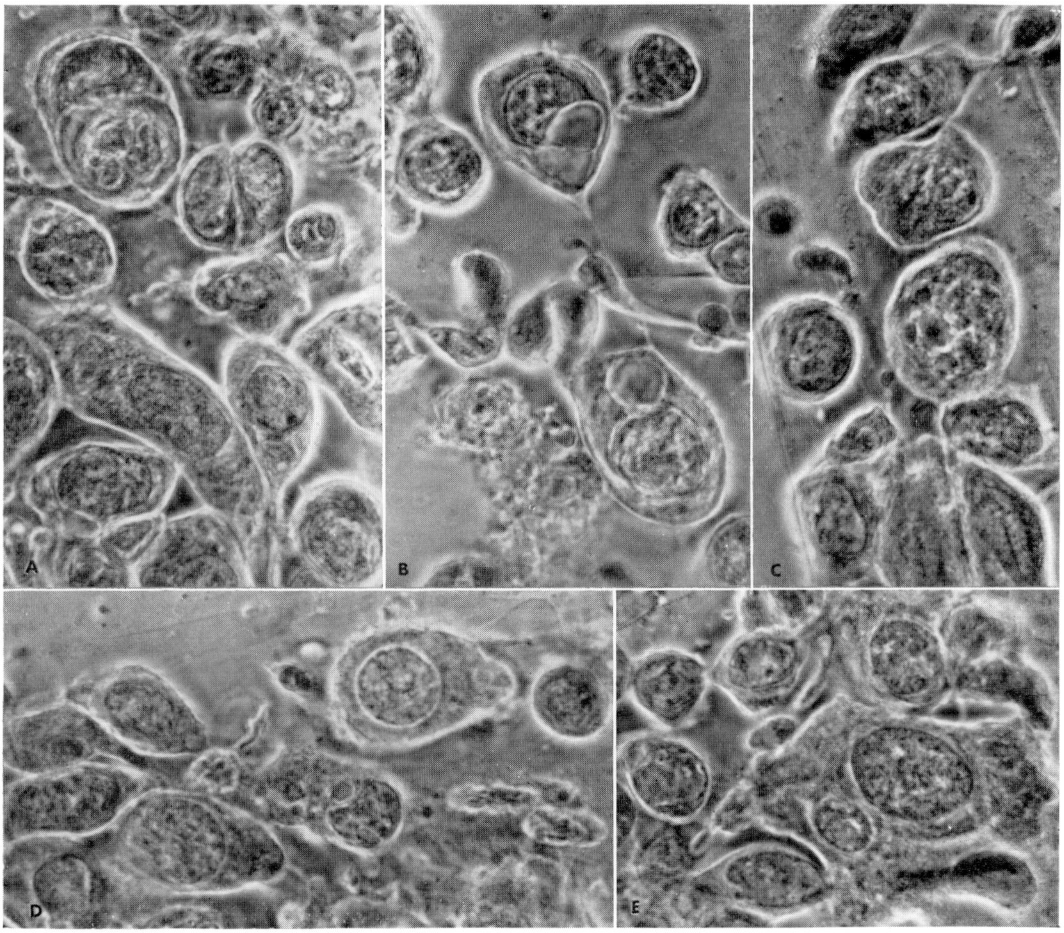

FIG. 168

Differentiated Malignant Cells of Squamous Type in a Cervical Smear

These fields demonstrate the characteristic features of differentiated malignant cells on phase contrast microscopy All the cells have strikingly abnormal nucleocytoplasmic ratios, with only a narrow rim of cytoplasm surrounding their enlarged nuclei. There is a marked degree of variation in nuclear size and shape. Compare the appearance of these cells with similar cells from the same specimen, stained by the Papanicolaou technique, illustrated in Figure 48. (×800.)

CHAPTER 26

THE ORGANISATION OF WORK IN THE LABORATORY

ALTHOUGH the detailed arrangement of work in any laboratory must depend on local circumstances and the preferences of the cytopathologist in overall charge, there are certain practical problems which are so frequently encountered that they warrant individual consideration.

THE RECEPTION OF SPECIMENS

When a specimen is received in the laboratory the first step is to check that it is correctly labelled and that this labelling corresponds with the identity of the patient described on the accompanying request form. Unidentified specimens must be discarded and a repeat specimen requested. After the identity of the specimen has been confirmed, it is entered in the day-book and the specimen and form are given a laboratory reference number. From that point onwards all slides prepared from the specimen or actually constituting the specimen (as is the case with cervical or vaginal smears) should be identified both by name and by laboratory reference number.

The next step is to ascertain from the request form exactly what type (or types) of investigation is required by the clinician, as this may affect the number of slides required and involve the use of a variety of staining techniques. Although most specimens may only require investigation for malignant cells and will be processed by the standard technique for the particular laboratory, considerable time is saved if any alternative or additional investigations are noted at this point, rather than when the slides reach the pathologist. Then, the gross appearance of the specimen (e.g., blood staining) and the volume of fluid specimens should be recorded. A note should also be made, in the case of unfixed specimens, of their time of arrival at the laboratory. Finally, any previous relevant cytology or pathology should be checked and the reports attached to the request form, so that all the information is readily available for the pathologist when he reports the specimen.

Number of slides

It is convenient and, occasionally, very useful to have spare slides available to cover diagnostic difficulties and technical errors. For this reason, when the clinician is making the smears (e.g., cervical specimens) two smears should be prepared from each patient. In routine screening of well women only one slide is stained. The second is kept in reserve and stained only if the first slide is technically unsatisfactory (e.g., because of drying distortion or scanty cellular material) or if any abnormality is noted on reading the first slide. However, if the smears come from symptomatic patients with suspicious symptoms such as intermenstrual, post-coital or post-menopausal bleeding, both slides should be stained initially as it will eventually be necessary to read all the cellular material available from these patients. Most other specimens (non-gynaecological) coming to the laboratory are from patients who are clinically suspicious. In this group four slides should be prepared and fixed (apart from any required for special stains). Two of these slides should be stained and screened and two held in reserve to cover diagnostic or technical difficulties.

SLIDE SCREENING

A methodical approach to slide screening ensures that all the cellular material is scrutinised. For this purpose it is useful to adopt certain conventions, which also simplify slide marking (q.v.). Everyone working in a particular laboratory should adhere to the same convention. A convenient one consists of placing the slide in the holder of the mechanical stage with the label at the right hand side and commencing screening at a right hand corner. The stage is moved so that each sweep across the smear is across the shorter diameter of the cover-slip. When a sweep is completed a cell is noted at the left hand margin of the field and the stage is then moved until this cell is at the right hand margin of the new field. The next sweep across the short diameter is then examined. In this way a series of overlapping sweeps across the smear is scanned. Initially there is a risk of failing to overlap consecutive sweeps, so that some areas of the smear are missed, or alternatively of overlapping fields so widely that slide reading is very slow. For these reasons trainees should become familiar with the technique while reading simple known smears, so that the method is automatically employed correctly by the time more difficult smears are encountered. For the beginner a simple way of checking on screening technique is to place a number (roughly 5-10) of tiny ink dots on the cover-slip before starting screening and note this number. As screening continues each dot is noted and the total provides a check on screening accuracy. (This 'dotting' should not be confused with slide marking.)

For screening a convenient low-power magnification is in the range ×125 to ×150. This can be achieved with ×10 objective and either ×10 or ×15 eyepieces, depending on whether the tube factor is 1 or 1.25. With this range for low-power work any doubtful cell or group must be examined at a high-power magnification of ×500 or ×600, achieved by changing from the ×10 objective to a ×40 objective. For slide marking (also for the initial examination of relevant pathological specimens) it is convenient to use a ×3.5 or ×4 objective. In routine work there is rarely any need to use an oil immersion objective (×100, giving a magnification of ×1,250 or ×1,500) although occasionally it may be valuable to demonstrate a point in teaching or to confirm an opinion of a cell. If the oil immersion objective is used, however, care should be taken that the observer is familiar with the appearance of benign cells and their nuclei at this order of magnification.

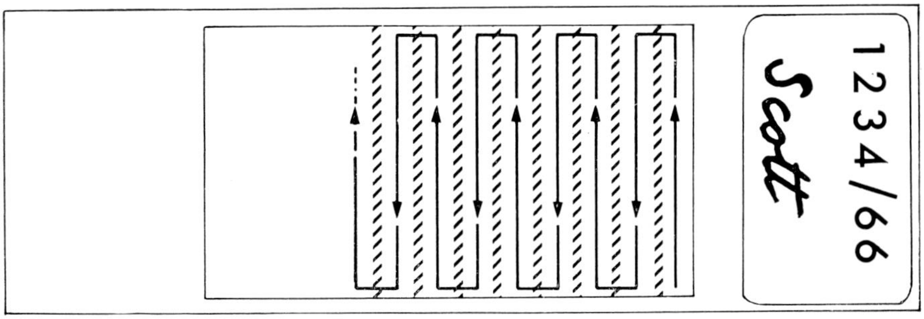

FIG. 169

The 'Screening' of a Smear

In order to examine all the cellular material a standardised pattern of screening should be adopted. By screening the smear in consecutive parallel sweeps, as illustrated, a series of overlapping bands, covering the whole area of the smear, is examined. (The shaded areas indicate the overlap between adjacent fields.)

SLIDE MARKING

It is essential for any member of the staff of a laboratory to be able to demonstrate a selected field to other members of staff. Various methods are available including the use of cross references on the micrometer scales, a graticule or marking objectives. A simple alternative where a laboratory has adopted a convention, as described above, is to use a single ink dot placed by general agreement in a fixed relationship to the selected group of cells. Again a convenient convention is always to place the dot above the relevant group. This is easily done at very low-power magnification (×3.5 or ×4 objective). All that is then necessary is for the person, to whom the slide is shown, to place the slide in the holder with the label to the right and move the stage until the dot is in the centre of the field. The selected group of cells should then lie immediately below the dot. This method avoids the use of any special equipment and also reduces the area of the smear obscured by the marking method to an absolute minimum.

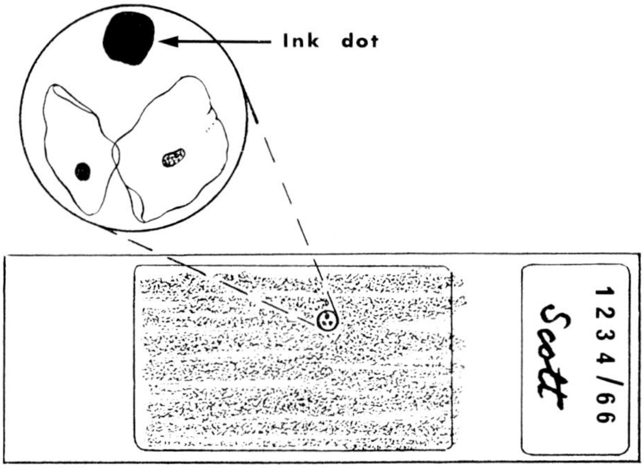

FIG. 170

A Simple Method of Marking a Field

By adopting certain conventions it is possible to identify a selected field accurately with a single ink dot. In this particular case the slide is initially inserted in the holder with the label at the right hand side and a dot is placed above the chosen group of cells. A second observer can easily find the group, by adhering to the right hand side position for the label when he places the slide in the holder, and then looking for a group of cells, which lies immediately below the ink dot.

REPORTING

Whatever the method of reporting specimens adopted by a laboratory it is essential to ensure that the report is readily understood by the clinician. The report on a specimen submitted for the investigation or detection of a malignant disorder can either be in the form of a numerical classification or grading on a previously known and mutually agreed scale or simply be a description. In practice a combination of the two methods is often employed, with a verbal description added to the numerical report. This is especially advisable in the case of smears showing abnormal features. In addition to reporting on the presence or absence of malignant cells in a specimen any other significant findings, which may prove to be relevant to the clinical management of the patient, should be included in the report.

The use of numerical classification or grading is a convenient method of standardising reporting in a laboratory. It can also be valuable for comparing results in different groups of patients. However, comparisons of this type are only valid if the classification is employed in a similar manner by all those responsible for the reports. The numerical classification or grading on the report to the clinician immediately summarises the essential feature of the report; this 'shorthand' communication can save time, but should not replace full reporting of abnormal specimens and joint discussion of management problems.

Descriptive reporting enables the cytopathologist to indicate exactly what abnormal features are present; moreover, it can often, by previous agreement, be combined with suggestions on further investigation. Its main hazard is that it may deteriorate to a point where a large proportion of specimens receive equivocal reports. This can lead to unnecessary anxiety and investigation for the patient and proves useless to the clinician. Although occasionally there are specimens on which it is impossible to reach a conclusion, especially with a single specimen, an attempt should be made to issue a definite report, if necessary after repeat specimens, whenever it is possible to do so.

The basis of most numerical methods of reporting specimens is the method originally described by Papanicolaou, but individual laboratories have modified the significance attached to the number in various ways. It is, therefore, always important to make sure that the particular significance of a classification or grading on a report from any area laboratory is locally known and understood. The Papanicolaou classification ranges from I to V, with the following significance:—

*Class I	Absence of atypical or abnormal cells.
Class II	Atypical cytology but no evidence of malignancy.
Class III	Cytology suggestive of, but not conclusive for, malignancy.
Class IV	Cytology strongly suggestive of malignancy.
Class V	Cytology conclusive for malignancy.

In verbal reporting the description should include a statement that malignant cells are or are not seen. However, in a few cases, when it is not possible to make such a statement categorically, abnormal features may be noted which arouse 'suspicion' in the laboratory. If possible it is better to describe these features and their significance, with a suggestion on further investigation rather than using the word 'suspicious' in a vacuum.

THE DIVISION OF WORK IN THE LABORATORY

The work of the laboratory can be divided into two main groups: specimens submitted primarily for the detection of malignant disorders and specimens submitted for other investigations, such as endocrine assessment. A further subdivision can be made of the first group into two categories: specimens from asymptomatic patients, who are being screened for the purpose of cancer detection, and specimens from symptomatic patients for cancer diagnosis.

* *Atlas of Exfoliative Cytology*, Papanicolaou, G.N. page 21, Cambridge, Mass. Published for the Commonwealth Fund, by Harvard University Press, 1954.

In order to reduce errors to a minimum it is helpful to organise a two-tier system of reporting of specimens submitted for the investigation of malignant disorders, so that specimens are screened and checked. However, the staff to whom the work is originally allocated can be varied according to the type of specimen and the availability of medical and technical staff. The cyto-pathologist should be responsible for the final reporting of all specimens from symptomatic patients with suspicious symptoms, and all abnormal specimens from asymptomatic patients, although preliminary screening may be done by technical or junior medical staff. The reporting of specimens submitted for the investigation of non-malignant diseases usually requires medical staff to correlate the laboratory findings with the clinical information. Whatever plan for reporting is adopted it is advisable for any abnormal or interesting specimens to be widely circulated in the laboratory to enlarge the experience of all the staff.

THE DISTRIBUTION OF WORK IN THE LABORATORY

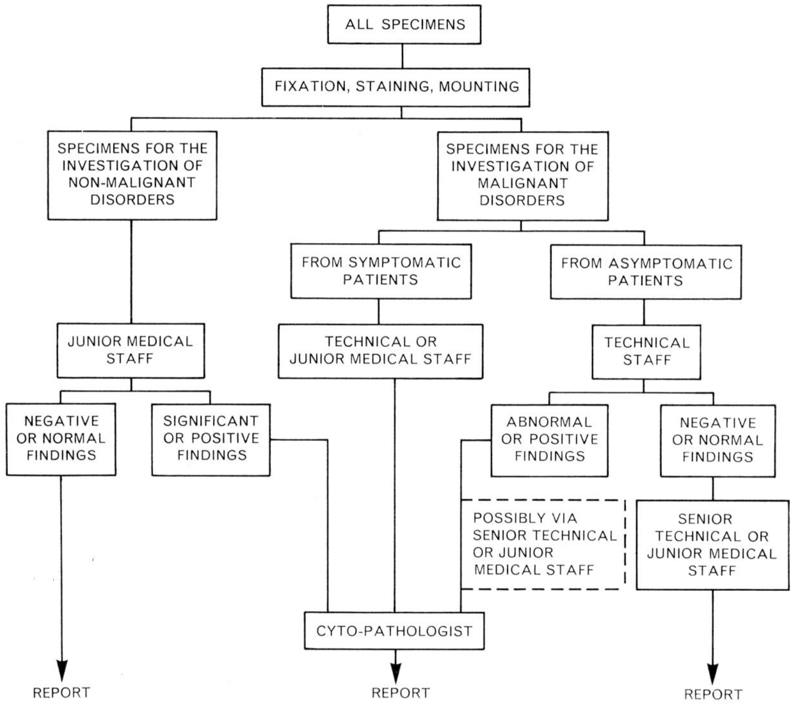

FIG. 171

SAFETY PRECAUTIONS

In view of the nature of many of the specimens arriving in the laboratory, certain precautions are advisable in specimen handling and disposal. The actual handling of specimens should be reduced to a minimum and disposable gloves should be available. As far as possible transparent disposable containers should be used for specimen collection, so that it is unnecessary to transfer the specimen to a fresh container in order to select suitable areas for smear preparation in such specimens as sputa. Receivers of antiseptic solution should be used as receptacles for all discarded specimens and containers. Facilities should be available for Mantoux tests and check radiographic examinations of staff handling fresh specimens.

CHAPTER 27

DIAGNOSTIC CYTOLOGY AND THE CARE OF THE PATIENT

PRIMUM NON NOCERE

CYTOLOGICAL investigation can contribute to the diagnosis of disease, to the control of treatment and to the follow-up of treated patients. However, if the patient is to derive the maximum benefit from diagnostic cytology, the cytopathologist must work in close co-operation with the clinician, so that the results from the cytology laboratory are fully understood and correlated with clinical findings and other investigations. Moreover, the cytopathologist must be prepared to view his results critically and accept that there are limitations to the method, while, at the same time, maintaining the highest standards of accuracy. Unless this standard of laboratory work is combined with properly planned further investigation of significant findings, patients can be exposed to two hazards, which are especially significant in relation to the problems of malignant disease. If 'negative' reports for malignant cells are allowed to counterbalance suspicion aroused by clinical evidence or by other investigations or if 'false negative' reports occur with significant frequency, then patients may be given a completely false sense of security. On the other hand, if single unconfirmed 'positive' reports are used alone as an indication for major surgery, or if 'false positive' reports occur, then the patient may be exposed unnecessarily to the risks of surgery. Both these problems are particularly liable to arise in screening projects. Although the investigation and treatment of any patient must be planned individually, certain problems commonly occur and a brief consideration of possible patterns of investigation may indicate some safeguards for the patient.

THE DETECTION OF CANCER IN ASYMPTOMATIC PATIENTS

At present large scale screening projects are applicable to the diagnosis of gynaecological carcinoma, especially of the cervix uteri, and bladder carcinoma.

Gynaecological patients

If patients who have had cervical smear tests for the detection of gynaecological cancer are free from significant or suspicious symptoms and there are no abnormal findings on pelvic examination, then the patient with a smear which is satisfactory and 'negative' for malignant cells (Class I or II) does not require further investigation for malignant disease at that time. It should, however, be explained to the patient that this 'negative' result is not a permanent guarantee, which allows her to ignore suspicious symptoms in the future. She should also be asked to return for a repeat smear after an interval of twelve months. Incidental non-malignant findings, such as trichomonal or yeast infection, which may give rise to symptoms, should be reported to the clinician, in these as in other patients, so that all relevant information is available to him. If the patient has a 'suspicious' (Class III) smear the first step is to repeat the smear, preferably after treatment of any local condition, such as a severe trichomonal infection, which might contribute to the abnormal appearance. Occasionally it may be necessary to obtain repeat smears over a period of time in this group of patients, because there is difficulty in clearing up an infection, or to allow healing to occur if, for example, the first smear has been taken shortly after a cervical cautery or operation, or if the patient is pregnant. If 'suspicious' (Class III) smears persist then further investigation by examination under anaesthesia, cone biopsy of the cervix and dilatation and curettage is indicated. If the patient has a 'positive' (Class IV or V) smear the first step is, again, to repeat the smear, and thus confirm the identity of the patient. If the repeat smear is again 'positive', the patient should be investigated by examination under anaesthesia, cone biopsy of the cervix and dilatation and curettage. In general these investigations should be performed as soon as possible, but, if the patient is pregnant, it may be decided, after a careful evaluation of the clinical and cytological findings, to postpone full investigation until after delivery. If the patient who has previously had a 'positive' smear is found to have a 'negative' smear when the test is repeated the possible explanations include a confusion of identity, a failure to obtain a satisfactory repeat smear (no endocervical cells seen) or a small lesion, which has been forcibly abraded on the first occasion or escaped the spatula on the second occasion. In most cases a further repeat smear will usually resolve the problem. It should be noted that smears should not be obtained immediately before biopsy, as the abnormal epithelium is readily detached, and this may mean that there is a complete loss of surface epithelium in the biopsy specimen.

Further treatment of these patients with 'suspicious' or 'positive' smears depends on the pathological report on the cone biopsy of the cervix and the endometrium. If the lesion is invasive this report is the essential factor in determining treatment. In pre-invasive lesions the pathological findings are combined with a consideration of the age, parity and wishes of the patient. Since the pathological report is the dominant factor in planning treatment it is of fundamental importance that the pathologist should investigate systematically and completely the whole specimen(s) sent to him, although this may involve him in the study of numerous blocks and sections.

Urological patients

The investigation by the cytological method alone, of patients in high risk groups from the dye and rubber industries, should be restricted to those patients who are completely symptom free. Symptomatic patients should be referred for full urological investigation. In the asymptomatic group those patients who receive a 'negative' report require no further investigation, but should be asked to submit a further specimen of urine in six months time and should also be warned to report any significant symptoms. Patients in whom 'suspicious' or 'positive' findings occur should be referred for full urological investigation, which will include repeat cytological examination.

THE DIAGNOSIS OF CANCER IN SYMPTOMATIC PATIENTS

The role of cytology in the investigation of patients with suspicious symptoms is somewhat different from the part it plays in the screening of well patients. In this group 'negative' findings do not allow the clinician to dismiss the patient, even when the specimen is satisfactory for cytological diagnosis. 'Positive' findings, however, should be checked and confirmed as carefully as in the asymptomatic group. Whenever possible a confirmatory 'positive' specimen should be obtained. In this group, too, the cytological diagnosis should be confirmed by a pathological diagnosis as the basis for major surgery or radiotherapy.

A problem which often arises in this group is the localisation of the lesion. The patient may have aroused suspicion by some generalised finding, such as a persistently elevated sedimentation rate or pyrexia, and cytological investigation may reveal a rather unexpected finding of malignant cells. Alternatively it may be difficult to locate the side on which a lesion is present. In these patients every effort should be directed to localising the lesion and to confirming pathologically that the lesion found is, in fact, the malignant one before carrying out major surgery or radiotherapy.

'FALSE NEGATIVE' AND 'FALSE POSITIVE' DIAGNOSES

These two possibilities give rise to some anxiety in any laboratory, but their most important aspect is their effect on the patient.

'False negative' reports

It is impossible to assess the frequency with which these occur since patients may have 'negative' cytological reports at one hospital and subsequently attend another hospital. Within a hospital or group of hospitals a cross check should be carried out between pathological and cytological reports. When a discrepancy is noted, the cytological slides should be reviewed, especially in lesions, such as carcinoma of cervix, in which a high degree of cytological accuracy can be achieved. In reviewing apparent 'false negative' results an attempt should be made to divide them into two groups; those in which the 'negative' report on the specimen was accurate, i.e., the lesion was inaccessible or failed to exfoliate cells, and those in which the 'negative' report on the specimen was inaccurate, since malignant cells were present, but were overlooked or unrecognised. This latter group is especially significant in the critical assessment of the work of the laboratory and every effort should be made to eliminate this group or reduce its size.

From the point of view of the patient both types of 'false negative' report are equally important. The only safeguards that can be provided are a high standard of accuracy and checking in the laboratory and an overall policy which does not allow a negative report to prevent further investigation of suspicious symptoms or signs.

'False positive' reports

In the laboratory these can be divided into two groups. In the first group the 'false positive' report is found to be inaccurate on reviewing the cytological specimen; it has arisen from a misinterpretation of the appearance of cellular or non-cellular material. The aim should be to reduce this first inaccurate group of 'false positive' reports to a minimum by ensuring that all 'positive' reports are checked by the cytopathologist. In the second group a review of the specimen confirms that malignant cells are present and that the actual report is accurate; the problem lies in finding the lesion.

For the patient both types of 'false positive' report require further investigation. In the first group where, on review, it is apparent that the laboratory has been inaccurate, the examination of further specimens, especially after treatment of any local condition which may have contributed to the problem, is usually sufficient to allay any doubts. However, in the second group, where review of the cytological smears confirms the previous 'positive' report, a rigorous search for the lesion is indicated. In the pathology laboratory the examination of all biopsy material available should be carried to completion. If this fails to indicate a source for the malignant cells, then it must be assumed that the malignant lesion is still present in the patient; often further cytological investigation will confirm this. Further biopsy should then be performed. However, provided major surgery or radiotherapy are only employed after biopsy confirmation of the cytological diagnosis, the 'false positive' report should not have involved the patient in unnecessary radical treatment.

SEROUS EFFUSIONS

One type of specimen requires further comment in relation to 'false negative' and 'false positive' results; this is the specimen of fluid aspirated from a serous body cavity. Since the finding of malignant cells in this type of specimen almost always indicates the presence of metastatic tumour deposits, it is apparent that an incorrect 'positive' report may lead to the discontinuation of investigation and even treatment in a patient who may not have secondary, or even primary, malignant disease. On the other hand an incorrect 'negative' report can cause little harm to the patient. For these reasons 'positive' reports on specimens of this type should be confined to those containing unequivocal malignant cells. This 'positive' report allows the clinician to start palliative therapy and it will also enable him to discuss treatment with the patient and his relatives, with a definite diagnosis in mind. Inevitably this conservative policy means that there will be a significant number of 'false negative' results, but these should be accepted by the laboratory and clinicians as they constitute less of a hazard to these patients than occasional 'false positive' results in this type of specimen.

BIBLIOGRAPHY

REFERENCES

INTRODUCTION

BEALE, L. S. (1860): *Arch. Med. London*, **2**, 44.
DONALDSON, R. (1863): *Am. J. med. Sci.*, **25**, 43.
DONNÉ, A. (1838): Quoted by PAPANICOLAOU, G. N., in *Ann. N.Y. Acad. Sci.* (1956), **63**, 1422.
DUDGEON, L. S., & BARRETT, N. R. (1934): *Br. J. Surg.*, **22**, 4.
DUDGEON, L. S., & PATRICK, C. V. A. (1927): *Br. J. Surg.*, **15**, 250.
PAPANICOLAOU, G. N. (1928): New cancer diagnosis. *Proceedings of the Third Race Betterment Conference*, p. 528.
PAPANICOLAOU, G. N., & TRAUT, H. F. (1943): *The Diagnosis of Uterine Cancer by the Vaginal Smear.* New York Commonwealth Fund.
POUCHET, F. A. (1847): *Théorie Positive de l'Ovulation Spontanée et de la Fécondation des Mammifères et de l'Espèce Humaine.* Paris: Baillère.
ROSENBACH, F. (1882): *Dt. med. Wschr.*, **33**, 452.
STOCKARD, R. C., & PAPANICOLAOU, G. N. (1917): *Am. J. Anat.*, **22**, 225.
WALSHE, W. H. (1846): *On the Nature and Treatment of Cancer.* London: Taylor & Walton.

FURTHER READING

GENERAL

Acta Cytologica (1962): Symposium on the comparative diagnostic accuracy, efficiency, and specificity of techniques for the detection of malignancy, **6**, 508.
BAMFORTH, J. (1966): *Cytological Diagnosis in Medical Practice.* London: Churchill.
BAMFORTH, J., & OSBORN, G. R. (1958): Diagnosis from cells. *J. clin. Path.*, **11**, 473.
BODDINGTON, M. M., & SPRIGGS, A. I. (1965): Cytolgoical diagnosis of cancer: its uses and limitations. *Br. med. J.*, **1**, 1523.
BRACHET, J., & MIRSKY, A. E. (1961): *The Cell*, Vols. I-V. New York: Academic Press.
CAPPELL, D. F. (1964): *Muir's Textbook of Pathology*, Eighth ed. London: Arnold.
CASPERSSON, T. D. (1950): *Cell Growth and Cell Function.* New York: W. W. Norton.
COWDRY, E. V. (1940): Properties of cancer cells. *Archs. Path.*, **30**, 1245.
COWDRY, E. V. (1955): *Cancer Cells.* Philadelphia: W. B. Saunders.
FOOT, N. C. (1950): Limitations of cytological diagnosis from pathological standpoint. *Illinois med. J.*, **98**, 284.
FROST, J. K. (1957): In *Gynaecologic and Obstetric Pathology, with Clinical and Endocrine Relations*, Sixth ed., p. 595. Ed. Novak, E. R., & Woodruff, J. D. Philadelphia: Saunders.
FROST, J. K. (1965): Cancer cells in circulating blood: a comparison with cytologic criteria for other sites. *Acta cytol.*, **9**, 83.
GRAHAM, R. M. (1958): Reliability of cytodiagnostic methods in cancer detection. *N.Y. med. J.*, **58**, 1265.
GRAHAM, R. M. (1963): *The Cytologic Diagnosis of Cancer*, Second ed. Philadelphia: Saunders.
HAAM, E. VON (1962): A comparative study of the accuracy of cancer cell detection by cytological methods. *Acta cytol.*, **6**, 508.
KOSS, L. G., & DURFEE, G. R. (1961): *Diagnostic Cytology and its Histopathologic Bases.* Philadelphia: Lippincott.
OSBORN, G. R. (1953): *Applied Cytology.* London: Butterworth.
PAPANICOLAOU, G. N. (1954): *Atlas of Exfoliative Cytology.* Cambridge, Mass.: Harvard University Press. Suppl. 1, 1956, Suppl. 2, 1960.
PAPANICOLAOU, G. N. (1955): The evolutionary dynamics and trends of exfoliative cytology. *Tex. Rep. Biol. Med.*, **134**, 901.
RAVEN, R. W. (1958): *Cancer.* Vol. 2, Part II. Pathology of malignant tumours. London: Butterworth.
ROBBINS, S. L. (1967): *Pathology.* Third ed. Philadelphia: W. B. Saunders.
SMITHERS, D. W. (1964): *On the Nature of Neoplasia in Man.* Edinburgh: Livingstone.
WIED, G. L. (1965): Quality control mechanisms for cytology programs. *Acta cytol*, **9**, 407.
WYNDER, E. L. (1955): Environmental factors in cervical cancer: an approach to its prevention. *Br. med. J.*, **1**, 743.

SECTION I

FEMALE GENITAL TRACT

Acta Cytologica (1957): Symposium on dyskaryosis, **1**, 23.
Acta Cytologica (1960): Symposium on endocervical adenocarcinoma, **4**, 271.
Acta Cytologica (1961): Symposia on probable or possible pre-malignant cervical lesions, **5**, 31 and 271.
Acta Cytologica (1962): Symposia on probable or possible malignant cervical lesions, **6**, 155.

BIBLIOGRAPHY

ANDERSON, A. F. (1957): Exfoliative cytology in the diagnosis of early cancer. *Jl R. Coll. Surg. Edin.*, **2,** 303.

ANDERSON, A. F. (1959): The management of the cytology test. *J. Obstet. Gynaec. Br. Emp.*, **66,** 239.

ANDERSON, A. F., GOMPEL, C., KOSS, L. G., LIMBURG, H., SCHULLER, E., WESPI, H. J. & YOUNGE, P. A. (1958): Carcinoma in situ of uterine cervix. *Acta Un. int. Cancr.*, **14,** 444.

ANDERSON, A. F., GRANT, M. P. S. & MCBRIDE, R. (1951): Diagnosis of cancer of the uterus. *Proc. R. Soc. Med.*, **44,** 308

ANDERSON, A. F., GRANT, M. P. S., MCBRYDE, R. M. & COCKBURN, M. K. (1953): The place of cervical smears in the diagnosis of early cervical cancer. *J. Obstet. Gynaec. Br. Emp.*, **60,** 345.

ASHLEY, D. J. B. (1966): The biological status of carcinoma in situ of the uterine cervix. *J. Obstet. Gynaec. Br. Commonw.*, **73,** 372.

ASHLEY, D. J. B. (1966): Evidence for the existence of two forms of cervical carcinoma. *J. Obstet. Gynaec. Br. Commonw.*, **73,** 382.

ATTWOOD, M. E. (1966): Cytology and the contraceptive pill. *J. Obst. Gynaec. Br. Commonw.*, **73,** 662.

AYRE, J. E. (1947): A disordered growth response to inflammation in the presence of oestrogen excess and nutritional deficiency; cytological, clinical, nutritional and pathologic studies. *Am. J. Obstet. Gynec.*, **54,** 363.

AYRE, J. E. (1947): Selective cytology smear for the diagnosis of cancer. *Am. J. Obstet. Gynec.*, **53,** 609.

AYRE, J. E. (1948): Diagnosis of preclinical cancer of the cervix: cervical cone knife; its use in patients with a positive vaginal smear. *J. Am. med. Ass.*, **138,** 11.

BECHTOLD, E., & REICHER, N. B. (1952): Relationship of Trichomonas infestations to false diagnoses of squamous carcinoma of cervix. *Cancer, N.Y.*, **5,** 442.

BODDINGTON, M. M., COWDELL, R. H., & SPRIGGS, A. I. (1960): Development of carcinoma of cervix uteri: observations resulting from cytologic examination of 10,000 cervical smears. *Br. J. Cancer*, **14,** 151.

BOYD, J. R., ROYLE, D., FIDLER, H. K., & BOYES, D. A. (1963): Cervical cancer control program in British Columbia. *Am. J. Obstet. Gynec.*, **85,** 328.

BOYES, D. A., FIDLER, H. K., & LOCK, D. R. (1962): Significance of in situ carcinoma of the uterine cervix. *Brit. med. J.*, **1,** 203.

British Medical Journal (1964): The elimination of cervical cancer, **2,** 1409.

BRYANS, F. E., BOYES, D. A., & FIDLER, H. K. (1964): The influence of a cytological screening program upon the incidence of squamous cell carcinoma of the cervix in British Columbia. *Am. J. Obstet. Gynec.*, **88,** 898.

CARSON, R. P., & GALL, E. A. (1954): Preinvasive carcinoma and precancerous metaplasia of cervix. *Am. J. Path.*, **30,** 15.

CHRISTOPHERSON, W. M., & BROGHAMER, W. L., Jun. (1962): A study of the reversibility of dysplasia of the uterine cervix. *Proceedings of the First International Congress of Exfoliative Cytology*, p. 269. Philadelphia: Lippincott.

CHRISTOPHERSON, W. M., & PARKER, J. E. (1961): A critical study of cervical biopsies including serial sectioning. *Cancer, N.Y.*, **14,** 213.

CHRISTOPHERSON, W. M., PARKER, J. E., & DRYE, J. C. (1962): Control of cervical cancer; report of a community cervical cancer control program. *J. Am. med. Ass.*, **2,** 178.

COMAN, D. R. (1944): Decreased material adhesiveness; property of cells from squamous cell carcinomas. *Cancer Res.*, **4,** 625.

CORSCADEN, J. A. (1964): *Gynaecologic Cancer.* Third ed. Baltimore: Williams & Wilkins.

DAVIS, H. J. (1962): The irrigation smear: accuracy in detection of cervical cancer. *Acta cytol.*, **6,** 459.

DAVIS, H. J. & JONES, H. W. (1966): Population screening for cancer of the cervix with irrigation smears. *Am. J. Obstet. Gynec.*, **96,** 605.

DUNN, J. E., Jun. (1953): Relationship between carcinoma in situ and invasive cervical carcinoma; consideration of contributions to problem to be made from population data. *Cancer, N.Y.*, **6,** 873.

FLUHMANN, C. F. (1961): *The Cervix Uteri and its Diseases.* Philadelphia: Saunders.

FOOTE, F. W., & STEWART, F. W. (1948): The anatomical distribution of intraepithelial epidermoid carcinoma of the cervix. *Cancer, N.Y.*, **1,** 431.

FORAKER, A. C. (1952): Analysis of hyperchromatism in the histologic diagnosis of carcinoma of the cervix uteri, with special reference to intraepithelial and invasive squamous cell carcinoma and squamous metaplasia. *Archs. Path.*, **53,** 250.

FORAKER, A. C., & REAGAN, J. W. (1959): Nuclear mass and allied phenomena in normal exocervical mucosa, squamous metaplasia, atypical hyperplasia, intraepithelial carcinoma and invasive squamous cell carcinoma of cervix. *Cancer, N.Y.*, **12,** 894.

FREMONT-SMITH, M., GRAHAM, R. M., & MEIGS, J. V. (1947): Vaginal smears as an aid in the diagnosis of early carcinoma of the cervix. *New Engl. J. Med.*, **237,** 302.

FREMONT-SMITH, M., GRAHAM, R. M., & MEIGS, J. V. (1948): Early diagnosis of cancer by study of exfoliated cells. *J. Am. med. Ass.*, **138,** 469.

FROST, J. K. (1962): Trichomonas vaginalis and cervical epithelial changes. *Ann. N.Y. Acad. Sci.*, **97,** 792.

GATES, O., & WARREN, S. (1949): *Handbook for the Diagnosis of Cancer of the Uterus by the Use of Vaginal Smears.* Third ed. Cambridge, Mass.: Harvard University Press.

GORE, H. & HERTIG, A. T. (1966): Carcinoma in situ of the endometrium. *Am. J. Obstet. Gynec.*, **94,** 134.

GOVAN, A. T. D., HAINES, R. M., LANGLEY, F. A., TAYLOR, C. W., & WOODCOCK, A. S. (1966): Changes in the epithelium of the cervix. *J. Obstet. Gynaec. Br. Commonw.*, **73,** 883.

GRAHAM, J. B., & GRAHAM, R. M. (1967): Cul-de-sac puncture in the diagnosis of early ovarian carcinoma. *J. Obstet. Gynaec. Br. Commonw.*, **74,** 371.

GRAHAM, J. B. & MEIGS, J. V. (1952): Earlier detection of recurrent cancer of the uterine cervix by vaginal smear. *Am. J. Obstet. Gynec.*, **64,** 908.

GRAHAM, R. M. (1953): The value of the vaginal smear to the gynaecologist. *J. Obstet. Gynaec. Br. Emp.*, **60,** 483.

GRAHAM, R. M. (1958): The diagnostic accuracy of vaginal smears for detection of endometrial cancer. *Acta cytol.*, **2,** 579.

GRAHAM, R. M., BARTELS, J., & GRAHAM, J. B. (1962): Screening for ovarian cancer by cul-de-sac aspiration. *Acta cytol.*, **6,** 492.
GRAHAM, R. M., & MEIGS, J. V. (1949): The value of the vaginal smear. *Am. J. Obstet. Gynec.*, **58,** 843.
GRAHAM, R. M., & VAN NICKERK, W. A. (1962): Vaginal cytology in cancer of the ovary. *Acta cytol.*, **6,** 496.
GRAY, L. A. (Ed.) (1964): *Dysplasia, Carcinoma in situ and Microinvasive Carcinoma of the Uterine Cervix.* Springfield, Ill.: Thomas.
GREENE, G. H. (1966): The significance of cervical carcinoma in situ. *Am. J. Obstet. Gynec.*, **94,** 1009.
GREENE, R. R., & PECKHAM, B. M. (1958): Preinvasive cancer of the cervix and pregnancy. *Am. J. Obstet. Gynec.*, **75,** 551.
GUSBERG, S. B. (1953): The relative efficiency of diagnostic techniques in the detection of early cervical cancer. *Am. J. Obstet. Gynec.*, **65,** 1073.
GUSBERG, S. B., MOORE, D. B., & MARTIN, F. (1954): Precursors of corpus cancer. II. A clinical and pathological study of adenomatous hyperplasia. *Am. J. Obstet. Gynec.*, **68,** 1472.
HAINES, M., & TAYLOR, C. W. (1962): *Gynaecological Pathology.* London: Churchill.
HECHT, E. L. (1952): The cytologic approach to uterine carcinoma. Detection, diagnosis and therapy. *Am. J. Obstet. Gynec.*, **64,** 81.
HECHT, E. L. (1953): The diagnostic accuracy of the endometrial aspiration smear. *Am. J. Obstet. Gynec.*, **66,** 1321.
HERTIG, A. T., & GORE, H. (1960): Tumors of the female sex organs. Part 2. Tumors of the vulva, vagina and uterus. Armed Forces Institute of Pathology. *Atlas of Tumor Pathology,* Section IX, Fasc. 33. Washington, D.C.
HERTIG, A. T., & YOUNGE, P. A. (1952): What is cancer in situ of the cervix? Is it the preinvasive form of true carcinoma? *Am. J. Obstet. Gynec.*, **64,** 807.
HUGHES, H. E., GORDON, A. M., & BARR, G. T. D. (1966): A clinical and laboratory study of trichomoniasis of the female genital tract. *J. Obstet. Gynaec. Br. Commonw.*, **73,** 821.
KASDON, C. S., & BAMFORD, S. B. (1962): *Atlas of in-situ Cytology.* Boston: Little, Brown.
KIRKLAND, J. (1963): Carcinoma in situ—diagnosis and prognosis. *J. Obstet Gynaec. Br. Emp.*, **70,** 232.
KOSS, L. G., & DURFEE, G. R. (1955): Cytological changes preceding appearance of in-situ carcinoma of uterine cervix. *Cancer, N.Y.* **8,** 295.
KOSS, L. G., & DURFEE, G. R. (1962): Cytologic diagnosis of endometrial carcinoma. Result of ten years' experience. *Acta cytol.*, **6,** 519.
KOSS, L. G., STEWART, F. W., FOOTE, F. W., JORDAN, M. J., BADER, G. M., & DAY, E. (1963): Some histological aspects of behaviour of epidermoid carcinoma in situ and related lesions of the uterine cervix, a long-term prospective study. *Cancer, N.Y.*, **16,** 1160.
KOSS, L. G., & WOLINSKA, W. H. (1959): Trichomonas vaginalis cervicitis and its relationship to cervical cancer; histocytological study. *Cancer, N.Y.* **12,** 1171.
KOTTMEIER, H. L. (1953): *Carcinoma of the Female Genitalia.* Baltimore: Williams & Wilkins.
Lancet (1957): Cytological detection of uterine cancer, **1,** 624.
LIU, W. (1959): *An Introduction to Gynaecological Exfoliative Cytology.* Springfield, Ill.: Thomas.
LONG, M. E., & TAYLOR, H. C., Jun. (1956): Nucleolar variability in human neoplastic cells. *Ann. N.Y. Acad. Sci.*, **63,** 1095.
MCGREGOR, J. E., & BAIRD, D. (1963): Detection of cervical carcinoma in the general population. *Br. med. J.*, **1,** 1631.
MCKAY, D. G., TERJANIAN, B., POSCHYACHINDA, D., YOUNGE, P. A., & HERTIG, A. T. (1959): Clinical and pathological significance of anaplasia (atypical hyperplasia) of cervix uteri. *Obstet. Gynec., N.Y.*, **13,** 2.
MCLAREN, H. C. (1963): *The Prevention of Cervical Cancer.* London: English Universities Press.
MCLAREN, H. C., & ATTWOOD, M. E. (1961): Methods of diagnosing cervical carcinoma in situ. *Br. med. J.*, **2,** 1111.
MCLAREN, H. C., TAYLOR, C. W., & ATTWOOD, M. E. (1956): The value of cytology in gynaecology. *J. Obstet. Gynec. Brit. Emp.*, **63,** 801.
MEIGS, J. V., GRAHAM, R. M., FREMONT-SMITH, M., KAPNICK, I., & RAWSON, R. W. (1943): Value of vaginal smear in diagnosis of uterine cancer. *Surgery, Gynec., Obstet.*, **77,** 449.
MUSSEY, E., & DECKER, D. (1967): Intraepithelial carcinoma of cervix in association with pregnancy. *Am. J. Obstet. Gynec.*, **97,** 30.
NESBITT, R. E. L., Jun., & HELLMANN, L. M. (1952): The histopathology and cytology of the cervix in pregnancy. *Surgery, Gynec., Obstet.*, **94,** 10.
NIEBURGS, H. E. (1956): Mass screening of the total female population of a county; prevalence and incidence rates. *Transactions of the First International Cancer Cytology Congress,* p. 327.
NOVAK, E. R. (1956): Postmenopausal endometrial hyperplasia. *Am. J. Obstet. Gynec.*, **71,** 1312.
NOVAK, E. R., & WOODRUFF, J. D. (1967): *Gynaecologic and Obstetric Pathology, with Clinical and Endocrine Relations.* Sixth ed. Philadelphia: W. B. Saunders.
PAPANICOLAOU, G. N. (1946): General survey of the vaginal smear and its use in research and diagnosis. *Am. J. Obstet. Gynec.*, **51,** 316.
PAPANICOLAOU, G. N. (1949): Cytologic diagnosis of uterine cancer by examination of vaginal and uterine secretions. *Am. J. clin. Path.*, **19,** 301.
PAPANICOLAOU, G. N. (1953): Observations on the origin and specific function of the histiocytes in the female genital tract. *Fert. Steril.*, **4,** 472.
PAPANICOLAOU, G. N., & TRAUT, H. F. (1943): Diagnosis of uterine cancer by the vaginal smear. New York: Commonwealth Fund.
PETERSEN, O. (1955): Precancerous changes of the cervical epithelium in relation to manifest cervical carcinoma; clinical and histological asoects. *Acta. radiol. Suppl.* 127.
PETERSEN, O., & WIKLUND, E. (1959): Further studies on the spontaneous course of cervical precancerous conditions. *Acta radiol. Suppl.* 188.

REAGAN, J. W., HICKS, D. J., & SCOTT, R. B. (1955): Atypical hyperplasia of uterine cervix. *Cancer, N.Y.*, **8**, 42.
REAGAN, J. W., & MOORE, R. D. (1952): Morphology of the malignant squamous cell. *Am. J. Path.*, **28**, 105.
REAGAN, J. W., SEIDEMANN, I. L., & SARACUSA, Y. (1953): Cellular morphology of carcinoma in situ and dysplasia or atypical hyperplasia of uterine cervix. *Cancer, N.Y.*, **6**, 224.
REAGAN, J. W., SEIDEMANN, I. B., & PATTEN, S. F. (1962): Developmental stages of in situ carcinoma in uterine cervix; an analytical study of the cells. *Acta cytol.*, **6**, 538.
RUSSELL, J. K. (1951): The dangerous vaginal pessary. *Br. med. J.*, **2**, 1595.
RUTLEDGE, C. E., Jun., CHRISTOPHERSON, W. M., & PARKER, J. E. (1962): Cervical dysplasia and carcinoma in pregnancy. *Obstet. Gynec., N.Y.*, **19**, 351.
SMOLKA, H., & SOOST, H.-J. (1965): *An Outline and Atlas of Gynaecological Cyto-diagnosis.* Second ed. London: Arnold.
SONG, Y. S. (1957): The significance of positive vaginal smears in extrauterine carcinoma. *Am. J. Obstet. Gynec.*, **73**, 341.
SPEERT, H. (1948): Corpus cancer. Clinical, pathological and aetiological aspects. *Cancer, N.Y.*, **1**, 584.
STERN, E., & DIXON, W. J. (1961): Cancer of the cervix. A biometric approach to aetiology. *Cancer, N.Y.*, **14**, 153.
STOCKS, P. (1955): Cancer of the uterine cervix and social conditions. *Br. J. Cancer*, **9**, 487.
TAYLOR, H. C., Jun. (1952): Controversial points in treatment of carcinoma of cervix. *Cancer, N.Y.*, **5**, 435.
TERRIS, M. (1962): Epidemiology of cervical cancer. *Ann. N.Y. Acad. Sci.*, **97**, 808.
Transactions of the international committee on histology terminology of the cervix (1961). *Proceedings of the First International Congress of Exfoliative Cytology.* Philadelphia: Lippincott.
UMIKER, W., & STEEN, M. (1953): Carcinoma of ovary with malignant cells in the vaginal smear of an asymptomatic patient. *Am. J. Obstet. Gynec.*, **66**, 674.
WACHTEL, E. (1964): *Exfoliative Cytology in Gynaecological Practice.* London: Butterworth.
WACHTEL, E., & PLESTER, J. A. (1952): The vaginal smear as an aid to the diagnosis of genital tract malignancy in women. *J. Obstet. Gynaec. Br. Emp.*, **59**, 323.
WALSH, H., HILDEBRANDT, R. J., & PRYSTOWSKY, H. (1965): Candidial vaginitis associated with the use of oral progestational agents. *Am. J. Obstet. Gynec.*, **93**, 904.
WAY, S. (1948): Primary carcinoma of vagina. *J. Obstet. Gynaec. Br. Emp.*, **55**, 739.
WAY, S. (1951): *Malignant Disease of the Female Genital Tract.* London: Churchill.
WAY, S. (1962): Exfoliative cytology, early carcinoma of the cervix, pregnancy and fertility. *Acta cytol.*, **6**, 547.
WAY, S. (1963): *Diagnosis of Early Carcinoma of the Cervix.* London: Churchill.
WAY, S. (1964): Microinvasive carcinoma of the cervix. *Acta cytol.*, **8**, 14.
WHEELER, J. D., & HERTIG, A. T. (1955): The pathologic anatomy of carcinoma of the uterus. 1. Squamous carcinoma of the cervix. *Am. J. clin. Path.*, **25**, 345.
WIED, G. L. (1955): Importance of site from which vaginal smears are taken. *Am. J. clin. Path.*, **25**, 742.
WIED, G. L. (1956): The potentialities of the smear technique for the differentiation of non-invasive and invasive cervical carcinoma. *Am. J. Obstet. Gynec.*, **71**, 793.
WIED, G. L., LEGORRETA, G., MOHR, D., & RAUZY, D. (1962): Cytology of invasive cervical carcinoma and carcinoma in situ. *Ann. N.Y. Acad. Sci.*, **97**, 759.
WORTH, A. J., BOYES, D. A., & FIDLER, H. K. (1967): The acceptance of the cervical cytology screening programme in the province of British Columbia. *J. Obst. Gynaec. Br. Commonw.*, **74**, 479.
WRIGLEY, A. J. (1960): A critical review of the value of vaginal cytology. *Practitioner*, **184**, 459.
WYNDER, D. L., CORNFIELD, J., SCHROFF, P. D., & DORAISWAMI, K. R. (1954): Study of environmental factors in carcinoma of cervix. *Am. J. Obstet. Gynec.*, **68**, 1016.
YOUNGE, P. A. (1965): The natural history of carcinoma in situ of the cervix uteri. *J. Obstet. Gynaec. Br. Emp.*, **72**, 9.

HORMONAL EFFECTS ON THE VAGINAL SMEAR

Acta Cytologica (1957): Symposium on androgenic effects on the vaginal epithelia, **1**, 70.
Acta Cytologica (1958): Symposium on the effects of exogenous oestrogens on the vaginal epithelia, **2**, 356.
Acta Cytologica (1959): Symposium on hormonal cytology during pregnancy and the post-partum period, **3**, 197.
Acta Cytologica (1960): Symposium on the effects of endogenous oestrogens on the vaginal epithelia, **4**, 13.
Acta Cytologica (1962): Symposium on the effects of progestational agents, **6**, 178 and 211.
DE ALLENDE, I. L. C., & ORIAS, O. (1950): *Cytology of the Human Vagina.* New York: Hoeber.
FERIN, J., & DEMOL, R. (1952): Étude comparée des différents techniques d'examen des frottis vaginaux au point de vue hormonal. *Annls Endocr.*, **13**, 527.
HAAM, E. VON (1961): The cytology of pregnancy. *Acta cytol.*, **5**, 320.
HUGHES, H. E., LORAINE, J. A., BELL, E. T., & LAYTON, R. (1964): Cytological observations, cervical mucus 'ferning' and hormone assays in early pregnancy. *Am. J. Obstet. Gynec.*, **90**, 1297.
LORAINE, J. A., & BELL, E. T. (1966): *The Clinical Application of Hormone Assay.* Second ed. Edinburgh: Livingstone.
NESBITT, R. E. L., Jun., GARCIA, R., & ROME, D. S. (1961): The prognostic value of vaginal cytology in pregnancy. *Obstet. Gynec., N.Y.*, **17**, 2.
OSMOND-CLARKE, F., & MURRAY, M. (1963): Vaginal cytology and recurrent abortions. *Br. med. J.*, **2**, 1174.
OSMOND-CLARKE, F., MURRAY, M., & WOOD, C. (1964): Endocrine cytology in pregnancy. Cytological changes before normal and premature labour. *J. Obstet. Gynec. Br. Commonw.*, **71**, 231.
PAPANICOLAOU, G. N. (1933): Existence of a 'post-menopause' sexual rhythm in women, as indiated by the study of the vaginal smear. *Anat. Rec.*, **55**, Suppl. 71.

PAPANICOLAOU, G. N. (1933): The sexual cycle in the human female as revealed by vaginal smears. *Am. J. Anat.*, **52,** Suppl. 519.
PAPANICOLAOU, G. N., & SHORR, E. (1936): Action of ovarian follicular hormone in menopause as indicated by vaginal smears. *Am. J. Obstet. Gynec.*, **31,** 806.
PAPANICOLAOU, G. N., TRAUT, H. F., & MARCHETTI, A. A. (1948): *The Epithelia of Woman's Reproductive Organs.* New York: Commonwealth Fund.
PIERCE, J. R. (1954): The vaginal smear as a guide to prognosis and treatment in threatened abortion. *Am. J. Obstet. Gynec.*, **67,** 47.
PIERCE, J. R. (1957): The abnormal smear in abortion. *Am. J. Obstet. Gynec.*, **74,** 119.
POUCHET, F. A. (1847): *Théorie Positive de l'Ovulation Spontanée et la Fécondation.* Paris: Baillière.
PUNDEL, J. P. (1950): *Les Frottis Vaginaux et Cervicaux.* Paris: Masson.
PUNDEL, J. P. (1952): *Les Frottis Vaginaux Endocriniens.* Paris: Masson.
PUNDEL, J. P. (1957): *Acquisitions Récentes en Cytologie Vaginale Hormonale.* Paris: Masson.
PUNDEL, J. P. (1966): *Précis de Colpocytologie Hormonale.* Paris: Masson.
PUNDEL, J. P., & VAN MEENSEL, F. (1951): *Gestation et Cytologie Vaginale.* Paris: Masson.
RAKOFF, A. E., WACHTEL, E., & WIED, G. (1958): Hormonal evaluation of patient through cytologic interpretation. *Acta Un. int. Cancr.*, **14,** 268.
RILEY, G. M., DONTAS, E., & GILL, B. (1955): Use of serial vaginal smears in detecting time of ovulation. *Fert. Steril.*, **6,** 86.
SOULE, S. D. (1964): The practical value of vaginal cytology in pregnancy. 1. Cytological prediction of the fate of early pregnancy. *Acta. cytol.*, **8,** 364.
STOCKARD, R. C., & PAPANICOLAOU, G. N. (1917): The existence of a typical oestrous cycle in the guinea pig, with a study of its histological and physiological changes. *Am. J. Anat.*, **22,** 225.
WACHTEL, E., & PLESTER, J. A. (1954): Hormonal assessment by vaginal cytology. *J. Obstet. Gynec. Br. Commonw.*, **61,** 155.
WIED, G. L. (1955): Importance of the site from which vaginal cytologic smears are taken. *Am. J. clin. Path.*, **25,** 742.
WIED, G. L. (1955): Suggested standard for karyopyknosis. Use in hormonal reading of vaginal smears. *Fert. Steril.*, **6,** 61.
WIED, G. L. (1957): Climacteric amenorrhoea. A cytohormonal test for differential diagnosis. *Obstet. Gynec. N.Y.*, **9,** 646.
WIED, G. L., DAVIS, M. I., FRANK, R., SEIGEL, P. B., MEIER, P., & ROSENTHAL, E. (1966): Statistical evaluation of effect of hormonal contracpetive on cytologic smear pattern. *Obstet. Gynec. N.Y.*, **27,** 327.
WOOD, C., OSMOND-CLARKE, F., & MURRAY, M. (1961): Vaginal cytology in pregnancy. *J. Obstet. Gynaec. Br. Commonw.*, **68,** 778.
YOUNG, S., BULBROOK, R. D., & GREENWOOD, F. C. (1957): The correlation between urinary oestrogen and vaginal cytology. *Lancet*, **1,** 350.

SECTION II

RESPIRATORY TRACT

BAMFORTH, J. (1946): The examination of the sputum and pleural fluid in the diagnosis of malignant diseases of the lung. *Thorax*, **1,** 118.
CHRISTOPHERSON, W. M. (1959): The cytologic examination of sputum and bronchial secretions. *J. Ky. St. med. Ass.*, November 1959.
CLERF, L. H., & HERBUT, P. A. (1950): The value of cytologic diagnosis of pulmonary malignancy. *Am. Rev. Tuberc. pulm. Dis.*, **61,** 60.
FARBER, S. M. (1961): Clinical appraisal of pulmonary cytology. *J. Am. med. Ass.*, **175,** 345.
FOOT, C. N. (1955): Cytologic diagnosis in suspected pulmonary cancer. *Am. J. clin. Path.*, **25,** 223.
GOWAR, F. J. S. (1943): Carcinoma of the lung: the value of sputum examination in diagnosis. *Br. J. Surg.*, **39,** 193.
GRUNZE, H. (1960): A critical review and evaluation of cytodiagnosis in chest diseases. *Acta cytol.*, **4,** 175.
KOSS, L. G. (1958): Cellular changes simulating bronchogenic carcinoma. *Acta Un. int. Cancr.*, **14,** 501.
KOSS, L. G., & RICHARDSON, H. L. (1955): Some pitfalls of diagnosis of lung cancer. *Cancer, N.Y.*, **8,** 5.
KOSS, L. G., MELAMED, M. R., & GOODNER, J. T. (1964): Pulmonary cytology. A brief survey of diagnostic results from July 1st 1952 to December 31st 1960. *Acta cytol.*, **8,** 104.
KUPAR, S. W. A. (1963): Exfoliative cytology in the diagnosis of cancer of the bronchus. *J. clin. Path.*, **16,** 399.
Lancet (1961): Detection of lung cancer, **1,** 1211.
NAIB, Z. M. (1959): Primary carcinoma of lung—cytological detection. Transactions of the Seventh Annual Meeting of the Inter-society Cytology Council, p. 294.
NAIB, Z. M. (1964): Pitfalls in the cytologic diagnosis of oat cell carcinoma of lung. *Acta cytol.*, **8,** 34.
PAPANICOLAOU, G. N., & CROMWELL, H. A. (1949): Diagnosis of cancer of the lung by the cytologic method. *Dis. Chest*, **15,** 412.
PHILPS, F. R. (1954): Adenocarcinoma cells in the sputum. *Br. J. Cancer*, **8,** 67.
PHILPS, F. R. (1954): Appearances which may cause confusion in the diagnosis of bronchial carcinoma from sputum examination. *Br. J. Cancer*, **8,** 422.

PHILPS, F. R. (1964): *A Short Manual of Respiratory Cytology*. London: Pitman.
UMIKER, W. (1952): Cytology in bronchogenic carcinoma. *Am. J. clin. Path.*, **22,** 558.
UMIKER, W. (1957): False negative reports in the cytologic diagnosis of cancer of the lung. *Am. J. clin. Path.*, **28,** 37.
WOOLNER, L. B., & MCDONALD, J. R. (1949): Cytologic diagnosis of bronchogenic carcinoma. *Am. J. clin. Path.*, **19,** 765.
WOOLNER, L. B., & SCHMIDT, H. W. (1950): Study of factors involved in exfoliation of malignant cells. *J. thorac. Surg.*, **20,** 125.

SECTION III

GASTRO-INTESTINAL TRACT

BODDINGTON, M. M., & SPRIGGS, A. I. (1959): The epithelial cells in megaloblastic anaemia. *J. clin. Path.*, **12,** 228.
GIBBS, D. D., & JENNINGS, D. (1961): Exfoliative gastric cytology. *Lancet*, **2,** 1413.
GRAHAM, R. M., & RHEAULT, M. H. (1954): Characteristic cellular changes in epithelial cells in pernicious anaemia. *J. Lab. clin. Med.*, **43,** 235.
HENNING, N., WITTE, S., & BRESSEL, D. (1964): The cytologic diagnosis of tumors of the upper gastrointestinal tract. (Esophagus, stomach, duodenum.) *Acta cytol.*, **8,** 121.
MACDONALD, W. C., BRANDBORG, L. L., TANIGUCHI, L., & RUBIN, C. E. (1963): Oesophageal exfoliative cytology: a neglected procedure. *Ann. intern. Med.*, **59,** 332.
MACDONALD, W. C., BRANDBORG, L. L., TANIGUCHI, L., & RUBIN, C. E. (1963): Gastric exfoliative cytology: an accurate and practical diagnostic procedure. *Lancet*, **2,** 83.
OAKLAND, D. J. (1964): The diagnosis of carcinoma of the colon by exfoliative cytology. *Proc. R. Soc. Med.*, **57,** 279.
RASKIN, H. F., KIRSNER, J. B., & PALMER, W. L. (1959): Rôle of exfoliative cytology in diagnosis of cancer of digestive tract. *J. Am. med. Ass.*, **169,** 789.
RASKIN, H. F., & PLETICKA, S. (1964): The cytologic diagnosis of cancer of the colon. *Acta. cytol.*, **8,** 131.
REECE, M. F., BOON, T. H., & SCHADE, R. O. K. (1961): Exfoliative cytology: its evaluation in the diagnosis of carcinoma of the stomach. *Lancet*, **2,** 1163.
RICHARDS, W. C. D., & SPRIGGS, A. I. (1961): The cytology of gastric mucosa. *J. clin. Path.*, **14,** 132.
SANDLER, H. C. (1962): Oral exfoliative cytology for detection of early mouth cancer. *Acta cytol.*, **6,** 355.
SANDLER, H. C. (1964): The cytologic diagnosis of tumours of the oral cavity. *Acta cytol.*, **8,** 114.
SCHADE, R. O. K. (1958): Exfoliative cytology of gastric carcinoma. *Br. med. J.*, **1,** 743.
SCHADE, R. O. K. (1959): A critical review of gastric cytology. *Acta cytol.*, **3,** 7.
SCHADE, R. O. K. (1960): *Gastric Cytology*. London: Arnold.
SEYBOLT, J. F., & PAPANICOLAOU, G. N. (1951): Cytology in the diagnosis of gastric cancer. *Cancer, N.Y.*, **4,** 286.
UMIKER, W. O., BOLT, R. J., HOEKZEMA, A. D., & POLLARD, H. M. (1958): Cytology in the diagnosis of gastric cancer. *Gastroenterology*, **34,** 859.

SECTION IV

URINARY TRACT

BAMFORTH, J. (1958): Cytological diagnosis of prostatic carcinoma. *Ann. R. Coll. Surg.*, **23,** 248.
BONSER, G. M., FAULDS, J. S., & STEWART, M. J. (1955): Occupational cancer: of the urinary bladder in dyestuffs operatives and of the lung in asbestos textile workers and iron-ore miners. *Am. J. clin. Path.*, **25,** 126.
British Medical Journal (1966): Bladder cancer in rubber industry, **1,** 329.
CRABBE, J. G. S. (1952): Exfoliative cytological control in occupational cancer of the bladder. *Br. med. J.*, **2,** 1072.
CRABBE, J. G. S. (1961): Cytology of voided urine with special reference to 'benign' papillomas and some of the problems encountered in the preparation of smears. *Acta cytol.*, **5,** 233.
CRABBE, J. G. S., CRESDEE, W. C., SCOTT, T. S., & WILLIAMS, M. H. (1956): The cytological diagnosis of bladder tumours amongst dyestuff workers. *Br. J. ind. Med.*, **13,** 270.
FERGUSSON, J. D., & GIBSON, E. C. (1956): Prostatic smear diagnosis. *Br. med. J.*, **1,** 822.
FOOT, N. C., PAPANICOLAOU, G. N., HOLMQUIST, N. D., & SEYBOLT, J. F. (1958): Exfoliative cytology of urinary sediments. Review of 2,829 cases. *Cancer, N.Y.*, **11,** 127.
MCFARLANE, W. E., CEELEN, C. H., & TAYLOR, J. N. (1964): Urine cytology after treatment of bladder tumours. *Acta cytol.*, **8,** 288.
MELAMED, M. R., KOSS, L. G., RICCI, A., & WHITMORE, W. F. (1960): Cytohistological observations on developing carcinoma of the urinary bladder in man. *Cancer, N.Y.*, **13,** 67.
PETERS, H. (1952): The cytologic interpretation of the prostatic smear. *Surgery, Gynec., Obstet.*, **94,** 69.

Solomon, C., Amelar, R. D., Hyman, R. M., Chaiban, R., & Europa, D. L. (1958): Exfoliative cytology of urinary tract: a new approach with reference to isolation of cancer cells and preparation of slides for study. *J. Urol.*, **80,** 374.
Umiker, W. (1962): Exfoliative cytology of papillomas and intraepithelial carcinomas of the urinary bladder. *Acta cytol.*, **6,** 255.
Umiker, W. (1964): Accuracy of cytologic diagnosis of cancer of the urinary tract. *Acta cytol.*, **8,** 186.

SECTION V

THE BREAST

Donné, A. (1845): *Cours de Microscopie Complémentaire des Études Médicales.* Paris: Baillière.
Godwin, J. T. (1964): Cytologic diagnosis of aspiration biopsies of solid or cystic tumours. *Acta cytol.*, **8,** 206.
Kjellgren, O. (1964): The cytologic diagnosis of cancer of the breast. *Acta Cytol.*, **8,** 216.
Papanicolaou, G. N., Bader, G. M., Holmquist, D. G., & Falk, E. A. (1956): Cytological evaluation of breast secretions. *Ann. N.Y. Acad. Sci.*, **63,** 1409.
Papanicolaou, G. N., Holmquist, D. G., Bader, G. M., & Falk, E. A. (1958): Exfoliative cytology of the human mammary gland and its value in the diagnosis of cancer and other diseases of the breast. *Cancer, N.Y.*, **11,** 377.
Robbins, G. F., & Brothers, J. A. (1954): Aspiration of breast cancer—dangers to the patient. *Cancer, N.Y.*, **7,** 774.
Saphir, O. (1950): Cytologic examination of breast secretion. *Am. J. clin. Path.*, **20,** 1001.
Saphir, O. (1954): Cytologic examination of breast secretion. *Proceedings National Cancer Conference*, **2,** 1238.

SECTION VI

THE BODY FLUIDS

Alexander, R. F., & Spriggs, A. I. (1966): The differential diagnosis of tumour cells in the circulating blood. *J. clin. Path.*, **13,** 414.
Foot, N. C. (1954): Identification of types and primary sites of metastatic tumours from exfoliated cells in serous fluids. *Am. J. Path.*, **30,** 661.
Foot, N. C. (1956): Identification of cells in exudates and effusates. *Ann. N.Y. Acad. Sci.*, **63,** 1324.
Foot, N. C. (1959): Identification of mesothelial cells in sediments of serous effusions. *Cancer, N.Y.*, **12,** 429.
Frost, J. K. (1965): Cancer cells in circulating blood; a comparison with cytologic criteria for other sites. *Acta cytol.*, **9,** 83.
Grunze, H. (1964): Comparative diagnostic accuracy, efficiency and specificity of cytologic technics used in the diagnosis of malignant neoplasms in serous effusions of pleural and pericardial cavities. *Acta cytol.*, **8,** 150.
Luse, S. A., & Reagan, J. W. (1954): A histocytological study of effusions. *Cancer, N.Y.*, **7,** 1155 and 1167.
Mavromattis, F. S. (1964): Some morphologic features of cells containing P.A.S. positive intracytoplasmic granules in smears of serous effusions. *Acta cytol.*, **8,** 426.
Naylor, B. (1964): The cytologic diagnosis of cerebrospinal fluid. *Acta cytol.*, **8,** 141.
Naylor, B., & Schmidt, R. W. (1964): The case for exfoliative cytology of serous effusions. *Lancet*, **1,** 711.
Saphir, O. (1949): Cytologic diagnosis of cancer from pleural and peritoneal fluids. *Am. J. clin. Path.*, **19,** 309.
Spriggs, A. I. (1954): Malignant cells in cerebrospinal fluid. *J. clin. Path.*, **7,** 122.
Spriggs, A. I. (1957): *The Cytology of Effusions in the Pleural, Pericardial and Peritoneal Cavities.* London: Heinemann.

SECTION VII

THE EFFECTS OF RADIATION

Agnew, A. M., Fidler, H. K., & Boyes, D. A. (1960): Evaluation of radiation response. *Am. J. Obstet. Gynec.*, **79,** 698.
Graham, J. B., & Graham, R. M. (1956): The sensitisation response in the vaginal smear. *Ann. N.Y. Acad. Sci.* **63,** 1458.
Graham, R. M. (1947): The effect of radiation on vaginal cells in cervical carcinoma. I. Description of cellular changes. II. The prognostic significance. *Surgery, Gynec., Obstet.*, **84,** 153 and 166.
Graham, R. M. (1957): Cytologic prognosis in cancer of cervix. *Proceedings of the Third National Cancer Conference*, p. 659.
Graham, R. M., & Graham, J. B. (1955): Cytological prognosis in cancer of the uterine cervix treated radiologically. *Cancer, N.Y.*, **8,** 59.
Rubio, C. A., Hertzberg, O., Kottmeier, H.-L., Olsson, E., & Zajicek, J. (1965): Sensitisation and radiation response in cases with carcinoma of uterine cervix. *Acta radiol. (Ther.)*, **3,** 241.
Wachtel, E. (1958): Simple cytologic test for cancer cure. *Br. med. J.*, **1,** 20.

APPENDIX

THE ORGANISATION OF A CYTOLOGY SERVICE

Acta Cytologica (1957): Symposium on fixation techniques, **1**, 62.
Acta Cytologica (1960): Symposium on the advantages and disadvantages of various techniques of obtaining material for examination, **4**, 221.
ANDERSON, A. F. (1956): The management of the cytology test. *J. Obstet. Gynec. Br. Emp.*, **66**, 239.
ANDERSON, A. F. (1965): Carcinoma in situ—before and after its interpretation. *J. Obstet. Gynaec. Br. Commonw.*, **72**, 140.
ANDERSON, A. F. (1965): Treatment and follow-up of non-invasive cancer of the uterine cervix. *J. Obstet. Gynaec. Br. Commonw.*, **72**, 172.
AYRE, J. E. (1947): Selective cytology smear for diagnosis of cancer. *Am. J. Obstet. Gynec.*, **53**, 609.
AYRE, J. E. (1948): Diagnosis of pre-clinical cancer of the cervix; cervical cone knife; its use in patients with a positive vaginal smear. *J. Am. med. Ass.*, **138**, 11.
BERTALANFFY, F. D. (1962): Evaluation of the acridine-orange fluorescence microscope method for cytodiagnosis of cancer. *Ann. N.Y. Acad. Sci.*, **93**, 715.
BERTALANFFY, L. VON, & BICKIS, I. (1956): Indication of cytoplasmic basophilia (ribonucleic acid) by fluorescence microscopy. *J. Histochem. Cytochem.*, **4**, 481.
BERTALANFFY, L. VON, MASIN, M., & MASIN, F. (1958): A new and rapid method for diagnosis of vaginal and cervical cancer by fluorescence microscopy. *Cancer, N.Y.*, **11**, 873.
CLYMAN, M. J. (1955): Cannula aspirators for vaginal, endocervical and endometrial cytology. *Obst. Gynec., N.Y.*, **6**, 258.
CONN, H. J. (1966): *Biological Stains.* Seventh ed. Edinburgh: Livingstone.
CULLING, C. F. A. (1963): *Handbook of Practical Histopathological Technique.* Second ed. London: Butterworth.
DAVIS, H. J. (1962): The irrigation smear, a cytologic method for mass population screening by mail. *Am. J. Obstet. Gynec.*, **84**, 1017.
DAVIS, H. J., & KURZ, L. (1962): Detection of preinvasive cancer by the irrigation smear technic. *Dan. med. Bull.*, **9**, 121.
FIDLER, H. K., BOYES, D. A., LOCK, D. R., & AUERSPERG, N. (1962): The cytology program in British Columbia. Part II. The operation of the cytology laboratory. *Can. med. Ass. J.*, **86**, 823.
GRUBB, C., & CRABBE, J. G. S. (1961): Fluorescence microscopy in exfoliative cytology. *Br. J. Cancer*, **3**, 483.
GUSBERG, S. B., & GRAHAM, D. A. (1950): The development of a vaginal cytology laboratory: its precision and significance. *Am. J. Obstet. Gynec.*, **59**, 1053.
HECHT, E. L. (1956): The endometrial aspiration smear: research status and clinical value. *Am. J. Obstet. Gynec.*, **71**, 819.
HIGGINS, J. A., & SMITH, J. P. (1963): Fixation and transport in a postal cytodiagnostic service. *J. clin. Path.*, **16**, 489.
LASH, A. F., & ANTONOW, A. M. (1953): The 'phase-contrast' microscope: use in the diagnosis of cervical and endometrial carcinoma. *Obstet. Gynec., N.Y.*, **2**, 584.
LOPES-CARDOZO, P. (1958): Advantages and disadvantages of the May-Grunwald Giemsa procedure in exfoliative cytology. *Acta cytol.*, **2**, 284.
MCGREGOR, J. E., FRASER, M. E., & MANN, E. M. F. (1966): The cytopipette in the diagnosis of early cervical carcinoma. *Lancet*, **1**, 252.
MAVROMATTIS, F. S. (1962): The identification of mesothelial cells by the technic of Mowry. *Acta cytol.*, **6**, 443.
MEEK, E. S. (1962): Application of the Feulgen reaction to smears in exfoliative cytology. *J. clin. Path.*, **15**, 387.
MOWRY, R. W. (1957): Alcian blue and alcian blue periodic acid-Schiff stains for carbohydrates. *Manual of Histological and Special Stain Technique*, p. 140. Washington, D.C.: Armed Forces Institute of Pathology.
PAPANICOLAOU, G. N. (1942): A new procedure for staining vaginal smears. *Science, N.Y.*, **95**, 438.
PAPANICOLAOU, G. N., & TRAUT, H. F. (1943): *Diagnosis of Uterine Cancer by the Vaginal Smear.* New York: Commonwealth Fund.
RICHART, R. R., & VAILLANT, H. W. (1965): Influence of cell collection techniques upon cytological diagnosis. *Cancer, N.Y.*, **11**, 1474.
SEAL, S. H. (1956): A method of concentrating cancer cells suspended in large quantities of fluid. *Cancer, N.Y.*, **9**, 866.
SELBACH, G. J., & BONDAR, M. (1962): Cytological filtration technics. *Acta cytol.*, **6**, 311.
SHORR, E. (1941): New technic for staining vaginal smears. III. A single differential stain. *Science, N.Y.*, **94**, 545.
SMITH, J. P., MUSKETT, J. M., & EASSON, E. C. (1965): Evolution of a community screening project. *Lancet*, **2**, 74.
STOCKWELL, R. E. (1945): Feulgen technique. *Stain Technol.*, **20**, 45.
UMIKER, W., & SOURENNE, R. (1961): A simple method for concentrating carcinoma cells in sputum. *Am. J. clin. Path.*, **35**, 411.
WACHTEL, E. (1965): The collection, preservation and staining of specimens for gynaecological cytology. *Broadsheet No.* 50. *Association of Clinical Pathologists*, London.
WIED, G. L. (1955): Importance of the site from which vaginal cytologic smears are taken. *Am. J. clin. Path.*, **25**, 742.
WIED, G. L. (1956): Phase-contrast microscopy, an office technique for pre-screening of cytologic vaginal smears. *Am. J. Obstet. Gynec.*, **71**, 806.
WIED, G. L. (1965): Quality control mechanisms for cytology programs. *Acta cytol.*, **9**, 407.
WIED, G. L., & MANGLANO, J. I. (1962): A comparative study of the Papanicolaou technic and the acridine-orange fluorescence method. *Acta cytol.*, **6**, 554.

EVALUATION OF RESULTS

Acta Cytologica (1964): Symposium on the comparative diagnostic accuracy, efficiency and specificity of technics for the detection of malignancy. **8,** 102 et seq.

BODDINGTON, M. M., & SPRIGGS, A. I. (1965): Cytological diagnosis of cancer: its uses and limitations. *Br. med. J.*, **1,** 1523.

BRYANS, F. E., BOYES, D. A., and FIDLER, H. K. (1964): The influence of a cytological screening program upon the incidence of squamous cell carcinoma of the cervix in British Columbia. *Am. J. Obst. Gynec.*, **88,** 898.

FIDLER, H. K., BOYES, D. A., AUERSPERG, N., & LOCK, D. R. (1962): The cytology program in British Columbia. Part I. An evaluation of cytology in the diagnosis of cancer and its application to the detection of carcinoma of the cervix. *Canad. med. Ass. J.*, **86,** 779.

GRUBB, C., & CRABBE, J. G. S. (1961): Fluorescence microscopy in exfoliative cytology. *Br. J. Cancer*, **15,** 483.

GUSBERG, S. B. (1953): The relative efficiency of diagnostic techniques in the detection of early cervical cancer. *Am. J. Obstet. Gynec.*, **65,** 1073.

HAAM, E. VON (1962): A comparative study of the accuracy of cancer cell detection by cytologic methods. *Acta cytol.*, **6,** 508.

RAVEN, R. W. (Ed.) (1958): *Cancer.* Vol. 3, Part VII. Cancer detection. London: Butterworth.

RICHARDSON, H. L., KOSS, L. G., & SIMON, T. R. (1955): An evaluation of the concomitant use of cytological and histological techniques in the recognition of cancer in exfoliated material from various sources. *Cancer, N.Y.* **8,** 948.

INDEX

Bold figures denote pages on which illustrations appear

A

Abnormal, cytoplasmic staining in trichomonal infection, **29**
— gastric columnar cells in megaloblastic anaemia, 129 et seq.
— squamous cells in carcinoma of cervix, **51**
— vaginal smears, in early pregnancy, **88, 89**
— — in late pregnancy, **90**
Abortion, vaginal smear in, 88, **89**
Acridine orange fluorescent technique, 218, **219**
Acute gastritis, 126
Adenocarcinoma, of bronchus, 117 et seq.
— of cervix, **42, 60, 61**
— of corpus uteri, 42, 62 et seq.
— of endometrium, 42, 62 et seq.
— of fallopian tube, 65
— of oesophagus, 132
— of ovary, **65, 66**
— — malignant cells in ascitic fluid, **167, 169, 223**
— of stomach, 134 et seq.
Albuminised slides, 211
Alimentary tract, collection of specimens from, 210, 211
Amenorrhoea, vaginal smear in, **88**
Anaemia, megaloblastic, 129 et seq.
Anaplastic metastatic carcinoma, 170, **171**
Androgens, effect of, on vaginal smear, 73
Anovulatory cycle, 73, 74
Anucleated keratinised squamous cells, in cervical and vaginal smears, 8, **9**
— in leukoplakia, **125**
— in sputum, **111**
Apparent nuclear enlargement, due to drying, 186, **187**
Asbestos bodies, in sputum, **107**
— Prussian blue reaction, 225
Asbestosis, 107
Ascitic fluid, histiocytes in, **160,** 161, **223, 224**
— lymphocytes in, **160,** 161
— maligant cells in, 165, **167, 169, 171**
— mesothelial cells in, 159, **160, 223, 224**
Aspirate, breast cyst, benign, 150
— — — ductal cells in, **151**
— — — foamy macrophages in, **151**
— — malignant, **156**
— bronchial, 208
— — advantages and disadvantages, 209
— — technique, 209
— endometrial, 62, 207
— vaginal pool, **201,** 203
Aspirates, from cysts, 150, **151, 156, 163,** 173
— gastric (*see Gastric washings*)
— serous fluids (*see Serous effusions*)
Aspiration smear technique, 155
Assessment of the vaginal smear, for hormonal studies, 71, 72
Atrophic post-menopausal vaginal smear, **82, 83**
Atypical, endometrial cells, **68**
— hyperplasia of the endometrium, 67
— post-menopausal vaginal smear, 83, **84**
— vaginal smear, in early pregnancy, **88, 89**
— — in late pregnancy, **90**
Ayre's spatula, **201, 202, 205, 206**

B

Bacilli, Gram positive, **33**
Basal squamous cells, in cervical smears, **9**
— in sputum, **102**
Basophil polymorphonuclear leucocytes in sputum, **97**
Benign, breast cysts, aspirates from, 150
— — ductal cells in aspirate, **151**
— — foamy macrophages in aspirate, **151**
— cells, in cervical smears, effects of radiation on, 176 et seq.
— columnar cells, in gastric washings, 122 et seq.
— — in sputum, 95, 96
— — denuded nuclei of, in gastric washings, **124**
— — — in sputum, **103**
— denuded nuclei in cervical smears, **25, 59**
— endocervical cells, 10 et seq.
— — inflammatory changes in, **23**
— endometrial cells, 12, **13, 64**
— — polypi, **28**
— mesothelial cells, 159 et seq.
— — changes in inflammation, 159
— — mitotic figures in, **160**
— multinucleated giant cells from renal pelvis, **141**
— ovarian cyst, histiocytes in aspirate from, **163**
— squamous cells, in cervical and vaginal smears, 6 et seq.
— — in gastric washings, 122, **123**
— — inflammatory changes in, 20 et seq.
— — in sputum, 102, **112**
— tadpole cells, **38, 183**
— transitional epithelial cells in urine, **140, 143**
'Benign' tumours of the urinary tract, 142
'Birds eye' appearance in malignant cells of squamous type, **53, 56**
Bladder, carcinoma of, 144 et seq.
— catheterisation, avoidance of, 211
— papilloma of, 142
Blood, 164, 173
Body cavities, effusions into, 158, 165 et seq.
Body fluids, 158
— cellular content, 159 et seq.
— inflammatory cells in, 161, **162**
— malignant cells in, 165 et seq.
— mesothelial cells in, 159, **160**
— non-malignant disorders, 158
— non-mesothelial cells in, 161, **162**
Borderline lesions, of cervix, 67
— of endometrium, **68**
Breast, cancer, 153 et seq.
— cystic carcinoma, **156**
— cysts, aspirates from, 150 et seq.
— — aspirates, malignant, **156**
— discharge from nipple, 150, **151**
— intraduct carcinoma of, 153
— non-malignant disorders, 150 et seq.
Bronchial, aspirates, 94, 208, 209
— — advantages and disadvantages, 209
— — ciliated columnar cells in, **95**
— — staining techniques for, 216, 222
— — technique, 209
— carcinoma, 110 et seq.
— — adenocarcinoma, 117 et seq.
— — oat cell type, 114 et seq.

— — large undifferentiated cell type, 114 et seq.
— — squamous cell type, 110 et seq.
— epithelial cells, **95, 96, 103**
— — prominent nucleoli in, **103**
— washings, 94, 209
— — advantages and disadvantages, 209
— — technique, 209

C

Cancer, cytological diagnosis of, 34 et seq.
— detection of, in asymptomatic patients, 235
— diagnosis of, in symptomatic patients, 236
— of breast, 153 et seq.
— of cervix uteri, 49 et seq.
— of endometrium, 62 et seq.
— of female genital tract, 37 et seq.
— of gastro-intestinal tract, 132 et seq.
— of ovary, **65, 66**
— of respiratory tract, 108 et seq.
— of urinary tract, 144 et seq.
— of vagina, 47, **48**
— of vulva, **46**
Candida albicans, in a cervical smear, **32**
— in sputum, **100**
Carbon pigment in histiocytes, **97**
Carbowax 1500, for postal specimens, 213
Carcinoma, of bladder, 144 et seq.
— of breast, 153 et seq.
— of bronchus, adenocarcinoma, 117 et seq.
— — oat cell type, 114 et seq.
— — squamous cell type, 110 et seq.
— of cervix, adenocarcinoma, **60, 61**
— — carcinoma-*in-situ*, 49, **53**
— — effects of radiotherapy, 176 et seq.
— — squamous cell carcinoma, 49 et seq.
— of endometrium, 62 et seq.
— of kidney, 147, **172**
— of larynx, 108, **109**
— of lower respiratory tract, 110 et seq.
— of oesophagus, 132
— of oral cavity, 132, **133**
— of ovary, **65, 66, 167, 169, 223**
— of prostate, 147
— of renal pelvis, 147
— of stomach, 134 et seq.
— of upper respiratory tract, 108
— of ureter, 147
— of vagina, 47, **48**
— of vulva, **46**
'Cart wheel' nuclei in plasma cells, 17
Cerebrospinal fluid, **164,** 173
— histiocytes in, **164**
— inflammatory cells in, **164**
— lymphocytes in, **164**
— polymorphonuclear leucocytes in, **164**
Cervical, mucus, 10
— polypi, 28
— scrape smear, 204
— — advantages, 204
— — cellular content, 7 et seq.
— — contamination, 188 et seq.
— — disadvantages, 205
— — equipment, **201, 202**
— — reporting, 233, 234
— — spermatozoa in, **18**
— — technique, **205**
— — *trichomonas vaginalis* in, **29, 30, 31**
Cervix uteri, adenocarcinoma of, **60, 61**
— carcinoma-*in-situ* of, 49 et seq.
— erosion of, 28
— leukoplakia of, 27
— squamous cell carcinoma of, 49 et seq.
— undifferentiated malignant cells in, 49 et seq., **60**
Childhood, vaginal smear in, 74
Chromatin, clumping of, 34 et seq.
Chronic atrophic gastritis, 126, **127**
— with intestinal metaplasia, **127**
Cilia, bronchial epithelial cells, **95**
— endocervical cells, 11, **12, 23**
'Clock face' nuclei in plasma cells, 17
Collection, of fluids, 212
— of specimens, 200 et seq.
— — alimentary tract, 210
— — female genital tract, 200 et seq.
— — respiratory tract, 208, 209
— — urinary tract, 211
— of vaginal smears by patient, 203, 204
Colon, carcinoma of, 137
— inflammation of, 128
— washings from, 122
— — technique, 211
Columnar cells, benign, 122 et seq.
— — endocervical, 10 et seq., **23, 25**
— — endometrial, 6, 12, **13, 64**
— — in bronchial aspirate, **95**
— — in gastric washings, 122 et seq.
— — in sputum, **96, 97**
— — of nasal origin, 94
— — of respiratory tract origin, 95 et seq.
Contaminants, extrinsic, 188 et seq.
— in cervical smears, 188 et seq.
— in sputum, **101**
— intrinsic, 192
Contamination, by 'floater', 190, **191**
— by glove powder, 188, **189**
— by lubricant, **188**
— by precipitate of stain, 188, **195**
'Cornified' cells, 7
Curschmann's spirals in sputum, **106**
Cysts, aspirates from, 150, **151,** 155, **156, 163,** 173
Cystadenocarcinoma of ovary, malignant cells, in ascitic fluid, **167**
— — in vaginal smear, **65**
Cystic carcinoma of breast, 155, **156**
Cytological diagnosis of cancer, 34 et seq.
Cytolysis, in cervical and vaginal smears, 24, **25,** 85, **86**
— in gastric washings, **124**
— in pregnancy, **59,** 85, **86**
Cytoplasmic changes, in inflammation, 20 et seq.
— — in endocervical cells, **23**
— — in squamous cells, **20, 21, 29**
— wrinkling, **7,** 72, 73, **78,** 85, **86**

D

'Deep' benign squamous cells in sputum, **102**
Degenerate malignant cells, in carcinoma of cervix, **51, 55**
— in squamous cell carcinoma of bronchus, **111, 112**
Degenerative changes, in mesothelial cells, 159,
— in squamous cells, 20, **21**
Denuded benign nuclei, in cervical smears, **25**
— in gastric washings, **124**
— of columnar cells in sputum, **103**
— of endocervical cells, **23**
— of parabasal cells in atrophy, **25**
— of squamous cells in pregnancy, **59,** 85, **86**
Denuded malignant nuclei, of gastric malignant cells, **135, 137**

— of oat cells, 114, **115**
— of squamous cells, in cervical smear, **59**
Detection of cancer in asymptomatic patients, 235
Diagnosis of cancer in symptomatic patients, 236
— 'false negative', 237
— 'false positive', 237
Diagnostic cytology and the care of the patient, 235
Diagnostic difficulties, 186
Differential diagnosis, practical problems in, 186
Differentiated malignant cells, in aspirate from breast cyst, **156**
— of adenocarcinomatous type, **42**
— — from bronchus, 117 et seq.
— — from cervix, **42, 60,** 61
— — from endometrium, 62 et seq.
— — from stomach, 134 et seq.
— — in fluids, 167 et seq.
— of squamous type, 37 et seq.
— — from bronchus, 110 et seq.
— — from carcinoma-*in-situ* of cervix, 37 et seq., 52 et seq.
— — from cervix, 37 et seq., 49 et seq.
— — from larynx, 108, **109**
— — from tongue, 133
— — from vagina and vulva, 46 et seq.
— — in fluids, 167 et seq., **170**
— — in sputum, 108 et seq.
— of transitional type from carcinoma of bladder, 144 et seq.
Discharge from the nipple, 150, 153
— ductal cells in, **151**
— foamy macrophages in, **151**
— undifferentiated malignant cells in, **154**
Distortion due to drying, 186, **187, 189**
Disturbances of the menstrual cycle, 79
Division of work in the laboratory, 233, 234
DNA, Feulgen reaction for, **226**
Döderleins bacilli, **25,** 85, **86**
Drying distortion, 186, **187, 189**
Ductal cells, in aspirate from benign breast cyst, **151**
— in discharge from the nipple, **151**
Dyskaryosis, 37, 44, **45, 48, 56, 58**
— in carcinoma-*in-situ* of cervix, **45,** 56, **58**
— in endocervical cells, **45**
— in intermediate cells in carcinoma-*in-situ* of cervix, **45**
— in parabasal cells in carcinoma-*in-situ* of cervix, **56, 58**
— in superficial cells in trichomonal infection, **44**
Dysplasia, 67

E

Early pregnancy, abnormal smears in, **88,** 89
Early puerperium, vaginal smear in, **91**
Effect, of androgens on the vaginal smear, 73
— of oestrogens on the vaginal smear, 72
— of 'Pill' on the vaginal smear, 79, **80**
— of progesterone on the vaginal smear, 73
— of progestogens on the vaginal smear, 79, **80**
Effects of radiation on cells, 176 et seq.
Effusions into body cavities, 158 et seq., 165 et seq.
Endocervical, adenocarcinoma, **42, 60, 61**
— cells, 10 et seq., **23, 25**
— — ciliated, **11, 12, 23**
— — cytoplasmic vacuolation in, **11, 23**
— — denuded nuclei, **23, 25**
— — dyskaryosis, **45**
— — 'Honeycomb' arrangement, 10, **11**
— — inflammatory changes in, **23, 195**
— — mucus secreting, 10 et seq.
— — multinucleation in, **12, 23**
— — nucleoli in, 10, **11**
— — 'Palisade' arrangement, 10, **11**
— — 'Picket fence' arrangement, 10, **11**
— — squamous metaplasia, **23, 26, 27**
— — terminal plates in, **11, 23**
— — with peaked nuclei, 10, **12**
— polyp, squamous metaplasia in, **45**
— smear, 207
— — advantage and disadvantage, 207
— — technique, 207
Endocrine assessment of the vaginal smear, 71
Endometrial, aspiration smear, 207
— — advantage and disadvantage, 207
— — technique, 207
— cells, 6, 12, **13, 64**
— — atypical, **68**
— hyperplasia, 67
— polypi, **28**
Endometrium, adenocarcinoma of, 62 et seq.
Eosinophil polymorphonuclear leucocytes in sputum, **97,** 222
Eosinophilic cytoplasmic staining, 70, **78,** 186, **187, 193**
Epithelial cells, changes in inflammation, 20 et seq.
— — of female genital tract, 6 et seq.
— — of gastro-intestinal tract, 122 et seq.
— — of respiratory tract, 102 et seq.
— — of urinary tract, 140 et seq.
— 'pearl' formation, by benign squamous cells, **194**
— — by malignant squamous cells, **194**
Erosion of cervix, 28
Erythrocytes, 6, **13, 14, 217**
— in body fluids, 161
Ethyl alcohol as fixative, 212, 213
Extrinsic contaminants, **188, 189**

F

Fallopian tube, adenocarcinoma of, 65
'False negative' diagnoses, 237
'False positive' diagnoses, 237
'False positive' diagnosis of malignant cells, **195**
Female genital tract, cancer of, 37 et seq.
— collection of specimens from, 200 et seq.
— inflammation, 20
— non-malignant disorders, 20
— normal smear, 6
Feulgen reaction for DNA, **226**
Fibre cells, malignant, 35, 37, **40, 50**
Fibroblast, **182**
Field marking, **232**
Fixation, 212, 213
— methods, 212, 213
— postal, 213
— routine, 213
— sputum, 213
Fixatives, 212, 213
'Floaters', 190, **191**
Fluid, ascitic, benign cells in, 158 et seq.
— — differentiated malignant cells in, 165, **167, 169, 171**
— — undifferentiated malignant cells in, 165, **171**
— cerebrospinal, **164,** 173
— from hydrocele, malignant cells in, **166**
— pleural, benign cells in, 158 et seq.
— — differentiated malignant cells in, 165, **168, 170, 172**
— — undifferentiated malignant cells in, 165, **172**

Fluids, body, 158 et seq.
— collection of, 212
Foamy macrophages, in aspirate from benign breast cyst, 150, **151**
— in discharge from nipple, 150, **151**
Folding of cytoplasm in squamous cells, **7**, 73, **78**, 85, **86**
Food contamination, of gastric aspirate, 122, **124**
— of sputum, **101**
Fungal infection, female genital tract, **32**
Fungi, in saliva, **100**
— in sputum, **100**

G

Gastric aspirate, food contamination of, 122, **124**
Gastric polypi, **128**
Gastric washings, 122, 210, 211
— benign columnar cells in, 122 et seq.
— benign squamous cells in, 122, **123**
— cytolysed columnar cells in, 122, **124**
— denuded nuclei, 122, **124**
— in chronic atrophic gastritis, 126, **127**
— in gastric adenocarcinoma, 134 et seq.
— in megaloblastic anaemia, 129 et seq.
— polypoid group of benign columnar cells in, **128**
— technique, 210, 211
Gastritis, acute, 126
— chronic atrophic, 126
Gastro-intestinal tract, cancer of, 132 et seq.
— non-malignant disorders of, 125 et seq.
— normal cellular content, 122 et seq.
Germinal cells, 26
Giant multinucleated, cells in urine, **141**
— histiocytes, 15, **16**, **84**, 177, **178**
Glove powder, contamination by, 188, **189**
Glycogen, 85, **86**, **92**
'Goblet' cells in sputum, **96**
Granules, keratohyaline, **6**
Granulosa-cell carcinoma of ovary, 65, **66**
'Ground glass' opacity, 186, **187**
Grouping of malignant cells, 35

H

Haemalum and eosin stain, **107**
Haematoxylin and eosin staining technique, 220, **221**
Haematoxylin, Harris', 216
Haemosiderin, Prussian blue reaction for, **105**, 107, **225**
Haemosiderosis, pulmonary, **105**
Histiocytes, 6, **15**, **16**, **63**, 75, **76**, **84**, 96, **97**, **98**, **105**, **115**, **160**, 161, **162**, **163**, **164**, 177, **178**, 192, **193**, **223**, **224**, **225**
— giant multinucleated, 15, **16**, **84**, 177, **178**
— — phagocytosis in, **16**
— in aspirate from benign ovarian cyst, **163**
— in body fluids, 160 et seq., **223**, **224**
— in cerebrospinal fluid, **164**
— in cervical smears, 6, **15**, **16**, **63**, 177, **178**, 192, **193**
— — with eosinophilic cytoplasmic staining, **193**
— in fluid from hydrocele, **162**
— in post-menopausal vaginal smear, **84**
— in post-radiation smear, 177, **178**
— in sputum, 96, **97**, **98**, **105**, **115**, **223**, 225
— in vaginal smears, 75, **76**, **84**
'Honeycomb' arrangement, endocervical cells, 10, **11**
Hormonal effects on the vaginal smear, 69 et seq.
Hydrocele fluid, histiocytes in, **162**
— malignant cells in, **166**
— mesothelial cells in, **162**
Hyperchromasia of nuclei, 34, 37 et seq., 110 et seq., 134 et seq., 145 et seq.
Hyperplasia, atypical of endometrium, 67

I

Infancy, vaginal smear in, 74
Infection, in cervical smear, monilial, **32**
— non-specific, **33**
— trichomonal, 29 et seq.
Inflammation, female genital tract, 20 et seq.
— of gastro-intestinal tract, 125 et seq.
— of respiratory tract, 102 et seq.
— of urinary tract, 142
Inflammatory cells, in body fluids, 161, **162**, **164**
Inflammatory changes, in endocervical cells, **23**
— in squamous cells, 20 et seq., **29**, **30**
Inflammatory conditions, specific, 29 et seq.
Ingested leucocytes in histiocytes, **16**
Inner layer basal cells (*see Basal cells*)
Inspissated mucus in sputum, **98**
Intermediate squamous cells, changes in inflammation, **21**, **22**, **29**, **30**
— denuded nuclei in cervical smear, **25**
— dyskaryosis, **45**
— in cervical smears, **6**, **7**, **21**, **22**, **25**, **29**, **30**
— in vaginal smears, 72, **76**, **77**, **78**, 80, 85, **86**, **92** et seq.
— of upper respiratory tract origin, **94**
Intestinal metaplasia, 126, **127**
Intraduct carcinoma of breast, 153
Intrinsic contaminants, 192

K

Karyolysis, **9**, 21, 29, **30**, 176
Karyopyknotic index, 71, 72, 181
Karyorrhexis, in cervical smears, **21**, **38**, **48**, **51**, 176
— in metaplastic cells in sputum, 104, **104**
Keratinised anucleated squamous cells, in cervical and vaginal smears, 8, **9**
— in sputum, **111**
Keratohyaline granules, **6**
Kidney, carcinoma of, 147, **172**

L

Lactobacillaceae, **25**, **33**, 85, **86**
Large undifferentiated malignant cells in sputum, 114, **116**
Laryngeal swab, 208
Larynx, cancer of, 108, **109**
Late pregnancy, abnormal smears in, **90**
Lateral vaginal wall smear, 204
— advantage and disadvantage, 204
— technique, 204
Leucocytes, in body fluids, 161, **162**
— in cerebrospinal fluid, **164**
— in cervical smears, **6**, 14, **15**
— in sputum, **97**, **98**
— in vaginal smears, 75, **76**, **79**
— ingested by histiocyte, **16**
— polymorphonuclear, **6**, 14, **15**
Leukoplakia, lingual, **125**
— of cervix, 27
— of the oral cavity, 125
Levine tube, 210
Lingual leukoplakia, **125**
Lower respiratory tract, cancer of, 110 et seq.
Lubricant, contamination by, **188**

Lymphocytes, 6, **15, 17, 98,** 105
— diagnostic confusion due to, **98,** 193
— in ascitic fluid, **160**
— in body fluids, 161
— in cerebrospinal fluid, **164**
— in cervical smear, 6, **15**
— in pleural fluid, **162**
— in sputum, **98,** 105
— in vaginal smear, **17**

M

Macrophages (*see also Histiocytes*)
— in aspirate from benign breast cyst, 150, **151**
— in discharge from nipple, 150, **151**
Malignant cell, characteristics (*see also individual sites*), 34 et seq.
— cytoplasm, 34, 35
— differentiated adenocarcinomatous, 42 et seq.
— differentiated squamous, 37 et seq.
— grouping, 35, 36
— nucleus, 34
— undifferentiated, **43**
Malignant cells, (*see also individual sites*)
— from body fluids, 165 et seq.
— from breast, 153 et seq.
— from female genital tract, 46 et seq.
— from gastro-intestinal tract, 132 et seq.
— from respiratory tract, 108 et seq.
— from urinary tract, 144 et seq.
— post-radiation, 176 et seq.
Malignant smear, overall appearance of, 36
Marking of slides, **232**
May-Grünwald-Giemsa stain, for sputum, **97**
— technique, 222, **223**
Megaloblastic anaemia, 129 et seq.
— abnormal gastric columnar cells in, **130, 131**
Menopause, vaginal smear at, 80, **81**
Menstrual cycle, disturbances of, 79
— normal vaginal smears, 74 et seq.
Menstrual phase, vaginal smear in, 75, **76**
Mesothelial cells, 159, **160, 162**
— ascitic fluid, **160, 223**
— benign, mitotic figures in, 159, **160**
— degenerative changes in, 159
— in fluid from hydrocele, **162**
— in pleural fluid, **162, 172**
— mitosis in, 159, **160**
— multinucleation in, 159, **160**
Mesothelioma, 165
Metaplasia, intestinal, 126, **127**
Metaplastic cells, in cervical smears, **26,** 27
— in sputum, karyorrhexis in, 104, **104**
Metastatic, anaplastic carcinoma in fluids, **171**
— carcinoma of female genital tract, 66
— carcinoma of respiratory tract, 120
— carcinoma of urinary tract, 147
— differentiated carcinoma in fluids, 165 et seq.
— tumours in body fluids, 165 et seq.
Methylene blue staining technique, 222, **223**
Micro-organisms, in saliva, 99
— in sputum, **99**
— simulating malignant nuclei in sputum, **99**
Millipore filter technique, 212
Mitotic figures, in benign mesothelial cells, 159, **160**
— in histiocytes, **160**
Monilia albicans, in a cervical smear, **32**
— in sputum, **100**
Monocytes, 105
Mucus, cervical, 10
— in sputum, **98, 106**
— inspissated in sputum, **98**
— secreting columnar cells in sputum, **96, 97**
— secreting endocervical cells, 10, **11, 12**
Multinucleated malignant squamous cells, in cervical smear, **49**
— in sputum, **112**
Multinucleation, in endocervical cells, **23**
— in histiocytes, 15, **16, 84, 177, 178**
— — in post-menopausal vaginal smear, **84**
— — in post-radiation smear, **177, 178**
— in mesothelial cells, 159, **160**
— in parabasal cells, **22**
— in squamous cells, in cervical smear, **49**
— — post-radiation, 176, **178, 182**
— — in sputum, **112**
— of giant cells in urine, **141**

N

Nasal columnar cells, 94
Nasal secretion, for eosinophils, 208
Nipple, discharge from, 150, 153
— discharge, ductal cells in, **151**
— — foamy macrophages in, **151**
— — malignant cells in, **154**
Non-cellular components of sputum, 98 et seq., **106, 107**
Non-malignant disorders, body fluids, 158 et seq.
— breast, 150 et seq.
— female genital tract, 20 et seq.
— gastro-intestinal tract, 125 et seq.
— respiratory tract, 102 et seq.
— urinary tract, 142 et seq.
Normal cellular content, body fluids, 158
— female genital tract smears, 6 et seq.
— gastro-intestinal tract specimens, 122 et seq.
— respiratory tract specimens, 95 et seq.
— urinary tract specimens, 140 et seq.
Normal vaginal smear in pregnancy, **86**
Nuclear changes, in inflammation, 20 et seq.
— in malignancy, 34, 35
Nucleoli, in benign cells, in bronchial epithelial cells, **103**
— — in endocervical cells, 10, **11**
— — in endometrial cells, 12, **13**
— — in gastric cells, 122 et seq.
— — in inflammation, 22
— — in megaloblastic anaemia, 129 et seq.
— in malignant cells, from adenocarcinoma, of bronchus, **117, 119**
— — — of cervix, **42, 60, 61**
— — — of endometrium, 62, **63**
— — — of stomach, 135 et seq.
— in undifferentiated malignant cells, **43,** 114, **116,** 134, **137**

O

Oat cells, in sputum, 114, **115**
Oesophageal washings, 122, 126, 132
— technique, 210
Oesophagus, cancer of, 132
— inflammation of, 126
Oestrogens, effect of, on vaginal smear, 72, **82**
Oral cavity, cancer of, 132
— inflammation of, 125
— leukoplakia of, **125**
— normal cellular content, 122

Oral specimens, collection, 210
Organisation of a cytology service, general principles, 198
Organisation of work in the laboratory, 230
Outer layer basal cells (*see Parabasal squamous cells*)
Ovarian tumours, malignant, 65 et seq., **167, 169, 171, 223**
Ovulatory cycle, 74 et seq.
— phase, vaginal smear in, **77**
'Oyster cells', **91**

P

'Palisade' arrangement, endocervical cells, 10, **11**
'Pap' cells, 102
Papanicolaou, classification for reporting, 233
— stain, technique, 215 et seq.
Papilloma of bladder, 142
Parabasal squamous cells, **8**
— changes in inflammation, 21, **22**
— denuded nuclei in cervical smear, 24, **25**
— dyskaryosis, **56, 58**
— multinucleation in, **22**
— vacuolation in, 8, **22**
Paranasal sinuses, cancer of, 108
Patient collection of vaginal smears, 203
— advantages, 203
— disadvantages, 204
— technique, 204
Peaked nuclei in endocervical cells, **12**
'Pearl' formation by squamous cells, **194**
Perinuclear vacuoles, 20, 22, **23,** 29, **30**
Periodic acid Schiff reaction, **224**
Peritoneal fluid (*see Ascitic fluid*)
Phagocytosis by histiocytes, 15, **16,** 96, **97,** 105
Phase contrast microscopy, 227 et seq.
'Pill', effect on vaginal smear, 79, **80**
Pipette, vaginal, **201,** 203
Plasma cells, 6, 14, **17, 127**
— diagnostic confusion due to, 193
Pleural fluid, histiocytes in, 160 et seq., **172**
— lymphocytes in, 160 et seq., **162**
— malignant cells in, **168, 170, 172**
— mesothelial cells in, 159, **160, 162, 172**
Pleomorphism of nuclei, **34**
Polyethylene glycol for postal specimens, 213
Polymorphonuclear leucocytes, **6,** 14, **15, 97, 98,** 222
Polyp, benign, **28**
— cervical, 28
— dyskaryosis of squamous metaplasia in, **45**
— endometrial, **28**
— gastric, **128**
— squamous metaplasia in, **45**
Post abortion, vaginal smear in, 88, **89**
Post-menopausal era, vaginal smear in, 80 et seq.
Post-menstrual phase, vaginal smear in, 75, **76**
Post-natal smears, **92**
Post-radiation smears, 176 et seq.
Practical problems in differential diagnosis, 186 et seq.
Precautions, safety, 234
'Pre-cornified' cells, 7
Pregnancy, cytolytic vaginal smear in, 85, **86**
— early, abnormal smears in, 88, **89**
— late, abnormal smears in, **90**
— normal vaginal smear in, 85, **86**
Pre-menstrual phase, vaginal smear in, 75, **79**
Preparation, of smears, 200 et seq.
— of specimens, 200 et seq.
Problems, diagnostic, of extrinsic origin, 186
— — of intrinsic origin, 192
Proliferative phase, vaginal smear in, 75, **77**
Progesterone, effect on the vaginal smear, 73
Progestogens, effect on the vaginal smear, 79, **80**
Prostate, carcinoma of, 147
Prussian blue reaction, for asbestosis, **107**
— for iron, method, **225**
Puerperium, vaginal smear in, **91, 92**
Pulmonary haemosiderosis, **105, 225**
Pyknosis, **6, 21, 104,** 176

R

Radiation, effects, 176
— response (R.R.), 181
Reception of specimens, 230
Rectal, polypi, 128
— washings, 122
— — technique, 211
Rectum, cancer of, 137
— inflammation of, 128
Renal carcinoma, 147
— metastatic cells in pleural fluid, **172**
Reporting, 232, 233
Reserve cells, 26
Respiratory tract, cancer of, 108 et seq.
— collection of specimens, 208, 209
— non-cellular components, 98
— non-malignant disorders, 102
— normal cellular patterns, 94
Routine fixation, 213

S

Safety precautions, 234
Screening of smears, **231**
Secretory phase, early, vaginal smear in, 75, **78**
— late, vaginal smear in, 75, **78**
Seminoma of testis, **166**
Sensitisation response (S.R.), 181
Serous effusions, malignant, 165 et seq.
— non-malignant, 158 et seq.
— reporting, 238
Shorr's staining technique, 221
Slide screening, **231**
Small, undifferentiated malignant cells (oat cells) in sputum, 114, **115**
Smears from female genital tract, collection, 200 et seq.
— technique, **206**
— — correct, **206**
— — incorrect, **206**
Specimens, collection of (*see also individual sites*), 200 et seq.
— preparation of, 200 et seq.
— reception of, 230
Spermatazoa, in cervical smear, 6, **18**
Sputum, malignant cells, differentiated, 110
— — — adenocarcinomatous, 117 et seq.
— — — squamous, 110 et seq.
— — undifferentiated, 110, 114
— — — 'oat cell' type, 114, **115**
— — — large, 114, **116**
— non-cellular components, 98 et seq., **106, 107**
— non-malignant disorders, 102 et seq.
— normal cellular content, 94 et seq.
— specimen preparation, 208, 209
— staining techniques, 215 et seq.
Squamo-columnar junction, 204, 205
'Squamoid' changes in transitional cell carcinoma of bladder, **146**

Squamous cells, benign, in endocrine assessment, 71, 72
— — in female genital tract, 6 et seq.
— — in gastro-intestinal tract, 122, **123**
— — in respiratory tract, 94, **102**
— — in urinary tract, 141, 143
— malignant, 34 et seq.
— — in body fluids, **170**
— — in female genital tract, 37 et seq.
— — in gastro-intestinal tract, 132, **133**
— — in respiratory tract, 108 et seq.
Squamous metaplasia, in cervical smear, **23, 26, 27**
— in endocervical polyp, **45**
— in sputum, **104**
Stain precipitate, contamination by, 188, **195**
Staining techniques, 215 et seq.
— acridine orange fluorescent, 218, **219**
— Feulgen reaction for DNA, **226**
— for bronchial aspirates, 222, **223**
— for endocrine cytology, 221
— for iron, **225**
— for serous fluids, **224**
— for sputum, 222, **223**
— haematoxylin and eosin, 220, **221**
— May-Grünwald-Giemsa, 222, **223**
— methylene blue, 222, **223**
— Papanicolaou, **6** et seq., 215 et seq.
— Periodic acid-Schiff reaction, **224**
— Prussian blue reaction, **225**
— Shorr's, 221
Stomach, acute gastritis, 126
— carcinoma, 134 et seq.
— chronic atrophic gastritis, 126
— megaloblastic anaemia, changes in, 129 et seq.
Superficial squamous cells, in female genital tract, **6**
— dyskaryosis, **44**
— of upper respiratory tract, **94**
Swab, laryngeal, 208

T

Tadpole cells, benign, 35, 37 et seq.
— — in female genital tract, **38, 183**
— malignant, 35, 37 et seq.
— — in female genital tract, **38, 48, 54, 183**
— — in sputum, **113**
Techniques, collection, 200 et seq.
— staining, 215 et seq.
Terminal plate, bronchial epithelial cells, **95**
— endocervical cells, 10, **11, 23**
Third-type malignant cells, 37, **41**
Tongue, squamous cell carcinoma of, 132, **133**
Touch smear method, 154
Transitional cell carcinoma of bladder, 145, **146**
— 'squamoid' changes in, **146**
Transitional epithelial cells, benign, in urine, **140, 143**
Trichomonas buccalis in sputum, 100
Trichomonas vaginalis infection in female genital tract, 29 et seq., **51**
Tumours, metastatic, in body fluids (*see also individual sites*), 165 et seq.

U

Undifferentiated malignant cells, characteristics, 43
— from breast, **154**
— in body fluids, 165 et seq.
— in female genital tract, 43, 59, **226**
— in gastro-intestinal tract, 134 et seq.
— in respiratory tract, 114 et seq.
— in urinary tract, 145, **146**
— post-radiotherapy, 180, **181**
Unsatisfactory smears, 186, **187, 188, 189**
Upper respiratory tract, cancer of, 108, **109**
— non-malignant disorders, **102**
Ureter, carcinoma of, 147
Ureteric catheterisation, 140, **141,** 211
Urinary tract, benign tumours of, 142
— cancer of, 144 et seq.
— collection of specimens, 211
— non-malignant disorders of, 142, **143**
— normal cellular patterns, **140, 141**
Urine, benign transitional cells in, **140**
— malignant cells in, 144 et seq.
— multinucleated giant cells in, **141**
— squamous epithelial cells in, 141
Uterus (*see Cervix and Endometrium*)

V

Vacuoles, perinuclear, 20, **22, 23,** 29, **30**
Vagina, carcinoma of, 47, **48**
Vaginal pool, aspirate, **201,** 203
Vaginal smear, assessment of, 71, 72
— at the menopause, 80, **81**
— atrophic post-menopausal, **83, 84**
— atypical late menopausal, **84**
— cellular content, 6 et seq.
— early secretory phase, 75, **78**
— effect of androgens on, 73
— effect of oestrogens on, 72, **82**
— effect of 'Pill' on, 79, **80**
— effect of progesterone on, 73
— effect of progestogens on, 79, **80**
— hormonal effects on, 69 et seq.
— in abortion, 88, **89**
— in amenorrhoea, **88**
— in childhood, 74
— in early puerperium, **91**
— in infancy, 74
— in late secretory phase, 75, **78**
— in ovulatory phase, 75, **77**
— in pregnancy, 85 et seq.
— in pre-menstrual phase, 75, **79**
— in puerperium, 85, **91, 92**
— menstrual phase, 75, **76**
— post-abortion, 88, **89**
— post-menopausal, 80 et seq.
— post-menstrual phase, 75, **76**
— post-radiation, 177 et seq.
— reporting, 233, 234
— *Trichomonas vaginalis* in, **29, 30, 31**
Vulva, carcinoma of, **46**
— leukoplakia of, 46

W

Washings, bronchial, 94, 209
— — advantages and disadvantages, 209
— — technique, 209
— gastro-intestinal tract, 122
— — technique, 210
Wrinkling of cytoplasm in squamous cells, **7,** 72, 73, **78,** 85, **86**